T0186318

Mathematical Principles
of
Remote Sensing:

Making Inferences from Noisy Data

by

Andrew S. Milman

Library of Congress Cataloging-in-Publication Data

Milman, Andrew S.
    Mathematical principles of remote sensing / Andrew S. Milman.
    p. cm.
    Includes bibliographical references.
    ISBN 1-57504-135-9
    1. Remote sensing--Mathematical models.   I. Title.
    G70.4.M53   1999
    621.36'78'015118--dc21

                                                99-14147
                                                CIP

    © 1999 by Sleeping Bear Press
    Ann Arbor Press is an imprint of Sleeping Bear Press

Mathematical Principles

of

Remote Sensing:

Making Inferences from Noisy Data

by

Andrew S. Milman

Library of Congress Cataloging-in-Publication Data

Milman, Andrew S.
   Mathematical principles of remote sensing / Andrew S. Milman.
   p. cm.
   Includes bibliographical references.
   ISBN 1-57504-135-9
   1. Remote sensing--Mathematical models.   I. Title.
   G70.4.M53   1999
   621.36'78'015118--dc21

                                          99-14147
                                          CIP

   © 1999 by Sleeping Bear Press
   Ann Arbor Press is an imprint of Sleeping Bear Press

Andy Milman earned his B.S. in mathematics from Columbia University and his Ph.D. in astronomy from the University of Maryland. His thesis topic was a survey of radio emission from interstellar dust clouds in the 115-GHz line of carbon monoxide. These observations showed their temperatures are about 9 K, making them the coldest physical objects in the universe. Ironically, new stars form in these interstellar dust clouds.

Dr. Milman has worked in many areas of remote sensing of the earth from space, primarily as a contractor to NASA. During this time, he realized the need for a systematic discussion of the problem that is central to this book: inferring the physical properties of a body by measuring the radiation it emits. With a few recent exceptions, all astronomical knowledge has been obtained in this way.

To Adrienne

To Adrienne

Acknowledgments

Quotations from the *Encyclopaedia Britannica* are from the Britannica CD 98® Multimedia Edition © 1994-1997. The title of each article is the name of the person being described.

The quotations that begin each chapter were taken from the *Furman University Mathematical Quotation Server*, edited by Mark R. Woodard, available at http://math.furman.edu/mqs.html, April 26, 1999.

Contents

Preliminary Remarks 1

1. Introduction..7
   1.1 Measurements and noise 10
   1.2 A matrix equation 11
   1.3 Automatic computing 13
   1.4 Noise 14
   1.5 Algorithms 19
   1.6 Systems of linear equations 21
   1.7 An example 22
   1.8 Nonlinear systems 25
   1.9 Iteration 26
   1.10 The rest of the book 27

2. Light and Atoms..29
   2.1 Light and radiometers 30
   2.2 Thermal radiation... 37
   2.3 Absorption and emission of radiation by gases 45
   2.4 Spectral lines.. 50
   2.5 Line broadening 58
   2.6 A remote-sensing example 62

3. Instruments and Noise...67
   3.1 Radiometers 67
   3.2 Noise 70
   3.3 Telescopes.. 74

4. Radiative Transfer...85
   4.1 Derivation 85
   4.2 Formal solution 87
   4.3 Isothermal atmosphere 88
   4.4 Inhomogeneous atmosphere 91
   4.5 Weighting functions... 92
   4.6 Resolution 96

5. Covariance Matrices...99
   5.1 Probability and moments 100
   5.2 Vectors and matrices.. 104
   5.3 Covariance matrices... 114

5.4 Eigenvalues and eigenvectors.. 119

5.5 Singular value decomposition.. 124

5.6 Empirical orthogonal functions 129

5.7 Non-negative definite matrices.. 132

5.8 Noise and the covariance matrix.. 133

5.9 Parallel component analysis 135

5.10 Finding eigenvectors.. 137

5.11 A canonical form for matrix equations... 138

5.12 Rank of the covariance matrix 141

6. Regression...**143**

7. Matrix Solution of Linear Equations **147**

7.1 Noise... 149

7.2 Retrieval.. 150

7.3 Least-squares inverse.. 152

7.4 An example and discussion 157

7.5 Singular value decomposition.. 159

7.6 A geometrical interpretation 160

7.7 Other matrix methods 163

7.8 Pseudoinverses.. 169

7.9 Extension to quadratic terms 170

7.10 Two approaches 174

8. Fourier Transforms **179**

8.1 δ-Functions 181

8.2 Fourier transforms and their properties 182

8.3 Fourier series 192

8.4 Discrete Fourier transform.. 198

8.5 The sampling theorem 199

8.6 Fast Fourier transform 203

8.7 Other transforms 205

8.8 A Fourier transform coloring book.. 207

9. Autocorrelation Functions and Spectra...**215**

9.1 Random variables 216

9.2 Estimating the spectrum 225

9.3 The structure function.. 230

9.4 Filters 233

9.5 Propagating waves 235

9.6 Separating waves from noiselike features 239

10. Integral Equations **247**

10.1 Hilbert spaces 247

10.2 Fredholm equations 249

10.3 Convolution integrals 251

10.4 Noise 253

10.5 Solution when noise is included 254

10.6 Existence, uniqueness, and stability 259

10.7 Eigenfunction expansion 262

10.8 Discussion 276

11. Iteration **279**

11.1 One-dimensional examples 280

11.2 Two perverse examples 286

11.3 Matrix iteration… 290

11.4 ….Is the same as the matrix-inverse solution 297

11.5 Integral equations 304

11.6 Nonlinear equations 310

11.7 Discussion 320

12. Resolution and Noise **323**

12.1 Matrix approach 324

12.2 Integral-equation approach 336

12.3 Discussion 341

13. Convolution and Images **343**

13.1 Deconvolution in the absence of noise 345

13.2 Deconvolution in the presence of noise 348

13.3 Numerical processing to improve the resolution 349

13.4 A Backus-Gilbert approach 353

13.5 Selecting a pattern 357

13.6 An integral-equation approach 360

14. Mathematical Appendix **367**

14.1 Constrained extrema 367

14.2 Vector derivatives of scalars 375

14.3 A matrix identity 378

14.4 Vector and matrix norms 380

14.5 Inequalities 381

14.6 Two useful relationships from statistics 382

14.7 Law of large numbers and the central limit theorem 383

14.8 Hilbert spaces 385

14.9 Orthogonal expansions 393

14.10 Orthonormalization 394

14.11 Summing series 394

14.12 Clenshaw's algorithm 395

14.13 Proof by induction 398

Since my own day on the Mississippi, cut-offs have been made at Hurricane Island, at Island 100, at Napoleon, Arkansas, at Walnut Bend, and at Council Bend. These shortened the river, in the aggregate, sixty-seven miles. In my own time a cut-off was made at American Bend, which shortened the river ten miles more.

Therefore, the Mississippi between Cairo and New Orleans was twelve hundred and fifteen miles long, one hundred and seventy-six years ago. It was eleven hundred and eighty after the cut-off of 1722. It was one thousand and forty after the American Bend cut-off. It has lost sixty-seven miles since. Consequently, its length is only nine hundred and seventy-three miles at present.

Now, if I wanted to be one of those ponderous scientific people, and "let on" to prove what had occurred in the remote past by what had occurred in a given time in the recent past, or what will occur in the far future by what has occurred in late years, what an opportunity is here! Geology never had such a chance, or such exact data to argue from!. Glacial epochs are great things, but they are vague—vague. Please observe:

In the space of one hundred and seventy-six years the Lower Mississippi has shortened itself two hundred and forty-two miles. That is an average of a trifle over one mile and a third per year. Therefore, any calm person, who is not blind or idiotic, can see that in the Old Oölithic Silurian Period, just a million years ago next November, the Lower Mississippi River was upward of one million three hundred thousand miles long, and stuck out over the Gulf of Mexico like a fishing-rod. And by the same token any person can see that seven hundred and forty-two years from now the Lower Mississippi will be only a mile and three-quarters long, and Cairo and New Orleans will have joined their streets together, and be plodding comfortably along under a single mayor and a mutual board of aldermen. There is something fascinating about science. One gets such wholesome returns of conjecture out of such a trifling investment of fact.

Mark Twain,
Life on the Mississippi

Preliminary Remarks

*In my opinion, a mathematician, in so far as he is
a mathematician, need not preoccupy himself with
philosophy—an opinion, moreover, which has
been expressed by many philosophers.*

Henri Lebesgue (1875-1941)

This is a book about remote sensing, the science of measuring the properties of objects we cannot touch, by measuring the radiation they emit or
scatter, or the acoustic waves they transmit. The quintessential problem
in remote sensing is that of the astronomer trying to learn about the stars in the
Galaxy. It will be a long time—if ever—before we can visit a star. Yet by
measuring the light a star emits, we can infer its temperature, composition,
and mass. Remote sensing applies these principles to studies of the earth.
Many parts of the earth are inaccessible to us, either for geographical or political reasons, so we observe the earth from satellites. Seismologists cannot
visit the center of the earth—except in fantasy—but they can learn about it by
measuring the time of arrival of seismic waves at different points on the surface of the earth.

* * *

One problem of being a scientist working in this field is the lack of a recognized name for the discipline. Although I have a Ph.D. in astronomy, I do not
call myself an astronomer because I do not study celestial objects. Neither am
I, by training or inclination, a meteorologist or a geologist. There is no de-
partment of remote sensing as such in any major university that I know of.
Each established scientific discipline, in addition to a body of factual knowledge, has, to some degree, a consistent view of its methods and subject matter.
Most of them have long histories; astronomy, for example, is probably the
second-oldest profession. Every astronomer who is at all philosophical has
considered the legacy left to us by Ptolemy, Copernicus, Kepler, and Newton.
Every physician is indebted to Galen, Vesalius, and Harvey.

Contrasted with this, remote sensing as a discipline has only a very short
history. While seismology dates back to the nineteenth century, remote sensing from space dates only from the advent of weather satellites in the 1960s. A
scientist who does remote sensing does not have a simple job title: I certainly
would not call myself a *remote sensor* (which might sound like someone who
passes on the fitness of distant movies for family viewing), the way someone
who does physics is a physicist or someone who studies the human mind is a
psychologist. When people ask one of my children, "What does your father

1

do?" the reply is likely to be, "He sits in an office and thinks." That is not a very informative answer, but there is no generally understood name for what I do.

If remote sensing has no name for the people who do it, it is not surprising that it does not seem to have an overall philosophy or tradition that is shared by most of the members of the profession. This is, in part, because we come from a wide variety of (recognized) disciplines: astronomy, meteorology, geology, geography, physics, electrical engineering, and many others.

Although they do not seem to be widely known as an organized body of knowledge, there are general methods for developing algorithms—numerical recipes for inferring the value of one physical variable from a measurement of another, related variable—for remote sensing. Even more important, there are methods for estimating how accurate these inferences will be, given a certain instrument and a certain physical situation. We can evaluate how well a certain sensor will work before we build it. This is invaluable when it comes to designing satellite sensors that view the earth from space, since they are very expensive, $100 million or more apiece[1], and it takes many years to design, build, and launch them. Also, by their very nature, such sensors cannot be fixed if they do not work the first time.[2] People working in the field, however, do not seem to share—or even be aware of—a consistent philosophy of remote sensing, a shared way of perceiving either the problems involved or the methods that can be used to solve them.

There is a lot of physics involved in remote sensing. Atomic and molecular physics are needed to understand the absorption coefficients of the gases in the atmosphere. Meteorologists tell us what parameters need to be measured. Climatologists tell us what accuracy we need to achieve. The instruments are always state-of-the-art, not in the least because of the severe constraints on size, weight, and power consumption that are imposed on a satellite. The challenge of building them is paramount to the engineers and engineering managers. Workers in each discipline usually see things from different perspectives and may discount the importance of work in areas that are not their own.

This book is not meant primarily as another discourse on the procedures of remote sensing—and much less on remote-sensing instruments. The real subject is the *philosophy* of remote sensing, the mathematical methods and ways of thinking that will help us to decide which problems can be solved and how best to solve them. The first chapter describes the central problem: any measurements we make are, to a greater or lesser extent, corrupted by errors. This is inescapable. There is no such thing as a perfect measurement. To my way of thinking, the core of all remote-sensing problems is that of understand-

---

[1] Plus launch costs.

[2] The Hubble space telescope may be the exception to this, but the cost of *repairing* it was greater than the combined cost of several weather satellites.

ing the effects of measurement errors and devising sensible ways to deal with them.

Although most of this book deals with the mathematics used in remote sensing, I have tried to keep the discussion as purely mathematical as possible so that the reader who is not familiar with a certain area of physics will not be put in the position of having to learn some (to him) irrelevant physics in order to understand the mathematical ideas being presented. I hope that the material in this book will be useful in many different physical situations.

The second and fourth chapters of this book, however, deal with light and radiative transfer. The goal here is to give the reader who is unfamiliar with remote sensing a feeling for how different processes on the earth affect the amount of light that reaches a satellite orbiting the earth. It is meant to give the reader a feeling for how remote sensing works; it is not meant to make him an expert in the area. I have tried to avoid details—partly, I confess, because I do not have the patience either to read up on them or write about them—in an effort to impart a flavor of the subject without confusing or boring the reader.

The rest of the chapters are entirely mathematical. While I always have a physical situation in mind, the mathematical treatment can be viewed as being completely abstract. My modest aim is to provide, by a systematic treatment of the mathematics involved in making inferences from imperfect data, a useful philosophy that can, at least in a small part, provide a philosophical basis for remote sensing as a discipline.

Most of the papers that I have encountered on this subject are, even for my taste, overly mathematical. I am entirely sympathetic with the scientist or engineer who either avoids reading them from lack of time or background, or who does wade through them, but resents having to do so. Therefore, I have tried, as far as possible, to keep the mathematics simple, while still providing enough rigor to show how the different aspects I discuss are interrelated and to prove some of the assertions that I base the treatment on. While it would be best for the reader to be familiar with linear algebra[3] and Fourier transforms, I have tried to provide enough background information so that those who are not familiar with them can understand the material here. Any reader who needs to is encouraged to consult a standard text on these subjects. Otherwise, no other mathematical background is needed beyond ordinary integral and differential calculus.

While my original intent in writing this book was to present the material on retrieving information from noisy data (as presented primarily in Chapter 7), the book has grown somewhat. I have included some topics that are directly relevant to remote sensing, like the questions of the relationship be-

---

[3] I do, however, expect the reader to understand the basics of determinants, matrix multiplication, matrix inverses, and the like, that are usually covered in high school or an introductory calculus sequence in college.

tween resolution and radiometric sensitivity, while others are more tangential, like the descriptions of Lagrange multipliers and Hilbert spaces. I have included these topics because they will enhance the reader's appreciation of some of the more central topics and make some of the derivations logically complete. They are topics that should be covered in graduate-level math courses but are often overlooked. Some readers may find them useful and—who knows?—even enjoy them.

* * *

This is what you will find in the rest of the book:

Chapter 1 introduces the mathematical problems involved in creating algorithms for making inferences from noisy data. It explains some of the notation and treats a very simple inversion problem: elaboration of this idea is the central theme of this book.

Chapter 2 describes some of the physical processes that make remote sensing possible. It describes how light interacts with matter, which is fundamental to the science of remote sensing.

Chapter 3 describes some fundamental properties of radiometers so that we can see how the instruments we use affect our measurements.

Chapter 4 is a short introduction to radiative transfer in the earth's atmosphere.

Chapter 5 reviews the properties of matrices and discusses covariance matrices and related topics in detail.

Chapter 6 treats regression very briefly and contains some comments on its use in remote sensing.

Chapter 7 gets to the heart of the matter: it discusses matrix methods that are used in inverting systems of linear equations. In particular, it concentrates on characterizing noise and finding optimal linear inversion algorithms. The focus here is finding algorithms that minimize the total error in the inversion. At the end, I show how we can extend these ideas to certain nonlinear problems.

Chapter 8 introduces Fourier transforms and discusses their properties; we need to use this material in the next chapter.

Chapter 9 discusses autocorrelation functions, statistical stationarity, and spectra of random processes.

Chapter 10 discusses integral equations, which provide an alternative viewpoint to the matrix approach to inverting systems of equations. I concentrate on the questions of the stability of the solutions and whether or not solutions exist. Fourier transforms play an important role in this topic.

Chapter 11 discusses iterative methods for inverting systems of linear or nonlinear equations. In it, I include some new results showing that, for systems of linear equations, matrix and iterative methods produce exactly the same results, and discuss some consequences of this.

Chapter 12 introduces the relationship between resolution and noise in the solution of systems of equations, and discusses alternative criteria that might be used to select an optimal algorithm. It discusses the method of Backus and Gilbert. The aim is to integrate their viewpoint into a more comprehensive theory of inversion methods.

Chapter 13 discusses methods for improving the resolution of certain images and fundamental limitations to what can be achieved.

Chapter 14 contains a miscellaneous collection of mathematical material, mostly material that is needed in other chapters. It includes a discussion of probability, Lagrange multipliers, some background mathematics (*e.g.*, a discussion of Hilbert spaces), and other topics.

* * *

This book undoubtedly contains some errors—or some unclear passages—despite my most diligent efforts to root them out. I plan to create a web page where I can post corrections. Anyone who finds an error, has a question, or wants to make a comment, should send me e-mail at *amilman@ieee.org*.

1. Introduction

It is well known that the man who first made public the theory of irrationals perished in a shipwreck in order that the inexpressible and unimaginable should ever remain veiled. And so the guilty man, who fortuitously touched on and revealed this aspect of living things, was taken to the place where he began and there is for ever beaten by the waves.

Proclus Diadochus (412 - 485)
Scholium to Book X of Euclid V.

Few of us who live in cities ever see the stars in a dark night sky. However, if you go out to the Grand Canyon in the middle of a cold winter night, you will see an overwhelming display of stars. There are thousands of tiny lights, some red, some blue, and some yellow or white. The regular movements of the stars, the irregular movements of the planets, the alternation of day and night, and the cycles of the seasons: these things gave our ancestors their first examples of an orderly, albeit mysterious, universe.

The stars are so remote. What are they? To someone who had no electric lights, but also no light pollution to obscure the starlight, these questions must have arisen daily (or rather, nightly). While speculations have come down to us from the earliest Greek philosophers, it was only in the nineteenth century that astronomers could provide some answers to the questions of how large stars are, how hot, and what they are made of. Aristotle taught that all heavenly bodies are made up of a substance (*quintessence*) that is different from those found on earth (earth, water, fire, and air). But the science of spectroscopy has shown us that the sun and stars are made up of the same elements that are found on the earth. People who live in cities rarely notice these things; only astronomers are much concerned about them.

We can only study stars from a distance, of course: even if we could get to the Sun, we could not touch it and survive for very long. All that we know about the stars, we know by studying the light they emit (see Figure 1-1). Astronomers have developed many instruments and theoretical tools for analyzing light that they have applied to studying stars and planets. Recently, we have started to use satellites to observe the earth from space, in the same way we have studied the heavens.

Remote sensing is the science of measuring the properties of objects that we cannot touch by measuring the amount of radiation they absorb, emit, or reflect at various wavelengths. Astronomy was the first remote-sensing sci-

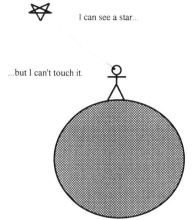

I can see a star...

...but I can't touch it.

Figure 1-1. Finding out about things we cannot reach.

ence. For example, we can measure the temperature of a star—or a hot piece of iron—by measuring its color. Red stars are relatively cool; yellow stars are hotter; and blue-white stars, the hottest. We can determine the chemical composition of a star from the wavelengths at which the coolest part of its atmosphere absorbs light.

On the earth, a similar situation applies. In places where it is convenient, we can measure the atmospheric temperature directly with a thermometer. But there many places where such measurements would not be convenient, or even possible. When we need to know the temperature in places we cannot reach directly, we use remote sensing. This also applies to many other properties of the earth's surface and atmosphere, but I shall use temperature as an example. While I can certainly measure the temperature outside my front door with a thermometer, I cannot get a global set of temperature measurements that way. There are too many places where no one lives to make measurements with a thermometer: the ocean surface, the middle of a desert, or the air one kilometer above the earth's surface. Furthermore, in order to understand and predict weather patterns, we need to know the temperature of the ocean surface everywhere, not just near the coasts or along common shipping lanes.

How do we measure temperatures of places are not accessible by normal means? The ocean surface is so vast that even the idea measuring its temperature everywhere with thermometers lies somewhere between the improbable and the impossible—we would need more than 700 *billion* buoys. Yet, if we are worried about the possibility of global climate change, knowledge of the sea surface temperature is of the utmost importance. And we need to know it everywhere, not just where people happen to live. As another example, we need to know the three-dimensional distribution of temperatures in the atmosphere if we want to forecast the weather accurately. How can we accomplish this?

Meteorologists make routine measurements of vertical temperature profiles (*i.e.*, the air temperature at each height above the surface) by means of *radiosondes*, which are instruments lifted by balloons high into the atmosphere. These instruments are launched twice a day from many places in the U.S. and other countries around the world; they measure tempera-

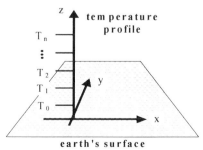

Figure 1-2. Vertical profiles.

ture, dew point, and other variables as they ascend and radio the data back to the surface. Such measurements are very expensive, however, and they are not made in areas where nobody lives. They are indispensable because they provide direct physical measurements of the properties of the atmosphere that are of interest to meteorologists, but they by no means provide adequate spatial or temporal coverage.

There are two areas where remotely sensed data are needed: weather forecasting and studies of global climate change. In each case, data are needed on a global basis on a time scale of a day or a few days. While there are many conventional meteorological measurements made in places where a lot of people live, most of the globe is uninhabited (remember that 70% is covered with water). Furthermore, no one lives in the stratosphere, so no direct measurements can be made there.

On the other hand, we know that changes in one part of the world affect the weather in other parts. The best-known example of this is probably the

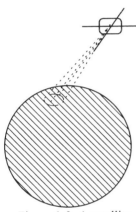

Figure 1-3. A satellite views the earth.

phenomenon called *El Niño*, which is a warming of the waters off the coast of Peru. From time to time— but usually in December, hence the name—the trade winds in the Equatorial Pacific stop blowing. This, in turn, causes a warming of the ocean surface in the Eastern Pacific. In 1982, this warming amounted to as much as 5° C above the average temperature, which is a huge effect. This localized warming near Peru affected the weather worldwide. One example of this is the heavy rains that occurred in Los Angeles in 1983. Another *El Niño*, in 1992, has led to even heavier rains in Southern California in the spring of 1993, even as I am writing this chapter.[1] In the preceding years, without *El Niño*, there was a drought there. If we want to understand the world's weather, we need to measure the meteorological conditions everywhere: over the oceans, in inaccessible mountain regions, and at altitudes high above the surface. However, for obvious practical reasons, there are very few conventional measurements made in these places.

We can make up for this lack of coverage by using measurements made from satellites (see Figure 1-3). Such measurements are indirect: we measure the intensity of infrared radiation emitted from the top of the atmosphere and make inferences about temperature profile[2] in the atmosphere below it. In addition to temperature profiles, we can measure the concentrations of water va-

---

[1] There was another strong *El Niño* in the winter of 1997-1998 that also brought torrential rains; as you can see, I am still writing this chapter.

[2] Here, and elsewhere, *profile* means a vertical profile in the atmosphere of temperature or some other quantity.

por and other gases remotely. But I have to emphasize that satellite instruments can only measure the infrared (or visible or microwave) radiance above the earth's atmosphere. These quantities are, in and of themselves, of no particular interest: their sole value lies in our ability to make inferences about the state of the atmosphere from them. This can be done with an algorithm that takes, as its input, the measured radiances and produces an estimate of the temperature profile in the atmosphere below it. That is the subject of this book.

Very similar mathematical problems arise in the field of seismology, where geophysicists want to determine the structure of the interior of the earth from measurements of the propagation of sound waves through it. Their problem is even more difficult than the meteorologists', who can at least make some direct measurements of the state of the atmosphere with instruments, say, borne by balloons. Geophysicists cannot bore even a single test hole 1000 km deep. However, since the speed of sound depends on the temperature, pressure, and density of the material through which it passes, measurements of the time it takes for a disturbance in one part of the earth to travel to the surface at many different places can be analyzed to give information about the structure of the earth's interior. The mathematics of this geophysical remote-sensing problem are very similar to those of the meteorological problem that I discuss throughout this book. In fact, some of the methods—a good example is the Backus-Gilbert method of solving integral equations—were developed by seismologists.

1.1 Measurements and noise

Every measurement contains some error: no measurement is perfect. I shall call anything that introduces a random error into a measurement a source of *noise*. The amount of noise and its nature will determine, in part, the usefulness of any measurement. I first encountered this problem when I was trying to develop an algorithm for determining temperature and water vapor profiles from satellite data. These data consisted of measurements made at several different wavelengths in the infrared part of the spectrum. I was somewhat surprised to find that there was no general discussion in the research literature of the effects of noise on the inferred temperature profiles, and I had to develop for myself many of the techniques I present in this book. I hope that this book will aid other researchers in understanding how to develop algorithms for making inferences from noisy measurements, both in the field of remote sensing and in other fields.

1.2 A matrix equation

The center of attention of this book will be a deceptively simple matrix equation

$$\mathbf{y} = \mathbf{A}\mathbf{x}. \tag{1}$$

Throughout this book I use bold lowercase letters to represent vectors, and bold uppercase letters to represent matrices.

There are three possibilities concerning equation (1):

1. If both $\mathbf{A}$ and $\mathbf{x}$ are known, it is trivial to find $\mathbf{y}$.

2. If we have many *pairs* of vectors $(\mathbf{x}_i, \mathbf{y}_i)$, we can determine the matrix $\mathbf{A}$. This process is called *linear regression*, and there are many books written about it. We should note that the $\mathbf{x}_i$'s or the $\mathbf{y}_i$'s will usually represent some measurements and therefore contain some noise, or errors in the measurements. If these errors are random and uncorrelated with the variables $\mathbf{x}$ and $\mathbf{y}$, however, their effect can be made vanishingly small by using enough pairs $(\mathbf{x}_i, \mathbf{y}_i)$ so that the noise averages out. Most books on linear regression do not mention noise at all.

3. The last possible application of equation (1) is when $\mathbf{A}$ and $\mathbf{y}$ are both known, and we must estimate the value of $\mathbf{x}$ that gave rise to the $\mathbf{y}$ we measured. In particular, I assume that $\mathbf{A}$ is known from an analysis of the physics of the problem under consideration, while $\mathbf{y}$ is a measured quantity (and therefore, contains some error). If the $\mathbf{y}$'s were known to be error-free, we could invert equation (1) with little trouble or thought, using pseudoinverses if necessary. The effects of noise, however, make the subject much more interesting. We will find that there are many possible inverses of $\mathbf{A}$, and that the noise will affect different ones differently. I shall show that the amount of noise in a measurement will, in large part, determine which inverse we should use.

1.2.1 Notation and nomenclature

A short digression is in order. We measure a value of a vector quantity $\mathbf{y}$ and want to infer the corresponding value of some other vector quantity $\mathbf{x}$. In the discussion that follows, we identify $\mathbf{x}$ with the vector (the notation will be explained shortly) $\mathrm{col}(x_1, x_2, \ldots)$. We assume that $\mathbf{A}$ is a known matrix. Here, as elsewhere, I want to avoid sounding pedantic, so from now on, I will refer to the process of finding the value of $\mathbf{x}$ by saying that we *retrieve* the value of $\mathbf{x}$, or simply that we retrieve $\mathbf{x}$. In this sense, "retrieve" is a synonym for "infer." Similarly, I will say that we measure $\mathbf{y}$, or refer to the measurement of $\mathbf{y}$. In other words, we shall discuss the *retrieval* of $\mathbf{x}$ from the *measured* value of $\mathbf{y}$.

In this sense, the symbols **x** and **y** may represent various physical quantities, not their values. We say equivalently that we *invert* a set of equations.

We need to distinguish the true value of **x** from the retrieved value, which I denote by $\hat{\mathbf{x}}$. Throughout this book, I represent a retrieved or estimated quantity with a superior hat,[3] and a measurement that includes noise, with a prime. The recipe for calculating $\hat{\mathbf{x}}$ from **y**′ is an *algorithm*: a prescribed sequence of arithmetic operations that will, after a finite number of operations, produce an answer. Equation (1) is linear; the inversion algorithm will also be linear. Even within this limited framework, there is a lot to be said about different retrieval methods, or different inversion algorithms. If we replace equation (1) with a nonlinear vector equation, we may want to find a linear inversion algorithm, or we may want to find a nonlinear one.

Depending on the system of notation, we could define vectors to be *row vectors*, in which case we must write **y** = **xA**, or we could define them a column vectors, and write **y** = **Ax**. I use the second system here, because it is more familiar, but column vectors are awkward to write, so I write the components in a row, with the notation col(x_1, ..., x_n) to indicate that it's a column vector.

1.2.2 The basic problem

Baldly put, the problem we need to solve is that we want to know the value of a vector **x** from measurement of a different vector **y**, where we know that they are related by equation (1). However, there is always an error in the measurement of **y**, so the measured value we have to work with is

$$\mathbf{y}' \;=\; \mathbf{y} + \mathbf{n} \;=\; \mathbf{Ax} + \mathbf{n} . \tag{2}$$

Here, **n** is a vector of random variables that changes with every measurement; we assume that it is uncorrelated with **y** and has zero mean. I shall generally assume that each component has the same variance σ_N^2. Furthermore, we have many measurements of different **y**'s coming into our computer every second and for years on end, so whatever we do to estimate the values of the **x**'s, we have to do automatically, without human intervention.

On the surface, this might look a lot like signal processing, but it is quite different. The basic difference is that in signal processing, there is a time-ordered series of different states of the device that receives the signal. For example, a digital transmission line produces a series of 1's and 0's which represents a message with some noise added. We use the time history to decode the message. Once we've decoded the message, we're done.

---

[3] A diacritical mark, not a piece of apparel.

In remote sensing, however, we measure a vector that represents several measurements made at one location and time. In general, the measurements at different times and places are uncorrelated, so each one has no history. All we have to work with is the measured value $\mathbf{y}'$ and, possibly, some ancillary information about the place where we are looking (*e.g.*, are we looking down at land or ocean?). Furthermore, the value of $\mathbf{y}$ is immaterial; we need to use our knowledge of the physics involved to infer the value of $\mathbf{x}$ from $\mathbf{y}'$. I have never found signal-processing theory useful for remote-sensing problems.

1.3 Automatic computing

There was a time when there were no computers—computations were always done by hand. A *computer* was a person who was employed to perform calculations for someone else. My 1905 *Webster's Imperial Dictionary* defines a computer as "one who computes; a reckoner; a calculator.[4]" In such an environment, one did not explore all of the possible numerical solutions to a problem. Life wasn't long enough. In the earliest days of satellite remote sensing, scientists tried inverse methods that were unacceptable because they were unstable. There is an infinite number of possible solutions to most such problems and people did not realize, at first, that they needed to provide a way to choose among them. Because solving matrix equations is so time-consuming without computers, no one had tried to apply a matrix inverse to thousands of measurements. When they tried it, they found that the straightforward approach didn't work.[5] Twomey, in the introduction to his book, *Introduction to the Mathematics of Inversion in Remote Sensing and Indirect Measurements* (1977), wrote

> Inversion problems have existed in various branches of physics and engineering for a long time, but in the past ten or fifteen years they have received far more attention than ever before. The reason, of course, was the arrival on the scene of large computers (which enabled hitherto abstract algebraic concepts such as the solution of linear systems of equations in many unknowns to be achieved arithmetically with real numbers) and the launching of earth-orbiting satellites....
>
> It was soon found that the dogmas of algebraic theory did not necessarily carry over into the realm of numerical inversion and some of the first numerical inversions gave results so bad—for example, in one early instance negative absolute temperatures—that the prospects for the remote sensing of atmospheric temperature profiles, water vapor concentration profiles, ozone profiles, and so on, for a time looked very bleak, attractive as they might have appeared when first conceived.
>
> The crux of the difficulty was that numerical inversions were producing results which were physically unacceptable but were mathematically acceptable (in the sense that had they existed they would have given measured values identical or almost identical with what was measured). There were in fact ambiguities—the computer was being "told" to find an $f(x)$ from a set of values for $g(y)$ at prescribed values of y, it was "told" what the mathematical process was which related $f(x)$ and $g(y)$, but it was not "told" that there were many sorts of

---

[4] Again, a person: no batteries were needed.

[5] If it had worked, I would never have written this book.

$f(x)$—highly oscillatory, negative in some places, or whatever—which, either because of the physical nature of $f(x)$ or because of the way in which direct measurements showed that $f(x)$ usually behaved, would be rejected as impossible or ridiculous by the recipient of the computer's "answer". And yet the computer was often blamed, even though it had done all that had been asked of it and produced an $f(x)$ which via the specified mathematical process led to values of $g(y)$ which were exceedingly close to the initial data for $g(y)$ supplied to the computer.

    For a time it was thought that precision and accuracy in the computer were the core of the problem and more accurate numerical procedures were sought and applied without success. Soon it was realized that the problem was not inaccuracy in calculations, but a fundamental ambiguity in the presence of inevitable experimental inaccuracy—sets of measured quantities differing only minutely from each other could correspond to unknown functions differing very greatly from each other. Thus in the presence of measurement errors (or even in more extreme cases computer roundoff error) there were many, indeed an infinity of possible "solutions". Ambiguity could only be removed if some grounds could be provided from outside the inversion problem, for selecting one of the possible solutions and rejecting the others. These grounds might be provided by the physical nature of the unknown function, by the likelihood that the unknown function be smooth or that it lie close to some known climatological average. It is important, however, to realize that most physical inversion problems are ambiguous—they do not possess a unique solution and the selection of a preferred unique solution from the infinity of possible solutions is an imposed additional condition. The reasonableness of the imposed condition will dictate whether or not the unique solution is also reasonable. There have been many advances in mathematical procedures for solving inversion problems but these do not remove the fundamental ambiguity (although in some instances they very effectively hide it). Our measurements reduce the number of possible solutions from infinity to a lesser but still infinite selection of possibilities. We rely on some other knowledge (or hope) to select from those a most likely or most acceptable candidate.

Twomey takes the first step, the realization that it is necessary to choose the best solution from the infinity of solutions that fit the measurements to within the errors in the measurement, but he does not, in my opinion, emphasize the effects of measurement noise sufficiently. In fact, *noise* does not even appear in the index. I believe that the noise has to be considered from the beginning, and any sensible inversion algorithm should be designed to minimize the effects of noise.

1.4 Noise

What do we mean by *noise*? I offer a very broad definition of noise, one that encompasses not only unwanted signals generated in the instrument that we are using to make a measurement, but also the effects that extraneous factors have on our measurements.

Suppose we had an instrument of some kind that measured a certain quantity m: we wish to infer the value of some other quantity q from this measurement. There are many other factors, however, that also affect the measurement: the quantities $r, s,...,t$ may also affect the value of m. I call *noise* the effect that all of the quantities, other than the quantity q that we specifically wish to measure, have on the measurement of the quantity m.

We can divide noise sources into two types. The first is internal, in that the error in the measurement arises inside the instrument that performed the

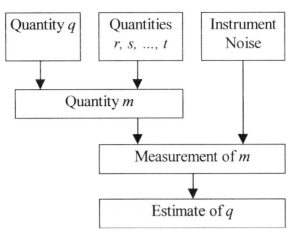

Figure 1-4. The effects of noise. The quantities q, r,
s,..., t all affect the measured value m.

measurement. For instance, an infrared or microwave radiometer will generate thermal noise[6] internally. Internal noise can usually be controlled by the design of the instrument; the noise in a radiometer, for example, might be reduced by cooling some critical components. There are, however, fundamental limits to as to how much the noise can be reduced; usually, instruments are designed so that the noise is not insignificant, but not an overwhelming problem, either. Sometimes it is not possible to reduce the noise to acceptable levels, in which case it might not be practical—or even possible—to build a particular instrument.

The other type of noise is external. It arises because there are other things, besides the quantity we want to retrieve, that affect the quantity we are measuring. I show this schematically in Figure 1-4, where we want to infer q from a measurement of m. The top box on the left contains the variable we want to retrieve, and the middle box contains other factors that affect m. The instrument noise affects the measurement of m, but not the value of m itself, while changes of r, s,..., t do affect the value of m. Finally, we estimate q from this the measured value of m. Note the distinction between the top middle and right boxes: the "other quantities" affect the true value of m; the noise only affects the measured value of m, not its true value.

1.4.1 An example

As a concrete example of these two kinds of noise, imagine that we have a mercury barometer: we determine the air pressure by measuring the height of a mercury column. This measurement itself has some error involved; this is what I meant by an internal error in the paragraph above. This noise arises

---

[6] The term *thermal noise* refers to radiation emitted by some part of the instrument that is performing the measurement. I discuss black body emission in Chapter 2.

from errors in measuring the height of the mercury column, perhaps because of errors in relating its height to a ruler or some other kind of measuring instrument. This error arises inside the barometer (which includes the ruler or whatever other device we use to measure the height of the mercury column). Perhaps we could reduce this error by substituting a laser interferometer for the ruler. Such errors affect the accuracy of the measurement, but they do not affect the actual height of the mercury column (see Figure 1-5).

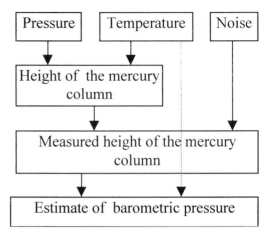

Figure 1-5. Effects of temperature and barometric pressure on the estimated pressure.

On the other hand, in addition to the barometric pressure, the ambient temperature might also affect the height of the mercury column. This is external noise, since it did not affect just the measurement, but it actually changed the height of the mercury column.

Well, suppose I know the error in measurement of the height h of the mercury column; call the error variance χ_h^2. (I will use the symbol χ^2 to represent an *error variance*; I will use σ^2 to represent the variance of a physical quantity.) By *error variance* I mean the following. Let $\hat{u}$ be the estimated value of the barometric pressure, and u its true value.[7] Suppose we make N independent measurements holding the pressure constant; the error variance is

$$\chi_x^2 = \frac{1}{N}\sum_{k=1}^{N}\left(\hat{u}_k - u_k\right)^2. \qquad (3)$$

Suppose that the functional relationship between h and the barometric pressure u is

$$h = cu, \qquad (4)$$

where c is a constant of proportionality that depends on the density of mercury relative to air. We might estimate the value of the barometric pressure $\hat{u} = h/c$. The error in $\hat{u}$ depends on the error in the measurement of h. From equa-

---

[7] Since we shall need to use the symbol p for probability distributions, I need to use a different symbol for pressure.

tion (4), $dh/du = c$, so the error (from now on *variance* is understood) in the air pressure, χ_u^2, is given by[8]

$$\chi_u^2 = \frac{1}{c^2} \chi_h^2. \qquad (5)$$

The propagation of errors in this fashion is discussed in innumerable textbooks. In principle, we can make the error in $\hat{u}$ as small as we want if we can make the error in the measurement of h small enough. What is "small enough" will depend on the value of the constant c.

We can also correct for errors like changes in the air temperature or the sun shining on the barometer and heating it directly. In this case, we need to know the mean temperature $<T>$ and the inherent variance of the temperature, σ_T^2. Suppose, for the moment, we do know how large σ_T^2 is.[9] Then, considering the temperature changes to be a source of noise when we want to know the barometric pressure, the height of the mercury column is related to the temperature by

$$h - h_o = \alpha_{Hg}\left(T - \langle T \rangle\right), \qquad (6)$$

where h_o is the height of the mercury column when the temperature has its average value $<T>$ and α_{Hg} is the thermal coefficient of expansion of mercury. The variance of the column height due to changes in air temperature is

$$\chi_u^2 = \alpha_{Hg}^2 \, \sigma_T^2. \qquad (7)$$

We could use a thermometer to measure the temperature and correct the height of the mercury column.

Note that we measure two quantities, temperature and column height, and infer the barometric pressure from them. There are two possible ways to view this process. One is that we use the temperature measurement, $\hat{T}$, to correct the measured column height, $\hat{h}$. We make a correction $\delta h = b\hat{T}$, were b is a constant, and estimate u from $\hat{u} = (\hat{h} + \delta h) / c$. Alternatively, we can view the process as simply being one of making two measurements and estimating $\hat{u} = (\hat{h} + b\hat{T})$. These two viewpoints obviously lead to exactly the same result. I shall always take the latter viewpoint, that we do not differentiate in the mathematical treatment between variables of direct interest to us (h) and variables (T) that interest us only because they affect the measurements (while the

---

[8] I discuss this in detail in Chapter 5.

[9] I am assuming here that temperature and pressure are statistically independent (although they are not, in reality).

ambient temperature might be just as important as the barometric pressure, the temperature of the barometer itself has no *meteorological* significance).

So now we are ready to calculate the errors? Not quite. In order to model the problem correctly, we need to know the *statistics* of the quantities involved. Assume that the temperature, barometric pressure, and so forth are all Gaussian[10] distributed random variables. That is, the probability of measuring a value for the temperature between T and $T+dT$ is

$$p(T)\,dT = \frac{1}{\sqrt{2\pi}\sigma} e^{-(T-<T>)^2/2\sigma^2}\,dT, \tag{8}$$

where $p(T)$ is a probability distribution and σ^2 is the variance. We also need to know the covariance of u and T, the true pressure and temperature, to specify the problem completely.

I should say a word about the variance here. Suppose we have an infinite set of temperature measurements taken where we have placed the barometer. Let T_k be the kth measurement of the temperature there. This set of measurements will have a *population mean* $<T>$ defined by

$$\langle T \rangle = \underset{N\to\infty}{Lim} \frac{1}{N} \sum_{k=1}^{N} T_k \tag{9}$$

and a population variance defined by

$$\text{var}(T) = \underset{N\to\infty}{Lim} \frac{1}{N} \sum_{k=1}^{N} \left(T_k - \langle T \rangle\right)^2 . \tag{10}$$

If we want to estimate the population variance based on a sample of the parent population, we use the sample variance[11]

$$\text{var}_{sample}(T) = \frac{1}{N-1} \sum_{k-1}^{N} \left(T_k - \langle T \rangle\right)^2 . \tag{11}$$

Throughout this book, I shall only deal with the population variance, rather than the sample variance, so the denominators will always contain N, rather than $N-1$. This makes the notation simpler, and, when the number of samples is large, the difference is immaterial, anyway.

---

[10] The *normal* distribution was discovered by Carl Friedrich Gauss (1777-1855), the German mathematician, and usually bears his name.

[11] Since we do not use the sample variance for anything, we do not need a special symbol for it.

1.4.2 Mean values

For many quantities, the value we call zero is purely arbitrary (there is a physical meaning to 0 degrees Kelvin, but not to 0 degrees Celsius). For the remainder of this book, for any variable u, it will be convenient to substitute u - $<u>$ for u in every equation. Therefore the numerical value of any variable will refer to the difference from its mean value. This simplifies the equations, but it does not make them less general. The reader will have to restore the mean values of the variables in any calculations he might do, and he should be careful in situations where the mean value changes from one set of calculations to another.

1.5 Algorithms

This isn't a detective novel, so I can reveal the ending at the beginning. It may help the reader if he knows what to look out for before he starts working on the balance of the material in this book.

Many people talk about algorithms, but I'm sure that many of them mean different things by that word. Let me propose my own definition here, as it applies to us. An *algorithm* is a mathematical recipe for estimating the value of a set of quantities $\mathbf{x} = \text{col}(x_1, \ldots, x_m)$, which are not directly measurable, from measurements of some other quantities $\mathbf{y} = \text{col}(y_1, \ldots, y_n)$.[12] Many, but no means all, algorithms are iterated, which means that you perform a calculation on the measured data to get an initial estimate of $\mathbf{x}$; then, using that result, perform another calculation to get an improved estimate of $\mathbf{x}$. Keep doing this a predetermined number of times, or until the sequence of estimates converges. Alternatively, an algorithm may simply perform a single calculation on the data that produces the final result. Either way, an algorithm needs to have two properties for it to be useful: 1) it must stop after a finite number of steps; and 2) it must converge to the correct solution in all but very rare instances.

1.5.1 Optimal algorithms

For any application, there will generally be more than one possible algorithm.[13] We would like, in general, to select the best possible algorithm. But in order to do this, we must have a criterion for deciding which is one is best. In most cases, we will use the one that produces the smallest r.m.s. error. While this is sensible, however, there are other possible criteria that could be used.

---

[12] Throughout this book, "col(p_1, …, p_m)" denotes a column vector whose elements are p_i. See Chapter 5.

[13] Unless there is *no* algorithm.

We might, for instance, want the algorithm that produces the *smoothest* solution (if the quantity we are estimating is a vector); in Chapter 12, we shall minimize the *spread* of the solution—how localized it is. So before you decide which is the best algorithm, you need to state explicitly what criterion you are using. Note that *criterion* is singular: there can only be one (at a time).

Next, which algorithm is best depends on the statistics of the independent variables. I will be much more specific later (in Chapter 7) about the detail, but suffice it to say for now that we need to know which are the most likely values of the independent variables $\mathbf{x}$ and which are the least likely. In general, an algorithm will not work equally well for all possible values of $\mathbf{x}$, and we want it to work best for the most common values. I will say more about this later, but for now let me note that the statistics of $\mathbf{x}$ might change with location or with time. This also might affect which algorithm that we choose.

Lastly, we would like to have a good physical and a mathematical model of the situation involved. With the mathematical model in hand, we can usually calculate what algorithm will work best and what the error will be when we use it. This gives us a systematic method for considering *all* possible algorithms (in a certain class) and selecting the best. In some cases, we have no adequate model (usually because we don't know the relevant physics very well) and we have to use statistical methods to find the algorithm. We can use pairs $(\mathbf{x}_i, \mathbf{y}_i)$ of measured values of the independent variable $\mathbf{x}$ and the dependent variable $\mathbf{y}$ (where i runs over the number of pairs) and use regression techniques to find the appropriate coefficients.[14] This latter approach has many pitfalls: the biggest one is that there will be errors in the measurements of both $\mathbf{x}$ and $\mathbf{y}$, and also, the measurements of $\mathbf{x}_i$'s will not coincide exactly with the measurements of the corresponding $\mathbf{y}_i$'s. All of these effects tend to blur the true relationship between $\mathbf{x}$ and $\mathbf{y}$. Also, the resulting algorithm will be rather sensitive to the statistical distribution of the components of $\mathbf{x}$. In particular, if we apply the algorithm to a set of data with different statistics, the result will be poor to truly awful. Whenever we can, we would prefer to rely on a good physical and mathematical model, where few of the constants involved will need to be determined from measurements.

Often, engineers derive algorithms intuitively: "I think that we should use this combination of the measured variables y_i and these coefficients...." Unless you are a magician or have a large collection of rabbits' feet, you can grope about forever without finding the optimal algorithm. And without a mathematical model, you can never prove that that particular algorithm is optimal. This book provides the mathematical tools and insights to evaluate an entire set of algorithms and pick the one that is optimal.

---

[14] See Chapter 6.

1.5.2 Pieces of information

We all learned in high-school algebra that, if there are n variables x_i, we need n simultaneous equations if we want to solve for these n variables. This is certainly true in remote sensing. Suppose that we have n physical quantities that influence our measurements of y_i, $i = 1, \ldots, m$. Then the problem is underdetermined unless $m \geq n$. But suppose that we only want to retrieve the value of only one of the x_i, say x_1 for the sake of discussion. If $m \geq 1$, we should be able to solve for one variable, x_1. No problem here.

But now we say, "let's solve for x_2 also." That's fine, assuming that $m \geq 2$. When we solve for x_2, we must remember that we have removed one degree of freedom from the set of measurements of the y's. This same consideration applies each time we want to retrieve another x_i. Therefore, we must be sure that we have enough *independent* measurements (y's) to account for each physical variable that affects the measurements. There may be a tendency to ignore this fact and try to write n algorithms to retrieve n variables from m independent measurements, where we treat each algorithm separately, and not realize that the procedure is invalid if $m < n$.

I will define exactly what I mean by *independent* measurements in Chapter 7. I'm sure that the reader has at least an intuitive idea of what this means. However, the y's we measure may not all be independent: in fact, they rarely are. Then if we simply count n x's and m y's (without regard to independence), we might be overestimating the number of independent measurements: in general, it will be smaller than m.

Let me put this another way. Given n independent variables and m dependent variables that we measure, we must analyze the single algorithm that retrieves all of the n independent variables simultaneously, rather than n independent algorithms that retrieve only one independent variable each.

1.6 Systems of linear equations

An astute reader may have objected by now that I am dealing with systems of linear equations, and that I am assuming that it can represent any physical remote-sensing problem. I know that most radiative transfer problems are inherently nonlinear, but there is no general inverse for a nonlinear set of equations. Systems of linear equations are difficult enough to deal with. I have several reasons for being satisfied, for the moment at least, with the linear form of equation (1).

As I will show in this book, I can devise an optimal linear retrieval algorithm for a linear process and determine what the errors will be in the retrievals. There is no way to perform a general nonlinear analysis of the errors involved. Except for linear problems, I know of no way to find what is provably an optimal solution.

Nonlinear algorithms might produce smaller errors overall than linear ones, but there may be situations where the errors in the nonlinear algorithms become enormous. Even if these situations are infrequent, they may make the retrieved variables untrustworthy. It is sometimes hard to know where a nonlinear algorithm will work well, and where it won't. In many cases, if we find that a linear algorithm is inadequate, a given nonlinear algorithm may only make matters worse. As we shall see, most nonlinear problems are solved by iteration, which may be hard to analyze completely. However, I deal with some iterated procedures in Chapter 11 that we can analyze completely, and I discuss a method for solving systems of quadratic equations at the end of Chapter 7.

Given an inherently nonlinear problem, we may be able to linearize it without making much of an error. Let $\mathbf{x}$ be the vector of n variables we wish to measure; let $\mathbf{y}$ be the vector of m measurements that we make. To be completely general, represent the relationship between $\mathbf{y}$ and $\mathbf{x}$ by

$$\mathbf{y} = f(\mathbf{x}), \tag{12}$$

where f is a nonlinear, vector function of $\mathbf{x}$. Now expand f in a Taylor series about $\langle \mathbf{x} \rangle$: let

$$\alpha_{ij} = \frac{\partial y_i}{\partial x_j}; \tag{13}$$

then

$$\mathbf{y} = \langle \mathbf{y} \rangle + \mathbf{A}\left(\mathbf{x} - \langle \mathbf{x} \rangle\right) + \cdots, \tag{14}$$

where the elements of the matrix $\mathbf{A}$ are α_{ij}. Setting $\langle \mathbf{y} \rangle$ and $\langle \mathbf{x} \rangle$ to zero, and neglecting the higher-order terms, this is just equation (1). In other words, we can linearize f, and we will find an inverse to this linearized equation. How well this will work will depend on just how nonlinear $f(\mathbf{x})$ is.

1.7 An example

I shall illustrate these concepts with a one-dimensional example. Suppose $y = \alpha x$. Given y, what is the value of x? Let n be the error in the measurement, and denote its variance by σ_N^2. Assuming that $<x> = <y> = <n> = 0$, we need to find a constant β that satisfies

$$\hat{x} = \beta(y + n). \tag{15}$$

The variance of $\hat{x}$ is

$$\mathrm{var}(\hat{x}) = \beta^2\left[\mathrm{var}(y) + \mathrm{var}(n)\right]$$

$$= \beta^2\left(\sigma_y^2 + \sigma_N^2\right), \tag{16}$$

where σ_y^2 is the variance of y. Also, assume that the covariance of y and n, $\mathrm{cov}(y,n) = 0$.[15] In the more general case where $y = f(x)$ is not linear, we can expand it in a Taylor series and keep only the first term:

$$y = f(x) \approx x\frac{\mathrm{d}f}{\mathrm{d}x} = \alpha x, \tag{17}$$

where $\alpha = \mathrm{d}f/\mathrm{d}x$ evaluated at $x = 0$ ($= <x>$). I shall take up the nonlinear case in later chapters.

Since $y = \alpha x$ and $<yn> = 0$

$$\chi_x^2 = \left\langle(\hat{x} - x)^2\right\rangle = \left\langle\left[\beta(y + n) - x\right]^2\right\rangle$$

$$= \beta^2\left(\left\langle y^2\right\rangle + 2\left\langle yn\right\rangle + \left\langle n^2\right\rangle\right) - 2\beta\left(\left\langle xy\right\rangle + \left\langle xn\right\rangle\right) + \left\langle x^2\right\rangle$$

$$= \beta^2\left[\alpha^2\sigma_x^2 + \sigma_N^2\right] - 2\alpha\beta\sigma_x^2 + \sigma_x^2 = \sigma_x^2(\alpha\beta - 1)^2 + \beta^2\sigma_N^2. \tag{18}$$

Therefore, the error in $\hat{x}$ is the sum of two terms: one is proportional to the noise σ_N^2; the other, to σ_x^2. These can be interpreted as follows.

The first term represents the error due to not inverting equation (1) exactly. This error is minimized by setting $\beta = 1/\alpha$. The second term represents the effect of noise. If $\beta > 1$, the noise will be amplified, while if $\beta < 1$, it will be diminished. We would like to make β as small as possible to minimize the effects of noise. Both of these conditions cannot be satisfied at once unless $\sigma_N^2 = 0$. To minimize $\chi_x^2 = \mathrm{var}(x - \hat{x})$, take the derivative of χ_x^2 with respect to β and set it equal to zero.

$$\frac{\mathrm{d}}{\mathrm{d}\beta}\mathrm{var}(\hat{x} - x) = 2\alpha(\alpha\beta - 1)\sigma_x^2 + 2\beta\sigma_N^2 \tag{19}$$

so the variance is minimum when

$$0 = 2\beta\sigma_N^2 + 2\alpha(\alpha\beta - 1)\sigma_x^2 \tag{20}$$

---

[15] I discuss variance and covariance in Chapter 5.

or

$$\beta = \frac{\alpha \sigma_x^2}{\sigma_N^2 + \alpha^2 \sigma_x^2}. \tag{21}$$

Note that $0 \le \beta \le 1$. The error is then [substitute equation (21) into (18)]

$$\chi_x^2 = \mathrm{var}(\hat{x} - x) = \frac{\sigma_N^2 \sigma_x^2}{\sigma_N^2 + \alpha^2 \sigma_x^2}. \tag{22}$$

When the noise becomes large, $\beta \to 0$ and $\chi_x^2 \to \sigma_x^2$; when the noise is small, $\beta \to 1/\alpha$ and $\chi_x^2 \to 0$. In the first case, we have no information about x, and we should always estimate x by $\hat{x} = <x>$, regardless of the data. Note that no sensible algorithm will have $\chi_x^2 > \sigma_x^2$ so this is the worst we can do. If $\sigma_N^2 = 0$, we have a perfect retrieval (needless to say, this happens very rarely).

In order to analyze this 1-dimensional problem, we need to know σ_x^2, σ_N^2 and $f(x)$ (and its derivative $\alpha = df/dx$). That β depends on σ_x^2 is quite important: depending on σ_x^2, the error χ_x^2 can be anywhere between σ_x^2 and 0. It is often very difficult to get statistical information about the quantities we want to measure (which is one reason why we need remote sensing in the first place), but some guess, at least, is needed before we can discuss the retrieval process intelligently.

To quantify how much the retrieval has improved the state of our knowledge of x, consider the *fraction of the variance explained*. If the variance of x is σ_x^2, and the error in $\hat{x}$ is χ_x^2, the part of the variance of x that we have not accounted for is χ_x^2. The part that we *have* explained is $\sigma_x^2 - \chi_x^2$. If we neglected the data and just set $\hat{x} = <x>$, the error would be σ_x^2; $\chi_x^2 > \sigma_x^2$ only if we are particularly foolish in our choice of an algorithm. Then the fraction of the variance that is explained

$$f_v = \frac{\sigma_x^2 - \chi_x^2}{\sigma_x^2} = \frac{\alpha^2 \sigma_x^2}{\sigma_N^2 + \alpha^2 \sigma_x^2} \tag{23}$$

[using equation (22)], is always ≥ 0. This represents the amount of information we have obtained about x from our retrieval. When $\sigma_N^2 = 0$, $f_v = 1$: in the absence of noise, we can explain everything. On the other hand,

$$\lim_{\sigma_N^2 \to \infty} f_v = 0. \tag{24}$$

That is, when the data are very noisy, we learn next to nothing.

1.8 Nonlinear systems

Equation (21) represents an optimal inverse for equation (1). Since β was chosen to minimize the error variance χ_x^2, no other *linear* algorithm can give a better result: a search for a better linear algorithm would be pointless. It may not be clear, until we have taken the time to understand the derivation, that there cannot be a better linear algorithm somewhere. I shall show that this is also true for multidimensional problems.

We have, so far, limited ourselves to using linear inversion algorithms. Such algorithms are obviously sufficient for truly linear problems. But for nonlinear problems, there is always the temptation to resort to nonlinear algorithms if the linear algorithm is perceived not to be adequate. There are two kinds of nonlinear problems and they are treated differently. For one kind, the fully nonlinear problem can be reduced to a linear one, while for the other kind, linearization can only be an approximation.

Suppose that the (nonlinear) function $f(\mathbf{x})$ can be written

$$f(x) = \operatorname{col}\left[f_1(x_1),\ldots,f_n(x_n)\right] \tag{25}$$

where each of the functions f_v depends only on the variable x_i. In this case, define

$$q_i = f_i(x_i); \tag{26}$$

and retrieve the values of $\mathbf{q}$. This linearizes the system of equations, although now the algorithm will depend on $<\mathbf{q}>$, since the variance of an ensemble of $\mathbf{q}$'s will depend on the mean value. In addition, the statistics of $\mathbf{q}$ will not be Gaussian, so the optimal algorithm could depend on moments of $\mathbf{q}$ higher than the second.

As an example, we might consider a problem where we know, from an analysis of the physics involved, that y is related to x by

$$y = ax + bx^2 \tag{27}$$

where a and b are constants other than zero. Although $<x> = 0$, $<y> = <b^2x^2> = b^2\operatorname{var}(x)$. This form is completely general; I can always define $y' = y + c$ to remove any explicit constant term.

Since

$$ax + bx^2 = b\left(x + \frac{a}{2b}\right)^2 - \frac{a^2}{4b}, \tag{28}$$

we can always transform the problem by defining

$$t = \left(x - \frac{b}{2a} \right)^2 \tag{29}$$

and

$$z = y + \frac{a^2}{4b}; \tag{30}$$

and using the relationship

$$z = bt. \tag{31}$$

Therefore, any quadratic problem in one variable can be transformed to a linear one, as long as we know the values of the constants a and b. Other forms can also be reduced to equivalent linear equations. I pursue this line of thought further in §7.9.

1.9 Iteration

During the 1970s, there were many articles about the accuracy of temperature profiles that could be obtained from certain satellite instruments. Often, writers would show a figure that compared a few temperature profiles retrieved from satellite data with corresponding radiosonde profiles. There would be little in the way of convincing statistics about the accuracy of the retrievals. One supposes that the statistics would not have looked very promising.

There was a lively discussion of different kinds of retrieval algorithms and which ones would give the best results. The late Henry Fleming of the National Oceanic and Atmospheric Administration (NOAA) published a short paper comparing what he called matrix algorithms with iterated algorithms.[16] In the first case, a matrix $\mathbf{M}$ would be chosen in some fashion, and the estimated values of the variables would be derived from $\hat{\mathbf{x}} = \mathbf{My}$. In the other case, one would employ equation (1) to find an iterated solution in the following manner.

1. Pick an initial estimate, perhaps $\mathbf{x}^{(0)} = <\mathbf{x}>$. Here, $\mathbf{x}^{(n)}$ refers to the nth estimated value.

2. Find $\mathbf{y}^{(n)} = \mathbf{Ax}^{(n)}$. Compare it with $\mathbf{y}'$, the measured value, and modify $\mathbf{x}^{(n)}$ appropriately to get $\mathbf{x}^{(n+1)}$. Repeat this process until $|\mathbf{y}' - \mathbf{y}^{(n)}| < \varepsilon$ for some positive constant ε. Unfortunately, there is no guarantee either that this process will converge, or that the solution is unique.

---

[16] See Fleming (1977).

Fleming showed that these two kinds of algorithms would give statistically identical results, in that the r.m.s. errors would be the same, even if the results for individual profiles would not be identical. It is possible to extend this work and show that the results for the two types of algorithm are identical for every profile. However, matrix algorithms are preferable for two reasons: firstly, they are much easier to compute, and secondly, they are amenable to a complete analysis.

1.10 The rest of the book

This chapter was meant to be an introduction to the material to follow and that will provide a framework for solving remote-sensing problems. Chapters 2 to 4 describe radiative transfer, and how the vertical temperature profile in the earth's atmosphere affects the radiation emitted into space. These chapters are included to introduce the people who are interested in remote sensing to the fundamentals of the subject and to provide a concrete physical setting for the material that follows. Many readers will want to have this physical context.

The following chapters, however, give a purely mathematical treatment of the subject. Except for a few places, this material can be understood without reference to the material in Chapters 2 to 4; readers who do not wish to learn about radiative transfer can skip them. In particular, no notational conventions are introduced in these chapters that are needed in later chapters (unless they are explained again). It is my hope that the mathematical development will be useful to workers in many other fields, and I do not want to force them to study a particular physical problem before they can make use of the mathematics I present here.

Stated quite abstractly, the problem is this. Given a set of pairs $\{(\mathbf{x}_i, \mathbf{y}_i)\}$, where $\mathbf{x}_i$ and $\mathbf{y}_i$ are the ith realization of the vector quantities $\mathbf{x}$ and $\mathbf{y}$, respectively, we want to infer the value of $\mathbf{x}_i$ from each measured value of $\mathbf{y}_i' = \mathbf{y}_i + \mathbf{n}$. We shall examine, at various times, the following possibilities:

1. Linear problem: $\mathbf{y} = \mathbf{A}\mathbf{x}$.

2. Linearized problem: $\mathbf{y} = f(\mathbf{x})$, where f is a vector function of $\mathbf{x}$, but it is approximated by $\mathbf{y} = \mathbf{A}\mathbf{x}$.

3. Nonlinear problem: $\mathbf{y} = f(\mathbf{x})$.

In addition, in Chapter 10 and elsewhere, we will consider an alternative form of the problem, where we investigate integral equations instead of matrix equations.

References

Fleming, H. E., 'Comparison of linear inversion methods by examination of the duality between iterative and inverse matrix methods,' in *Inversion Methods in Atmospheric Remote*

Sounding, A workshop held at Langley Research Center, Hampton, Virginia, December 15-17, 1976, Edited by A. Deepak, NASA CP-004 (1977).

Twomey, S., *Introduction to the Mathematics of Inversion in Remote Sensing and Indirect Measurements*, Elsevier Scientific Pub. Co., Amsterdam (1977).

2. Light and Atoms

*... that, in a few years, all great physical constants will
have been approximately estimated, and that the only oc-
cupation which will be left to men of science will be to
carry these measurements to another place of decimals.*

James Clerk Maxwell (1813-1879)

This chapter presents an overview of the physical processes by which light interacts with matter.[1] There could be no remote sensing without this knowledge, since it forms the fundamental link between the radiation we sense and the physical phenomena we want to measure. While the rest of the book is a rather general inquiry into how we can make inferences from noisy data, this chapter and the next two provide a direct physical and mathematical background to a certain class of remote-sensing problems—those that deal, primarily, with remote sensing of the earth from space.[2]

I wrote this chapter primarily to give a concrete, physical example of how remote sensing works: how we can measure the properties of the earth by measuring the light it emits. To do the topic justice would require a book in itself—a very large book. It would take several chapters just to describe how we calculate the absorption coefficients of different gases; how to solve different kinds of radiative-transfer problems; how to calculate the emissivity or radar cross-section of the ocean surface; and so forth. But the focus of this book is really the mathematical—and perhaps, the philosophical—basis of remote sensing. In order to do the mathematics justice, I have decided to keep the treatment of the physics of the interaction of electromagnetic radiation with matter to this single chapter, followed by short chapters on radiometers and radiative transfer. Some readers will want this introduction so that they can keep a physical situation in mind; or because they are curious about how the microwave radiation a planet emits is related to the temperature of its oceans or the wind speed. Others will be unsatisfied by this chapter because they want to know all of the details that I have left out. The material here is not required for an understanding of the mathematics in the rest of the book.

---

[1] The classic work on the subject is Chandrasekhar (1960). See also Liou (1980) or Ulaby *et al.* (1981).

[2] The reader who is interested in remote sensing of the earth from space should read the special issue of the *IEEE Transactions on Geoscience and Remote Sensing*, **14** (July, 1998). This is a special issue devoted to the instruments on the EOS AM-1 Platform, which contains a suite of visible and infrared instruments. It discusses both the instruments and some of the physical principles that will be used to interpret the data.

However, the reader should also realize that all remote sensing depends, ultimately, on understanding how light interacts with matter. I discuss the physics here, some principles of radiometers in Chapter 3, and the mathematics of radiative transfer calculations in Chapter 4. In particular, I hope that the reader will take note of how an understanding of radiative processes is related to the Second Law of thermodynamics. This apparently simple law, that heat always flows spontaneously from a hot to a cold object, leads directly to Kirchoff's law (§2.2.2), which could reasonably be called the fundamental law of remote sensing.

There is a lot of physics that is involved in making the connection between the physical variables that concern us—*e.g.*, the temperature of the ocean surface—and the radiant intensities that we measure from space. We shall see bits of atomic physics, thermodynamics, E&M theory, wave physics, and so forth, being used to deal with the physics part of the problem. In general in the physical sciences, we use our knowledge of physics to write an appropriate set of equations, along with boundary conditions and whatnot, that describe a problem. Then we can forget the physics for a while[3] and concentrate on solving the math problems involved. This chapter is on the physics; the rest of the book concerns solving the math problems.

Since remote sensing is the science of determining the properties of a body—most notably, the earth—from measurements of the light[4] it emits, this chapter describes the ways that the physical properties of the earth's atmosphere and surface affect the emission and absorption of light. In the atmosphere, the main absorbers[5] are molecules of O_2, H_2O, O_3, and CO_2. I shall describe very briefly how absorption coefficients are controlled by the temperature and pressure of a gas, as well as emission and reflection from a surface.

2.1 Light and radiometers

In remote sensing, we use reflected, scattered, or emitted radiation to study the earth. Except for sensing with radar (*i.e.*, an instrument that emits a pulse of electromagnetic radiation and measures the amount reflected back from a target), we are measuring the intensity of light emitted by the earth (primarily in the microwave and infrared parts of the spectrum) or reflected sunlight (in the visual part). A device that measures radiant intensity is called a *radiometer*; I discuss some aspects of radiometers in Chapter 3. Because the physical conditions in the atmosphere and on the surface have different effects at different

---

[3] Not completely; all of our thinking should be guided by physical principles.

[4] By *light*, I mean electromagnetic radiation of any wavelength: this includes ultraviolet, visible, and infrared light, as well as radio waves.

[5] *Of electromagnetic radiation* will be understood from now on.

wavelengths, it is common to employ radiometers that make measurements at several different wavelengths simultaneously. For example, one might use an infrared radiometer that makes simultaneous measurements of light passed through several different filters, each one of which passes light in a band centered around a different wavelength. We speak of making measurements with different *channels*, or bands of wavelengths. I shall return to this later.

2.1.1 Matter and radiation

There are five ways that electromagnetic waves (for our purposes, microwaves, infrared radiation, and visible light, in order of increasing frequency) interact with matter. Light that is incident on a body (the atmosphere or surface of a planet, for example) can be *reflected, absorbed, transmitted,* or *scattered.* The relative importance of these mechanisms depends, in large part, on whether we are talking about gases, liquids, or solids. On the earth we encounter all three: the atmosphere, the ocean, and land surfaces. Although other configurations are possible, we will be concerned almost exclusively with the situation where there is a surface that separates a volume of air from a volume of soil or water. In particular, I assume in this chapter that we are in a satellite looking down through the atmosphere to the land or water surface of the earth. At some wavelengths, the atmosphere absorbs very little radiation, so we can see all the way to the surface from space and measure the properties of the surface. At other wavelengths, or under different conditions, the atmosphere absorbs all of the radiation that emerges from the earth below it: in turn, it emits radiation, some of which will reach our spacecraft. In this case, we can measure the properties of the atmosphere (but not of the surface below it).

In addition to the processes mentioned above, matter *emits* radiation. At a given wavelength, the amount of light emitted by a body depends both on the temperature of the object and, in the case of a solid or liquid surface, the *emissivity* of the surface. I will define emissivity later, but for the moment, think of it as being an efficiency, between 0 and 1, that represents the fraction of the energy that a body *does* emit, relative to the maximum amount it *could* emit at a given temperature. The emissivity of a surface depends primarily on three factors: the complex *permittivity* of the material (which, in turn, depends in part on the wavelength); the roughness of the surface; and the angle that an emitted ray makes with the normal to the surface. It also depends on the polarization of the radiation.

Radiative transfer problems arose in the study of stars. As I mentioned in the last chapter, since stars are so inaccessible, all we can know about them is what we can determine from analyses of the light they emit. There are two components to this light: a continuum, where the light is spread out more or less evenly over the spectrum, and which is determined primarily by the temperature of the star; and absorption or emission lines at discrete frequencies

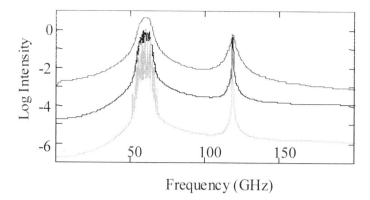

Figure 2-1. Oxygen (O_2) absorption coefficient. Note the single line at 118 GHz, and multiple lines between 50 and 60 GHz. T = 260 K; the pressure is 1000 mb (top line), 100 mb (middle); and 10 mb (bottom line).

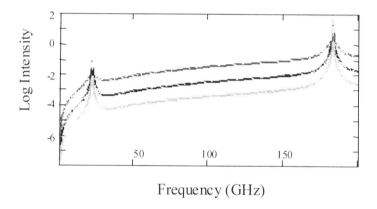

Figure 2-2. Water vapor absorption coefficient for T = 260 K, pressures of 1000 mb (top line); 100 mb (middle line); and 10 mb (bottom line).

due to the presence of different elements (and, in the case of cooler stars, some molecules). Each element or molecule has its own set of frequencies where it can emit or absorb light. Nineteenth-century astronomers were able to determine the chemical composition of the sun, and later of other stars, by comparing the wavelengths of absorption lines in the solar spectrum with the wavelengths of light emitted in the laboratory by different elements, most notably sodium. The element helium, in fact, was first discovered by its spectral lines in the sun; only later was it identified on the earth. The continuum and line components arise from different mechanisms and may provide different kinds of information to scientists involved in remote sensing.

The earth's atmosphere has different properties in the three parts of the spectrum that are of interest to remote sensing. I shall discuss each of them very briefly.

2.1.2 Microwave radiation

In the microwave part of the spectrum, frequencies between about 1 GHz (30 cm wavelength) and 200 GHz (0.15 cm), the dry component of the atmosphere is transparent, except that molecular oxygen absorbs microwaves at frequencies between about 50 and 60 GHz and at a small range of frequencies around 118 GHz (see Figure 2-1). The proportion of oxygen in the atmosphere is constant for all practical purposes; the amount of absorption depends only on the temperature and pressure of the air at a given height. We can use measurements of microwave emission in the 50-60 GHz band to infer vertical temperature profiles in the atmosphere.

Hydrometeors—atmospheric water droplets and ice particles in all of their different forms—absorb and scatter microwaves at all wavelengths. The absorption coefficients of water droplets and ice particles are very small at low frequencies, but increase with increasing frequency. The amount of water in the atmosphere and the form it is in varies greatly with time and place, so we can use microwaves to study the properties and concentrations of water in the atmosphere. Water vapor has prominent absorption lines at 22 GHz and 183 GHz (see Figure 2-2). At frequencies between 1 and 10 GHz or so, clouds have little effect on microwave radiation, so microwave instruments can be used to look through clouds at the earth's surface at these wavelengths. This is often an important consideration, as we cannot see through clouds with visible or infrared radiation. At higher frequencies, microwave radiometers are used to sense the state of the atmosphere.

Land and ocean surfaces reflect and emit microwaves. The emissivity of a surface depends primarily on the roughness of the surface and its complex permittivity. Water, for instance, has an emissivity around one-half; land has a much higher emissivity. This is because water has a very large permittivity, while rocks have a much smaller permittivity. Therefore, at microwave frequencies, land is much brighter than water, even if they have the same physical temperature.

Because microwave sensors measure radiation emitted by the earth—or backscattered radiation from a man-made source, for radar—rather than reflected solar radiation,[6] they work equally well at night and during the day.

2.1.3 Infrared radiation

Atmospheric gases have very large absorption coefficients in many parts of the infrared part of the spectrum (1 to 12 micron wavelengths). The three principal absorbers in the infrared part of the spectrum are water vapor, carbon dioxide, and ozone. The absorption is due to rotational and vibrational

---

[6] As in the visible part of the spectrum.

transitions in these molecules, which produce thousands of individual absorption lines. In addition, water vapor produces continuum absorption: *i.e.*, absorption that varies smoothly over a very large range of wavelengths. The source of this continuum absorption is not well understood, although it is usually attributed to the formation of water polymers (many water molecules stuck loosely together) in the atmosphere. In addition, hydrometeors absorb and scatter infrared radiation. Infrared sensors cannot see through any but the very thinnest clouds. I show the transmittance of the atmosphere—the fraction of light that can escape from the surface to space—in Figures 2-3 to 2-10, for wavelengths between 0.5 and 20 μm. This is the spectral region where the black-body curve is maximum for temperatures that obtain on the earth (see Figure 2-12); these figures show how CO_2, ozone, and water vapor contribute to warming the earth via the greenhouse effect.

The emissivity of land and water surfaces at the longer infrared wavelengths is, in general, quite high. Differences in the amount of radiation emitted from the surface depend almost entirely on differences in temperature. One important use of infrared sensors is making routine measurements of the temperature of the ocean surface by measuring the amount of infrared radiation it emits. At shorter infrared wavelengths (closer to 1 micron), different kinds of surfaces have different reflectivities, just as they do at visible wavelengths.

2.1.4 Visible radiation

At the temperatures encountered on the surface of the earth (outside of active volcanoes, anyway) matter is too cold to emit visible radiation. The gases in the atmosphere are completely transparent to visible radiation (wavelengths roughly 0.5 to 1.0 μm), although there is some scattering (that is why the sky is blue). Clouds, as we know, scatter and absorb visible radiation, as do aerosols in the atmosphere. We can use visible light to measure the presence of clouds from space and to map different kinds of surfaces—land, water, and sea ice—by measuring their different colors. Since visible light does not penetrate clouds—that is why they look dark—we cannot see the surface below the clouds with instruments that sense visible light. Naturally, sensors that measure visible radiation are useful only during the day—or rather, on the daylight side of the earth.

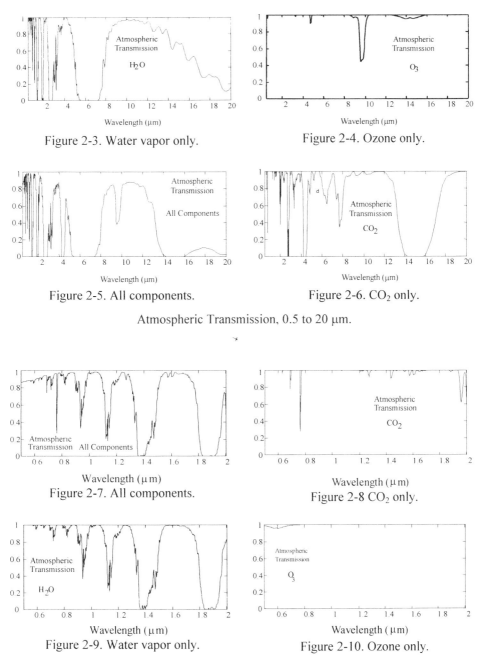

Figure 2-3. Water vapor only.

Figure 2-4. Ozone only.

Figure 2-5. All components.

Figure 2-6. CO_2 only.

Atmospheric Transmission, 0.5 to 20 μm.

Figure 2-7. All components.

Figure 2-8 CO_2 only.

Figure 2-9. Water vapor only.

Figure 2-10. Ozone only.

Atmospheric Transmission, 0.5 to 2 μm.

Table 2-1. Some Physical Constants.

Planck's constant	h	6.62620×10^{-27} erg s
Boltzmann's constant	k	1.3802×10^{-16} erg deg^{-1}
Speed of light	c	2.9979250×10^{10} cm s^{-1}
Electron mass	m_e	9.11×10^{-28} g
Proton mass	m_p	1.672661×10^{-24} g
Rydberg constant for ^{1}H	R_H	109677.576 cm^{-1}
Rydberg constant for infinite mass	R_∞	109737.312 cm^{-1}
Stefan-Boltzmann constant	σ	5.6696×10^{-5} erg cm^{-2} s^{-1} K^{-4}

2.1.5 The nature of light

In the following sections, I will describe some of the fundamental concepts pertaining to the ways that electromagnetic radiation interacts with matter. Although there are plenty of mathematical descriptions of the different concepts and phenomena, I do not make any particular effort to be complete.

I shall say very little about the physical nature of light and the physics of electromagnetic fields. While these topics are important, I feel that an adequate understanding of how remote sensing works can be achieved without knowing anything about the nature of light itself, except for knowing that light can be polarized. The reader is undoubtedly aware that electromagnetic fields can be described by a set of equations called Maxwell's equations. Considering that they involve vector calculus, they are not very complicated. It is not uncommon for a book like the present one to give the reader a whole chapter or more devoted to the derivation of the properties of light from these equations. I shall not present such a discussion here, for fear that it would bore, rather than enlighten, the reader. Even more to the point, it would bore me.

One physical principle that I shall use in several places is the Second Law of thermo
dynamics, which says that heat can only travel spontaneously from a hotter body to a colder one. This law has been established through countless experiments and observations, including many everyday observations; *e.g.*, you put ice in your drink to make the liquid colder; as a consequence the ice melts. You have never seen the ice cube get colder while the drink spontaneously gets hotter: heat always flows from the drink to the ice cube, never the other way. A violation of this law would be a surprise indeed. We shall use this law several times to help us understand the nature of thermal radiation.

One consequence of this law, along with the law that energy is conserved, is that it is impossible to build a perpetual-motion machine. Any time there is something moving and we extract energy from its motion, that system must

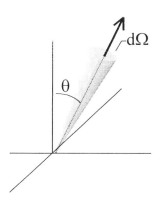

Figure 2-11. Solid angle dΩ.

eventually run down unless energy is also put back. There are systems where no energy is lost by a moving body, so it can move forever without violating this principle. The moon revolving around the earth is an example of this (although in reality, some energy is lost due to tidal friction, so the earth-moon system can only last for several billion years, not forever). In several places, I shall use the observation that perpetual-motion machines are impossible to explain how radiation interacts with matter. As an aside, applications of thermodynamic principles often lead to much simpler analyses than do applications of electromagnetic theories derived from Maxwell's equations.

2.1.6 Intensity

Intensity is the rate at which electromagnetic radiation transports energy from one place to another: it is somewhat analogous to the current in a wire. Consider a pencil-beam of radiation passing through an area A with the radiation being emitted into an infinitesimal solid angle dΩ. The radiation makes an angle θ with the normal to the area A; the *projected area* through which the radiation passes is $A\cos\theta$. The intensity is the amount of energy passing through a unit area per second per steradian:

$$I_\lambda = \frac{E_\lambda}{A \cos\theta \, \Delta\Omega}, \tag{1}$$

where E_λ is the amount of energy with wavelength λ passing through that area per second; $\Delta\Omega$ is the solid angle (see Figure 2-11).

2.2 Thermal radiation

Almost any physical object will absorb some of the light that falls upon it, reflecting the rest. Only a perfect mirror would absorb no light. Since light is a form of energy, and it turns into heat when it is absorbed, a body that absorbs light must either emit light, lose heat by conduction, or have its temperature increase. Consider a planet in orbit around the Sun: it absorbs sunlight, yet it reaches an equilibrium temperature that is far lower than the temperature of the Sun. Since it is surrounded by a vacuum, it cannot lose energy by conduction. Therefore, it must emit light also. Because it is cooler than the Sun, it emits mainly infrared radiation.

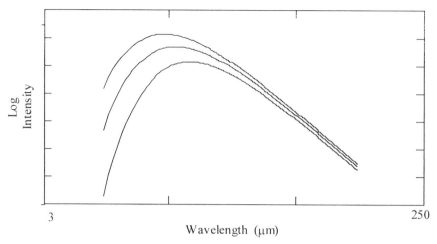

Figure 2-12. The Planck black-body intensity for temperatures of 320 K, 260 K, and 200 K (top to bottom curves).

We do not always see the radiation emitted by the objects around us. Room-temperature objects only emit radiation in the microwave and infrared parts of the spectrum: only very hot objects, like a star or the tungsten filament in a lamp, emit visible light. The wavelength at which the emitted radiation is maximum depends on the temperature of the object, with hotter objects emitting light at shorter wavelengths (see §2.2.4). However, all material bodies do emit electromagnetic radiation.

2.2.1 Planck radiation law

Imagine an object made out of a material that absorbs all of the radiation that falls on it. Such an idealized object is called a *black body*, since it reflects none of the light incident on it. A black body will emit light according to the law[7]

$$B(\lambda, T)\,d\lambda \;=\; \frac{2hc^2}{\lambda^5}\,\frac{1}{e^{hc/\lambda k T} - 1}\,d\lambda, \tag{2}$$

where h is Planck's constant; c, the speed of light; k, Boltzmann's constant; λ, the wavelength of the emitted radiation; and T, the temperature in kelvins. A black body with a temperature T will emit $B(\lambda,T)d\lambda$ erg s^{-1} cm^{-1} between the wavelengths λ and $\lambda + d\lambda$. This relationship was derived by Max Planck in 1900, who needed to assume that light is quantized in order to derive the correct law of black-body radiation. This was the beginning of the quantum me-

---

[7] Max Planck (1858-1947) was one of the founders of quantum mechanics.

chanics. The quantum hypothesis states that light consists of particles,[8] or quanta, and that each quantum of light has an energy $E = h\nu$, where $\nu = c / \lambda$ is the frequency. I show the intensity for temperatures of 200 K, 260 K, and 320 K, in Figure 2-12.

2.2.2 Kirchoff's law

A body that absorbs radiation must also emit radiation: it is this physical fact, more than any other, that makes remote sensing possible. Without it, we would be not be able to make any inferences about a body by measuring the amount of light it absorbs or emits. A theorem, published by Kirchoff[9] in 1859, showed that any body must be capable of absorbing light in proportion to its ability to emit light. His argument rests on the Second Law of thermo-dynamics.

Consider a body immersed in a radiation field. If it does not *emit* radiation while it keeps *absorbing* radiation from its surroundings, its temperature will increase without limit. The emission coefficient ε_λ is defined by

$$\delta I_{e,\lambda} = \varepsilon_\lambda \delta\sigma \tag{3}$$

where $\delta I_{e,\lambda}$ is the amount of radiation emitted from a surface area $\delta\sigma$ of the body at wavelength λ. The amount of light absorbed is $\delta I_{a,\lambda} = \kappa_\lambda \delta\sigma$, where κ_λ is the absorption coefficient of the surface. The net change in intensity is

$$\delta I_\lambda = (\kappa_\lambda - \varepsilon_\lambda)\delta\sigma. \tag{4}$$

In equilibrium, $\delta I_\lambda = 0$. This can only happen if the ratio $\varepsilon_\lambda / \kappa_\lambda$ is a constant that depends only on the wavelength λ and the temperature T, but not on the intensity of the radiation field or the nature of the body. There must be a universal law such that, for any body,

$$\frac{\varepsilon_\lambda}{\kappa_\lambda} = J(\lambda, T) \tag{5}$$

where $J(\lambda, T)$ is some universal function that had yet to be determined. It took physicists forty years to discover what form $J(\lambda, T)$ takes. The result was the Planck's radiation law, which he derived by assuming that light, at least as interacts with matter, is quantized. As Pais (1982, p 364) puts it: "Gustav

---

[8] We now call them *photons*.

[9] Gustav Robert Kirchoff (1824-1887) was a German physicist; he published the laws pertaining to electrical circuits that also bear his name in 1845. There are laws in two different branches of physics that bear his name.

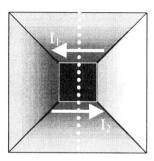

Figure 2-13. A black-body enclosure in normal matter with one wall
(on right) made of an (imaginary)
"extraordinary" substance (see
text).

Kirchoff from Heidelberg… proved a theorem and posed a challenge: The response to Kirchoff's challenge led to the discovery of quantum theory."

The more modern statement of *Kirchoff's law* is that

$$\frac{\varepsilon_\lambda}{\kappa_\lambda} \;=\; B(\lambda, T)\,, \tag{6}$$

where $B(\lambda, T)$ is Planck's radiation law (see §2.2.1). One consequence of equation (6) is that no material can emit radiation at wavelengths where it does not
absorb radiation.

Note that Kirchoff's law has to be obeyed separately at each wavelength.
Suppose there was a substance—call it *Substance E* for *extraordinary*—that,
unlike ordinary matter, had an absorption coefficient $\varepsilon_1 > \kappa_1 B(\lambda_1, T)$ at some
wavelength λ_1, and $\varepsilon_2 < \kappa_2 B(\lambda_2, T)$ at some other wavelength λ_2. Make a cavity
in ordinary matter at temperature T: in equilibrium, the radiation field inside
the cavity will be $B(\lambda, T)$ at every wavelength λ. Line one wall of the cavity
with Substance E (the right-hand wall in Figure 2-13). At wavelength λ_1 there
is a net flow of energy away from the wall; at λ_2, the flow is toward the wall.
Now put a separation down the middle, as shown by the white dotted line. If
the partition is made of a substance that is transparent at wavelength λ_1 but a
reflector at λ_2,[10] the wall made from Substance E will emit more light than it
can absorb, so its temperature $T_E < T$. If the partition has the opposite property, we will have $T_E > T$. But either situation would violate the Second Law.

---

[10] Substances that transmit light at one wavelength and reflect it at another wavelength are
common.

Although I discussed Kirchoff's law in terms of a solid body, exactly the same reasoning applies to a gas: the ratio of the emission and absorption coefficients still must equal $B(\lambda, T_{gas})$.

2.2.3 Raleigh-Jeans approximation

There is an approximation to Planck's radiation law that is useful at microwave frequencies. First, write the Planck radiation law as a function of frequency,

$$B(\nu, T) = \frac{2h\nu^3}{c^2} \frac{1}{e^{h\nu/kT} - 1} \tag{7}$$

[to get this, substitute $\nu = c / \lambda$ and remember that $d\nu = -(c / \lambda^2) d\lambda$ Note that equation (2) represents the energy emitted per unit wavelength, while this one represents the energy emitted per unit frequency.] In the limit $h\nu \ll kT$,

$$e^{h\nu/kT} - 1 \cong h\nu / kT, \tag{8}$$

so

$$B(\nu, T) \cong \frac{2\nu^2}{c^2} kT. \tag{9}$$

This is called the *Raleigh-Jeans* approximation. It is very useful to radio astronomers or those of us who do microwave remote sensing of the earth because, at low frequencies, we can measure the intensity in units of temperature. For a layer of gas that absorbs all of the radiation that is incident on it, the intensity (in temperature units) is just equal to the temperature of the gas. The *brightness temperature* of radiation of a certain intensity is the temperature of the black body that would emit the same intensity. It is also called *apparent temperature*, since the source being measured looks like a black body with that temperature. The brightness temperature is defined by

$$T_B = \frac{c^2}{k\nu^2} I_\nu, \tag{10}$$

where I_ν is the intensity at frequency ν. The factor of 2 does not appear in this definition because any physically realizable antenna can measure only one polarization of radiation. For unpolarized radiation, the antenna measures only one-half of the intensity. Therefore, we multiply by 2 to account for the intensity with the orthogonal polarization. For black-body radiation, the two polarizations are equal; this is assumed in equation (10).

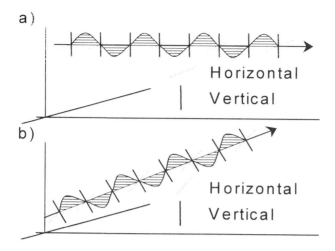

Figure 2-14. Vertical and horizontal polarizations for a) horizontal propagation; and b), propagation at 45° incidence angle.

2.2.4 Stefan-Boltzmann and Wien's laws

The total amount of radiation emitted by a black body per unit area is found from integrating equation (2):

$$B_{tot} = \frac{\sigma}{\pi} T^4, \qquad (11)$$

where σ is the Stefan-Boltzmann constant.[11] Note that a small increase in temperature can produce a large increase in the rate of radiative cooling. The peak of the black-body radiation curve is found by differentiating equation (2):

$$\lambda_{max} = \frac{0.28979}{T}, \qquad (12)$$

where λ_{max} is in cm. This is known as the *Wien displacement law*.[12] As the temperature increases, λ_{max} decreases. A hot piece of iron, for example, will emit mostly in the red part of the spectrum; a hotter carbon arc, in the blue;

---

[11] Joseph Stefan (1835-1893) was the Austrian physicist who derived this expression empirically; later, Boltzman derived it theoretically. Ludwig Eduard Boltzmann (1844-1906), also an Austrian physicist, is most famous for his papers on statistical mechanics. He showed in a paper published in 1870 that the second law of thermodynamics can be explained by applying classical mechanics and the laws of probability to the motions of atoms.

[12] Wilhelm Carl Werner Otto Fritz Franz Wien (1864-1928) won the Nobel Prize in Physics in 1911 for his work on black-body radiation.

and an extremely hot star will emit most of its light in the ultraviolet. At the other extreme, a room-temperature object emits radiation only at microwave or far-infrared frequencies.

2.2.5 Polarization

Electromagnetic radiation is a transverse wave; *i.e.*, the electric field oscillates in a direction that is perpendicular to the direction of motion of the wave. Suppose that a wave is propagating in a horizontal direction. The electric field can oscillate in a direction parallel to the ground, or a direction perpendicular to the ground. These are two states of linear polarization: they are called (reasonably enough) *horizontal* and *vertical* polarizations, respectively. Sometimes they are called parallel and perpendicular polarizations. Even if the wave is propagating toward or away from the surface, there is still a horizontal direction parallel to the surface: we still define horizontally polarized radiation as oscillating in this direction. Vertically polarized radiation is oscillating in the direction perpendicular to this, although it is not now perpendicular to the surface (see Figure 2-14). Note that we have described the state of the polarization with reference to a surface, which will usually be taken as being the surface of the earth—or of the ocean. If the radiation is emitted in the direction normal to the surface, horizontal and vertical polarizations are the same.[13]

Radiation from a true black body would be unpolarized. However, because the emissivity of most surfaces is different for horizontal and vertical polarizations, radiation in the atmosphere may be polarized. This can be quite important at microwave frequencies where the atmosphere is optically thin, so emission and reflection from the surface make a large contribution to the radiation emitted from the top of the atmosphere, but it is of little importance in the infrared.

2.2.6 Reflection and emission

A gas can only emit and absorb radiation, but a liquid or solid (think here of the surface of the earth) can also reflect light. We characterize properties of a surface by its emission and reflection coefficients. Let I_0 be the intensity incident on a surface; the material emits an intensity $I_{e,p}$ and reflects an intensity $I_{r,p}$ given by[14]

$$I_{e,p} = e_p \, B(\lambda, T) \tag{13}$$

---

[13] Light can also be elliptically or circularly polarized, but that does not concern us here.

[14] Throughout this chapter, I shall use ε to denote the emissivity of a gas, and e, the emissivity of a solid surface.

and

$$I_{r,p} = r_p \, I_0 \tag{14}$$

(see Figure 2-15). The emissivity e_p and the reflectivity r_p (the subscript p refers to the polarization) are related by the condition

$$e_p + r_p = 1. \tag{15}$$

We can prove this the same way that Kirchoff's law (see §0.2.1) is proved. Let us perform the following *gedanken* experiment.

Suppose that we have a perfectly absorbing material with a cavity on the inside. This is called a *black-body enclosure;* its walls absorb any radiation that falls on them and emit radiation according to the Planck radiation law. Let the temperature of the walls be T. In equilibrium, the cavity will contain an isotropic radiation field given by equation (2). We now introduce an object into this cavity. It will absorb and emit radiation until it reaches the temperature T (we assume that the heat capacity of the walls of the black-body enclosure is so large that T cannot change appreciably due to our introduction of the object into the cavity). In equilibrium, the surface of the object must absorb and emit radiation at equal rates and the radiation field must again be isotropic.

I can prove this as follows. Suppose that the radiation field were not isotropic. Then I could take two radiation detectors—devices that absorb radiation and produce electric currents—and aim them in different directions, one toward the maximum of the radiation field, the other, toward its minimum. The two detectors would produce different voltages, so I could extract energy from this radiation field by making use of the current caused the difference between the two voltages. If I could do this, however, I would have built a perpetual motion machine, something that we know cannot exist. Therefore, the radiation field inside a black-body enclosure in equilibrium must be isotropic.

Consider the surface of the object. The radiation incident on it is $B(\lambda,T)$; of this, a fraction r_p is reflected, so a fraction $e_p = (1 - r_p)$ must be absorbed. The object must emit radiation just as fast is it absorbs it if it is equilibrium, so it must emit an amount of radiation $(1 - r_p)B(\lambda,T) = e_p B(\lambda,T)$, thereby proving equation (9). The reflection

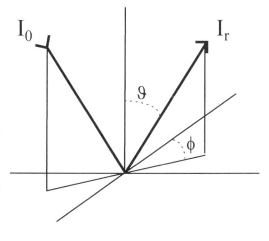

Figure 2-15. Incident and reflected radiation.

and emission coefficients both depend on wavelength, and this balance must hold separately for each wavelength λ.

The emission coefficient of a material depends on four things: the complex permittivity of the material; the polarization of the radiation; the roughness of the surface; and the angle at which the radiation is being emitted (see Figure 2-15). For a smooth dielectric surface, the reflectivity is given, as a function of angle and permittivity, by the *Fresnel reflection coefficients* . The emission coefficients for vertically and horizontally polarized radiation are[15]

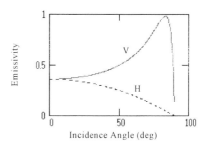

Figure 2-16. Emissivity of seawater for vertical and horizontal polarizations, including the effects of surface roughness.

$$r_{\mathrm{v}} = \left(\frac{\cos\theta - \sqrt{\eta - \sin^2\theta}}{\cos\theta + \sqrt{\eta - \sin^2\theta}} \right)^2 \qquad (16)$$

and

$$r_{\mathrm{h}} = \left(\frac{\eta\cos\theta - \sqrt{\eta - \sin^2\theta}}{\eta\cos\theta + \sqrt{\eta - \sin^2\theta}} \right)^2 \qquad (17)$$

where θ is the incidence angle and η, the complex permittivity (or dielectric constant). Also, r_{v} and r_{h} are the reflectivities for vertically and horizontally polarized radiation. This only applies to smooth surfaces; calculating reflection coefficients for randomly rough surfaces—like the ocean surface—is much more complicated. The emissivity of seawater is shown in Figure 2-16.

2.3 Absorption and emission of radiation by gases

The amount of radiation absorbed or emitted by a gas is determined by the absorption coefficient κ_λ, which depends on wavelength and the chemical composition of the gas. Usually, κ_λ is expressed in terms of the amount of light absorbed per unit mass of absorber and per unit path length. Each atom or molecule has its own set of wavelengths at which it can emit or absorb radiation.

Suppose that there is a plane-parallel layer of some gas with a uniform density ρ and thickness d. The sides of the layer are parallel to the x-y plane. The density ρ is zero for $z < 0$ or $z > d$, and constant for $0 \le z \le d$. Light with

---

[15] See Born and Wolf (1980); Jackson (1967); Ulaby *et al.* (1981, pp 73-74).

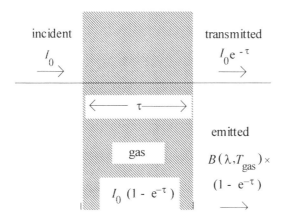

Figure 2-17. Absorption and emission by a gas with temperature T_{gas}.

an intensity $I_{0,\lambda}$ is incident on the layer (see Figure 2-17). Some of the incident radiation will be absorbed, and some, transmitted; we ignore scattering here. Let $I_{t,\lambda}$ be the transmitted intensity; the amount absorbed will be $I_{a,\lambda} = I_{0,\lambda} - I_{t,\lambda}$.

The *optical depth* is defined to be

$$\tau_\lambda(z) = \rho \int_0^z \kappa_\lambda(t)\, dt, \tag{18}$$

where ρ is the density, which is assumed to be constant and z is the path length. We assume explicitly that the absorption coefficient may change with z. The optical thickness of the layer is

$$\tau_\lambda(d) = \rho \int_0^d \kappa_\lambda(t)\, dt. \tag{19}$$

The differential form of this expression is

$$d\tau_\lambda = \rho \kappa_\lambda\, dz. \tag{20}$$

A gas that has a small optical depth τ is said to be *optically thin*; one that has a large optical depth is said to be *optically thick*. We can see through a gas if it is optically thin; it is opaque if it is optically thick. The same gas may be optically thick at some wavelengths and optically thin at others.

When this is integrated through the layer, the amount of light transmitted (ignore emission for the moment) is

$$I_{t,\lambda} = I_{0,\lambda} e^{-\tau_\lambda}. \tag{21}$$

Since the absorption coefficient κ_λ depends on wavelength, so does the optical depth: I write it as τ_λ to denote the wavelength dependence explicitly.

We wish to know the intensity of the radiation the gas emits parallel to the z-axis. If the layer of gas with a uniform temperature T and an optical thickness[16] $\tau_{\lambda,\text{gas}}$, it will emit an intensity I_λ in a direction perpendicular to the x-y plane (see Figure 2-17). I shall show presently that

$$I_\lambda = B(\lambda, T)\left(1 - e^{-\tau_{\lambda,\text{gas}}}\right). \tag{22}$$

Assume that there is no radiation incident from behind the gas. At each position z in the gas, the corresponding optical depth is $\tau_\lambda(z)$. The change in intensity there is the amount of radiation emitted in an infinitesimal layer of the gas with thickness δZ minus the amount absorbed in that layer. The total change in the intensity (emission minus absorption) is

$$\delta I_\lambda(\tau_\lambda) = \varepsilon_\lambda \rho \delta z - I_\lambda(\tau_\lambda)\kappa_\lambda \rho \delta z$$

$$= \left[\varepsilon_\lambda - I_\lambda(\tau_\lambda)\kappa_\lambda\right]\rho\delta z. \tag{23}$$

Here, $I_\lambda(\tau_\lambda)$ means the radiation with wavelength λ at the level where the optical depth is τ_λ. It **is** often convenient to change the independent variable from z to τ_λ, although at different wavelengths, τ_λ corresponds to different values of z.

Using equation (6), we can rewrite this as

$$I_\lambda = B(\lambda, T)\left(1 - e^{-\tau_{\lambda,\text{gas}}}\right). \tag{24}$$

where I have written the Planck function simply as B_λ with the temperature dependence being understood. Since $k_\lambda \rho \delta z = \delta \tau_\lambda$,

$$\frac{dI_\lambda(\tau_\lambda)}{d\tau_\lambda} = B_\lambda - I_\lambda(\tau_\lambda)$$

or

$$\frac{dI_\lambda(\tau_\lambda)}{d\tau_\lambda} + I_\lambda(\tau_\lambda) = B_\lambda. \tag{25}$$

This is the Equation of Radiative Transfer in its differential form. It is fundamental to the study of the radiative properties of gases that we need to under-

---

[16] I will use the term optical depth to refer to a distance within a gas; the term optical thickness refers to its total optical depth.

stand in order to interpret remote-sensing data properly. The equation shown here applies to an isothermal gas; we will consider a more general case later.

In order to solve this differential equation and prove equation (22), assume that the intensity is given by

$$I_\lambda(\tau_\lambda) = \alpha(1 - e^{-\tau_\lambda}). \tag{26}$$

Taking the derivative and using equation (25),

$$\frac{dI_\lambda}{d\tau_\lambda} = \alpha e^{-\tau_\lambda} = B(\lambda, T) - I_\lambda. \tag{27}$$

Given the boundary condition that there is no radiation incident on the layer from outside it—$I_0 = 0$ in Figure 2-17—the solution is simply $\alpha = \beta_\lambda$, thus proving equation (22).

It is instructive to look at two limits. If τ is very large, $(1 - e^{-\tau}) \cong 1$, so the maximum intensity is $B(\lambda, T)$. When τ is $<< 1$, $(1 - e^{-\tau}) \cong \tau$ and the intensity is $\tau B(\lambda, T)$. Note in particular that the gas cannot emit *more* radiation than a black body at the same temperature.

Suppose we are looking down at the earth from space. At wavelengths where the atmosphere is optically thin, we see down to the surface. Where it is optically thick, we see down only a small distance. Let the optical depth at the top of the atmosphere be zero; optical depth increases with decreasing altitude. Roughly speaking, we can see down through the atmosphere to the level where $\tau \approx 1$, assuming that we do not reach the surface before this happens. We can measure the vertical temperature profile in the atmosphere by using one channel (*i.e.*, band of wavelengths) where τ is very small; another, where τ is larger; ...; and one where τ is so large that we see only a small distance down into the atmosphere. Then, within limits, we can infer the temperature of each level from the intensity of the radiation measured in the appropriate channel. This is done both in the infrared and the microwave parts of the spectrum. Since the absorption coefficient depends on the wavelength λ, the same volume of gas may be optically thick at one wavelength and optically thin at another wavelength.

Exercise 2-1: Is our atmosphere optically thin or thick at visible wavelengths? Infrared wavelengths? Why?

Exercise 2-2: Suppose there is a black-body enclosure. If there is a gas inside that has zero optical depth, the radiation field is isotropic and has a value $B(\lambda, T)$, where T is the temperature of the walls of the cavity. If the cavity is filled with a gas that has a moderate optical depth and is in equilibrium with the enclosure, what is the intensity inside the cavity? Why? What happens if the optical depth of the gas is very large?

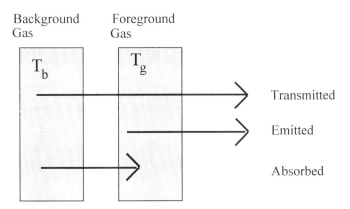

Figure 2-18. Background and foreground gases. Arrows represent radiation transmitted, emitted, and absorbed by the foreground gas.

Exercise 2-3: Why can't we use the visible part of the spectrum to measure vertical temperature profiles in the earth's atmosphere?

Now suppose that there is a plane layer of gas with optical thickness τ_g and temperature T_g; behind it, is another (background) gas that is optically thick and has a uniform temperature T_b (see Figure 2-18). The background gas emits radiation with an intensity $B(\lambda, T_b)$. The foreground gas will emit radiation and absorb radiation from the gas behind it. Using equation (22), the intensity emitted by the foreground gas is $I_{\lambda, g} = B(\lambda, T_g)(1 - \exp\{-\tau_g\})$. The radiation from the background gas is $B(\lambda, T_b)$, and it is attenuated by a factor $\exp\{-\tau_g\}$ as it passes through the foreground gas. Consequently the net intensity will be

$$I_\lambda = B(\lambda, T_b)e^{-\tau_g} + \left(1 - e^{-\tau_g}\right)B(\lambda, T_g)$$

$$I_\lambda = B(\lambda, T_b)e^{-\tau_g} + \left(1 - e^{-\tau_g}\right)B(\lambda, T_g) \tag{28}$$

(From now on, I will not denote the wavelength dependence of τ explicitly.) The first term on the right of the first equation is the intensity emitted by the background gas and absorbed by the foreground gas; the second term, the intensity emitted by the foreground gas. These terms are rearranged in the second equation to show the effect of the foreground gas. The foreground gas will increase or decrease the intensity, depending on whether or not it is hotter than the background gas. If $T_g = T_b$, there will be no effect whatsoever, regardless of the optical depth of the foreground gas. The role of the background gas could be taken by the surface of the earth. If the surface has a high emissivity—as it does in the infrared part of the spectrum—and the temperature of the atmosphere is approximately that of the surface, the atmosphere near the surface will have little effect on the brightness upwelling from near the surface. For this reason, it is often difficult to measure the temperature of the atmosphere near the surface with infrared remote-sensing data.

If the background gas is a star like the sun, it will emit light like a black body at a temperature of 5600 K or so. Its spectrum will be very broad; in other words, it emits a continuum spectrum. (This black-body intensity could also be produced by the hot filament in an incandescent light bulb or some other artificial source.) If $T_b < T_g$, a gas like hydrogen, which absorbs and emits light only at certain discrete frequencies, will produce absorption lines at these frequencies. If $T_b > T_g$, it will produce emission lines.

2.4 Spectral lines

Remote sensing is only possible because matter emits and absorbs light.[17] Because this process involves absorption by atoms and molecules, quantum mechanics is an important part of remote sensing. The amount of radiation a gas absorbs at each wavelength depends on the atomic[18] makeup of the gas; the temperature; and the pressure of the gas. I shall give a brief description of the processes involved in this section.

The formation of absorption lines is one of the macroscopic consequences of quantum mechanics. In fact, how the spectra of hydrogen and other elements are formed was one of the problems that led to the formulation of quantum mechanics in the early twentieth century. Although there is no significant amount of atomic hydrogen in the atmosphere, it is a very useful atom for explaining spectral lines because it is the simplest atom there is. Because hydrogen is the most abundant element in the universe, and the most abundant element in any star, its spectrum was observed in the sun in the nineteenth century.

A hydrogen atom consists of nothing more than a single proton with a single electron orbiting around it. Protons and electrons have positive and negative charges, respectively, that are equal in magnitude; the mass of a proton is about 1800 times that of an electron. Early pictures of atomic structure likened it to a star with a single planet in orbit around it. Therefore, the nucleus[19] contains most of an atom's mass, and remains almost stationary as the electrons orbit about it, just as the sun contains most of the mass in the solar system. This analogy with a solar system (where the force of gravity holds

---

[17] As we have seen, any gas that absorbs light must also emit light, and *vice versa.* From now on, I will discuss absorption with the tacit recognition that emission is also included.

[18] Both atoms and molecules have spectral lines. To avoid awkward locutions, I shall assume that "atom" includes molecules as well as atoms proper.

[19] The nucleus contains both protons and neutrons, which are neutral particles with about the same mass as a proton.

things together, as opposed to electrostatic forces in an atom) lent some insight into the nature of emission of light by atoms. However, there are fundamental differences between an atom and a solar system (to say the least). Basically, the main problem with the solar-system analogy is that the electron, being in orbit around the nucleus, is accelerating constantly and should therefore emit radiation continuously, losing energy in the process. If this actually happened, the electron would fall into the nucleus in a very brief time. Of course this doesn't happen; to explain *why* it doesn't happen requires quantum mechanics.

The analogy with a solar system also provides the concept of the energy of an atom. Consider a proton with an electron circling around it in an orbit with a radius *r*. Speaking in terms of classical mechanics, the electrostatic attraction between the proton and the electron keep them from flying apart, while the momentum of the electron keeps it from falling into smaller and smaller orbits until it crashes into the proton (*cf.* the astronaut hero in an out-of-control spaceship headed straight toward the Sun!). Clearly, if we do work on the electron (somehow) to move it farther from the proton, we have added energy to the system. If the electron moves closer to the proton, it can release some energy. Although you cannot push an electron with a subminiature rocket motor, an atom can change its energy by emitting or absorbing light. When an atom emits light, an electron moves from an orbit with a large radius to one with a smaller radius, and *vice versa*.

2.4.1 Energy levels

The basic problem for physicists that studied the emission and absorption of light by atoms was that each kind of atom absorbs and emits light only at certain discrete wavelengths, not at every wavelength imaginable. Furthermore, an atom that emits light at a certain wavelength also can absorb light at that wavelength, and *vice versa*. This leads to the notion that the energy in an atom must be *quantized*; i.e., there are only certain values that the energy of an atom can have. We call these different *energy levels.* These different energy levels are described by a set of four quantum numbers; the energy of the atom depends primarily on the principal quantum number *n* (see Figure 2-19). We shall only discuss the hydrogen atom, and we shall ignore the other quantum numbers. The lowest energy level is $n = 1$; the next is $n = 2$, ... The spacing between adjacent levels gets smaller and smaller as *n* increases.

The energy of level *n* in the hydrogen atom is

$$E_n = -\frac{R}{n^2} \tag{29}$$

where *R* is the Rydberg constant defined by

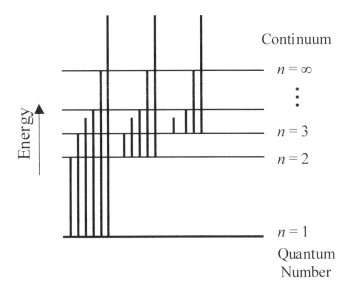

Figure 2-19. Energy levels in the hydrogen atom.

$$R_{\text{H}} = \frac{2\pi^2 m e^4}{ch^3},\tag{30}$$

and m and e are the mass and charge of an electron.[20] Equation (29) also shows why the energy levels start with 1, not 0. If we start with an atom in its lowest—or ground—state, and add an amount of energy

$$\Delta E_{n,1} = E_n - E_1,\tag{31}$$

representing the difference in energy between the level n and level 1, the atom becomes excited.[21] The electron has jumped to a higher orbit, but it is still bound to the nucleus. We can add an amount of energy corresponding to any value of n. In the limit as $n \to \infty$, $\Delta E_{n,1}$ approaches a limit called the *ionization potential* of the atom, denoted by $\Delta E_{\infty,1}$. As long as the amount of energy added is less than this limit, the electron remains bound to the nucleus.

However, if we add an amount of energy $\Delta E > \Delta E_{\infty,1}$, the electron will no longer bound to the nucleus and the atom is then said to be *ionized*. When the atom is ionized, the electron and proton are moving apart so fast that the electrostatic force between them cannot reunite them. As long as $\Delta E > \Delta E_{\infty,1}$, the atom can absorb any amount of energy, not just some specific quantized

---

[20] The Rydberg constant depends on the ratio of the mass of an electron to the mass of the nucleus. Since the ratio $m_p / m_e \approx 1800$ this effect is small: R_∞, the Rydberg constant for an infinitely massive nucleus, is only slightly larger (see Table 2-1).

[21] It is not the *electron* that becomes excited; the energy is a property of the entire atom.

amounts. It is called a *continuum*, since there is a continuous range of energy levels, compared to the separated energy levels when the energy added is below this limit. These ideas are illustrated in Figure 2-19.

Since the energy of an atom in a given state—or energy level—is potential energy, we are at liberty to set the zero point anywhere we want. It is convenient to set the zero of potential energy to the level where an atom is just ionized, so the energy corresponding to any level $n < \infty$ is negative. Since the zero point of potential energy is arbitrary, only the *difference* between energy levels has a physical meaning. Putting $n = \infty$ into equation (29) shows that the ionization limit does indeed correspond to an energy $= 0$.

The energy of a photon emitted when an atom jumps from the (upper) energy level E_u to the (lower) level E_l is

$$\Delta E_{ul} = R\left(\frac{1}{n_l^2} - \frac{1}{n_u^2}\right). \tag{32}$$

When an atom emits light, the energy lost to the atom is equal to the energy of the light emitted. Because the energy levels in an atom are quantized, the amount of energy emitted by an atom is also quantized. We call a quantum of light a *photon*. The energy of a photon is

$$E = h\nu = \frac{hc}{\lambda}, \tag{33}$$

where h is Planck's constant; c, the speed of light; ν, the frequency; and λ, the wavelength, of the photon.

This relationship is a necessary consequence of several experiments that showed, at the beginning of the twentieth century, that light behaves like little particles, thereby reviving a controversy over the nature of light that had been going on, at least, since the days of Isaac Newton. Newton believed that light is made up of corpuscles, or tiny particles. His contemporary, the Dutch physicist Christian Huygens, believed that light is a wave. At the time, there was no way to resolve the question, but in 1801, Thomas Young demonstrated that two beams of light can interfere. Since interference is a wave phenomenon, this seemed to show that light is definitely a wave. However, some experiments in the early twentieth century seemed to demonstrate the particle-like nature of light. Einstein's explanation of the photoelectric effect in 1905 and Planck's explanation of black-body radiation in 1901 both interpreted light as being quantized, thus reviving the corpuscle theory. Today we treat light as having both wave and corpuscular properties, but not *being* either one or the other: this is called the wave-particle duality of light.

Anyway, because the energy levels in the hydrogen atom are quantized, the amount of energy absorbed or emitted by a hydrogen atom is quantized. Because of the relationship between wavelength and energy, light is absorbed or emitted only at discrete wavelengths. Real atoms and molecules can have many, many different energy levels and different spectral lines corresponding to jumps from one level to another. The spectrum of even the simplest of atoms contains a multitude of different lines. The relative strengths of these lines depend on the pressure, the intensity of the ambient radiation field (and also, in some cases, the strength of the ambient magnetic field), and the temperature of the gas, as well as the quantum-mechanical structure of the atom. While the absolute strengths of the lines of an atom will depend on all of these factors, the relative strengths of the lines often depend primarily on the temperature. For example, astronomers can measure the temperature of the atmosphere of a star by measuring the relative intensities of different lines.

2.4.2 Absorption

Let us explore a bit further a particularly simple atom, a hypothetical one with only two levels. Let the energies be E_l and E_u. Aside from collisions, there are three radiative processes that allow the atom to jump from one level to another: absorption of a photon, spontaneous emission of a photon, and stimulated emission. The properties of the atom determine the rates of these processes.[22]

The absorption coefficient of an atom—or molecule—is proportional to the number of atoms in the lower state and to the transition probability $\mathcal{B}_{lu}$. That is,

$$\kappa = N_l B_{lu},\tag{34}$$

where N_l is the number of atoms (per unit volume) in the lower state. $\mathcal{B}_{lu}$ does not change with the physical conditions.

The amount of light emitted by a gas is proportional to [see equation (22)]

$$B(\lambda, T)\left(1 - e^{-N_l B_{lu}}\right) \approx B(\lambda, T)\, N_l\, B_{lu}.\tag{35}$$

In cases where the temperature is known, we can infer the density of the atomic species from a measurement of the intensity of the emitted radiation. If the density is known, we can infer the temperature from the same kind of measurement. If neither the temperature nor the density is known, we might be able to infer both of them from two or more measurement of the intensities at wavelengths corresponding to different transitions in the same atom.

Exercise 2-4: What circumstances would hinder our ability to measure both temperature and density? Assuming the relationships given above between optical depth and the number of atoms, derive a formal solution in terms of

---

[22] See Mihalas (1970, pp 81-85).

the intensities that would be measured. Assume also that the absorption coefficient does not change over the bandwidth of each measurement, although it will be different for different bands.

Exercise 2-5: What happens when the absorption coefficient changes within each band?

2.4.3 Spontaneous and stimulated emission

Spontaneous emission happens when an atom jumps spontaneously from a higher state to a lower one, emitting a photon. Stimulated emission, as the name implies, arises when a photon of just the right energy passes an excited atom and induces a transition from the higher to the lower state.[23] The atom emits another photon with the same wavelength and moving in the same direction. The emitted photon is coherent with the incident one—the emitted photon has the same phase as the one that stimulated it. It is as though the presence of the light near the atom encouraged it to make this transition, just as the presence of food makes us salivate. That stimulated emission should occur is quite a puzzlement, as there is no classical phenomenon that would predict its existence. However, as I will show below, stimulated emission must occur for the Second Law of thermodynamics to be satisfied.

Let I_{lu} be the intensity of radiation with a wavelength corresponding to the transition between states l and u in an atom, where $E_l < E_u$.[24] The rate of absorption of photons is proportional to the number of atoms in the lower state, l. Let $\mathscr{B}_{lu}$ be the absorption coefficient; let N_l and N_u be the number of atoms in each state; the total number of atoms, $N_l + N_u$, is fixed. Then the rate of absorption of photons is

$$\frac{\mathrm{d}N_l}{\mathrm{d}t} = \mathrm{B}_{lu} I_{lu} N_l. \tag{36}$$

Let $\mathscr{A}_{ul}$ be the rate of spontaneous emission of photons from the higher state, u; let $\mathscr{B}_{ul}$ be the rate of stimulated emission from state u. The difference between these two terms is that the former rate applies to an atom regardless of its surroundings, while the latter rate is proportional to the ambient intensity I_{lu}. The rate of stimulated emissions is $N_{stim} = N_u \mathscr{B}_{ul} I_{lu}$. The total emission rate (spontaneous plus stimulated) is

---

[23] This was first demonstrated by Einstein in 1917. The first practical use of stimulated emission was in the *maser*, invented by Charles Townes in 1953, which achieved amplification by stimulated emission of microwaves. Of course, the laser also operates on this principle.

[24] I ignore collisions here; assume that the gas is very tenuous.

$$\frac{dN_u}{dt} = N_u(A_{ul} + I_{lu} B_{ul}).$$ (37)

Suppose that the gas is in equilibrium with a radiation field $B(\lambda,T)$. In a steady state, the number atoms leaving the lower state must equal the number leaving the upper state. Combining the last two equations,

$$\frac{N_l}{N_u} = \frac{B_{ul} I_{lu} + A_{ul}}{B_{lu} I_{lu}}.$$ (38)

The probability that an atom will be in a state with energy E_u is $\exp\{-E_u/kT)$. This is proved in any book on statistical mechanics, but it is beyond the scope of this chapter. The relative populations of the two states is

$$\frac{N_u}{N_l} = e^{(E_l - E_u)/kT}.$$ (39)

Using equation (33),

$$\frac{E_u - E_l}{kT} = \frac{hc}{\lambda kT}.$$ (40)

Therefore, combining equations (37) and (38),

$$\mathscr{A}_{ul} = \left(\mathscr{B}_{lu} e^{hc/\lambda kT} - \mathscr{B}_{ul}\right) B(\lambda,T).$$ (41)

Since the constants $\mathscr{A}_{ul}$, $\mathscr{B}_{ul}$, and $\mathscr{B}_{lu}$, depend only on the structure of the atom, and not on temperature or other aspects of the environment,[25]

$$\mathscr{B}_{lu} = \mathscr{B}_{ul}.$$ (42)

Then

$$\mathscr{A}_{ul} = \left(e^{hc/\lambda kT} - 1\right)\mathscr{B}_{ul} \left[\frac{2hc^2}{\lambda^5\left(e^{hc/\lambda kT} - 1\right)}\right].$$ (43)

So, using the Plank black-body law [equation (2)],

$$\mathscr{A}_{ul} = \frac{2hc^2}{\lambda^5} \mathscr{B}_{ul}.$$ (44)

---

[25] Strictly speaking, this equation should read $g_{lu}\mathscr{B}_{lu} = g_{ul}\mathscr{B}_{ul}$, where g_{lu} and g_{ul} are the statistical weights or the lower and upper levels. For the present purposes, we can take them to be unity.

These coefficients are called the Einstein coefficients; they depend only on the nature of the atom and not on the temperature, the ambient radiant intensity, or any other aspect of the atom's environment. Therefore, equations (42) and (44) are true in general. In particular, note that this argument shows that stimulated emission *must* exist; *i.e.*, $\mathscr{B}_{ul} \neq 0$. This proved the existence of stimulated emission long before it had been observed in the laboratory; today, of course, it is the basis of the laser (**L**ight **A**mplification by **S**timulated **E**mission of **R**adiation) industry. Stimulated emission is necessary, as I have already mentioned, to keep the Second Law of thermodynamics from being violated, because without it, atomic absorption and emission processes could not be in balance with emission from a black body. If there is an enclosure with perfectly black walls with temperature T, then the radiation field inside the enclosure must be equal to the Planck function $B(\lambda, T)$. But now introduce a gas into that enclosure. Without stimulated emission, there would be a continuous transfer of energy either to or from the gas. This, as we know, is impossible.

2.4.4 Collisions

Just as a transition between two levels in an atom can occur with the simultaneous absorption or emission of a photon, it can also occur when the atom collides with another particle. Let E_i and E_f be the initial and final energies of the atom. During the collision, the energy of the atom changes by

$$\Delta E_{coll} = E_f - E_i; \qquad (45)$$

since energy is conserved, the sum of the kinetic energies of the atom and the particle it collided with changes by $-\Delta E_{coll}$. I use the notation E_i and E_f because the energy of the atom might increase or decrease during a collision. The probability that a collision will cause an atom to jump from one level to another—the collisional transition rate—depends on the temperature; the number of collisions in a unit volume of a gas is also proportional to the product of the densities of the two gases.

The relative populations of the different levels—the number of atoms in each level—depend, for a given atom, on the ambient pressure, temperature, and radiation field. On a microscopic level, the populations are affected by collision rates and the ambient intensity. On the other hand, the relative population in each level also affects the absorption coefficient: therefore, the absorption coefficient of a gas depends in a complicated way on the temperature and pressure. Changes in these variables affect the amount of light emitted or absorbed by a gas.

2.5 Line broadening

The study of atomic and molecular absorption lines started with the astro-physicists who were interpreting the spectral lines that they measured in stars;[26] later, people who study radiative transfer in the Earth's atmosphere also studied these processes.[27]

The lines observed with a spectrograph are not infinitely narrow, as the previous discussion might suggest. Spectral lines in the atmosphere have a finite width because of two broadening mechanisms, Doppler[28] broadening and pressure broadening. Because these mechanisms are affected differently by changes in temperature and pressure, absorption lines in the earth's atmosphere are much broader near the surface of the earth than at higher altitudes. This has a substantial effect on the radiance emitted at the top of the atmosphere. Let us examine each one in turn.

2.5.1 Doppler broadening

Doppler broadening is caused by the random motions of the particles of a gas toward or away from the observer; these radial motions cause changes in the wavelength of the emitted or absorbed radiation. Since there is a distribution of radial velocities among the particles of a gas, there will be a distribution of wavelengths where light is absorbed or emitted.

If ν_0 is the frequency of the light emitted or absorbed by an atom or molecule that is at rest, an observer moving with speed u sees a frequency

$$\nu = \nu_0 - \Delta\nu = \nu_0\left(1 - \frac{u}{c}\right), \tag{46}$$

where $\Delta\nu = \nu_0 u / c$. This equation is interpreted as having $u > 0$ for an atom that is receding; $u < 0$, for one that is approaching. In other words, $u = ds/dt$, where s is the distance from the observer to the atom. Therefore, atoms will appear to have a distribution of frequencies, centered on ν_0, where they absorb light. This means that a spectral line is not a δ-function; it has a substantial width due, in part, to Doppler broadening.

Temperature is a measure of the average kinetic energy of a gas particle. For a gas with temperature T, the average kinetic energy of a particle is

$$E(T) = kT, \tag{47}$$

---

[26] See, for example, Cowley (1970) or Mihalas (1970).

[27] See Liou (1992).

[28] Christian Doppler (1803-1853) was an Austrian physicist.

where k is Boltzmann's constant. The probability of finding a particle with an energy between E and $E + \Delta E$ is

$$p(E) = \frac{1}{kT} e^{-E/kT}. \tag{48}$$

The kinetic energy is $E = \frac{1}{2}mu^2$, where u is the speed of a particle; we can find the distribution of velocities of the atoms from the distributions of the energies. The probability that the radial velocity will be between u and $u + \Delta u$ is[29]

$$p(u) = \sqrt{\frac{m}{2\pi kT}} \, e^{-\frac{mu^2}{2kT}}. \tag{49}$$

I shall derive this formula in §14.1.1.

Given the previous two equations, we can see that a spectral line that is broadened by atomic or molecular motions has a shape given by

$$\alpha(v) = \alpha_o \, e^{-[(v-v_0)/\Delta v_D]^2}, \tag{50}$$

where the absorption coefficient α depends explicitly on frequency. The line width parameter is

$$\Delta v_D = 3.581 \cdot 10^{-7} \sqrt{T/m} \tag{51}$$

where T is the temperature in kelvins and m, the atomic or molecular weight.[30] Therefore, gases at higher temperatures or smaller particle masses have broader lines.

2.5.2 Pressure broadening

In additional to Doppler broadening, spectral lines are also broadened by collisions. This can be explained in terms of the *uncertainty principle*, which limits how much we can know about the state of a particle.

Heisenberg's[31] uncertainty principle applies to spectral line shapes as follows. Suppose there is an atom with two energy levels, denoted again by u and l, with u having the higher energy. Now the atom, if it is in level u, can emit a photon and jump to level l; if it is in level l, it can absorb a photon and jump to level u. However, collisions can also cause an atom to move from one state

---

[29] See, for example, Cowley (1970); Mihalas (1970); Liou (1992); or Peach (1975).

[30] See Gaut (1968).

[31] Werner Heisenberg (1901-1976), a German physicist, developed a matrix method of quantum mechanics. He published his work on the uncertainty principle in 1927.

to another, with the difference in the energy of the atom before and after the collision being equal to the difference in the kinetic energy of the two particles that collided. If an atom is in state u, it will stay there until either the atom emits a photon or it collides with another particle. Let Δt be the mean time an atom can remain in that state before one of these things happens. We call Δt the *lifetime* of that state. If E_{ul} is the difference in energy between levels u and l,[32] it has some uncertainty ΔE_{ul} due to the finite lifetime of the higher state. Because of the uncertainty principle, the uncertainty in energy is related to the lifetime of the higher state by

$$\Delta E_{ul} \geq \frac{h}{4\pi\Delta t} . \tag{52}$$

The uncertainty principle applies to any pair of conjugate variables; *i.e.*, quantities whose product has the dimension of action. Such pairs of quantities are energy and time; or position and momentum. In the later case, if x is the position of a particle along a certain axis, and $p = mv$ is its momentum in that direction, the uncertainties Δp and Δx obey the relationship

$$\Delta x\, \Delta p \geq \frac{h}{4\pi} . \tag{53}$$

The rate of collisional excitation and de-excitation of molecules in the air depends on the temperature and pressure of the air. While the temperature in the troposphere does not change much, compared with a typical temperature of 300 K, the pressure does (*e.g.*, between the surface and an altitude of 10 km). The higher the pressure is, the more frequent collisions will be, and consequently, the shorter the lifetime of any state of an atom. According to the uncertainty principle, this means that there is greater uncertainty in the energy of that state. This uncertainty broadens the spectral lines. Since increased pressure causes an increase in the line width, this effect is called *pressure broadening*. While this argument about the cause of pressure broadening is undoubtedly correct in its broad features, there is some question as to exactly what the pressure-broadened line shape should be; this is a topic of ongoing research.

The time that an atom spends in an excited state, before it emits a photon, is[33]

$$\Delta t \;=\; \frac{1}{4NvC}, \tag{54}$$

---

[32] Note that I have changed the notation slightly, representing the energy difference now by E instead of ΔE.

[33] See Lipson and Lipson (1969). There is also a new edition published in 1995.

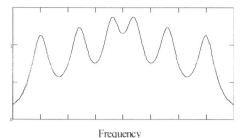

Figure 2-20a. Absorption coefficient at high pressure.

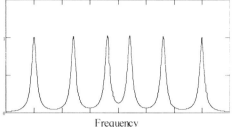

b. Absorption coefficient at lower pressure.

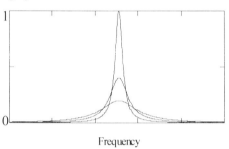

c. The Lorentz profile for high (solid line) to low (dashed line) pressures. Absorption is in arbitrary units.

where N is the number density of the gas; v the speed of a molecule; and C, the collision cross-section. Then, in the absence of Doppler broadening, the line shape is given by

$$\alpha(v) = \frac{2\,\Delta t}{\left[1 + (v - v_0)^2 (\Delta t)^2\right]}.$$ (55)

This is called the *Lorentz*[34] line shape. I illustrate this in Figure 2-20. Panels *a* and *b* show the absorption coefficient of a hypothetical gas with several lines at high and low pressure, respectively. Note how the lines blend at higher pressure. Panel *c* shows the profile of a single line at different pressures.

In general, Doppler broadening is more important at low pressures, where collisions are infrequent, while pressure broadening is more important at higher pressures. At intermediate pressures, both broadening mechanisms may contribute significantly to the line shape.

Exercise 2-6: Would Doppler broadening be more important for a hydrogen molecule or a xenon atom? Why?

---

[34] Hendrik Antoon Lorentz (1853-1928) was a Dutch physicist whose work on moving electromagnetic fields was a precursor of the special theory of relativity.

In general, atomic or molecular lines can be quite narrow in the strato-sphere or the lower troposphere, while they are much, much broader near the surface. In the infrared part of the spectrum, there are literally thousands of lines due to ozone, carbon dioxide, water vapor, and other molecules. At the surface, these lines blend together almost into a continuum.

We have already seen the absorption coefficients for oxygen (Figure 2-1) and for water vapor (Figure 2-2), in the microwave part of the spectrum. Note the complicated lines in the oxygen spectrum between 50 and 60 GHz. Fig-ures 2-3 to 2-10 show the absorption by the important molecules in the atmo-sphere in the range from 0.5 to 20 μm Spectra for water vapor, carbon diox-ide, and ozone are shown separately. The ordinate in these plots is the trans-mission $e^{-\tau}$, where τ is the optical depth of the species being plotted. A value of 0 corresponds to complete absorption, while a value of 1 corresponds to complete transmission.

The foregoing discussion was meant to give a physical feeling for how changes in variables like temperature and pressure affect the amount of light emitted by a gas, so that the reader will develop a feeling for the processes that make remote sensing possible. If these changes did not affect the amount of light we measure, we would be able to infer nothing about the temperature of a gas from measurements of the amount of radiation it emits. In order to understand how light interacts with matter, we need a detailed understanding of the quantum mechanical phenomena that determine such things as energy levels of atoms and absorption coefficients. This knowledge of atomic and molecular physics is one of the tools that is needed in the study of remote sensing. Fortunately, most of the relevant physical parameters—the energy levels and transition probabilities of different atoms and molecules, for exam-ple—have been determined quite well by spectroscopists, although there are some notable exceptions.

2.6 A remote-sensing example

The simplest remote-sensing problem is that of measuring the temperature of a black body (*e.g.*, of a star, which is a pretty good black body). Imagine that I have two photometers, one of which is sensitive in the red part of the spec-trum (wavelength λ_1) and one in the blue (wavelength λ_2). Since a black body emits a spectrum with a shape given by the Planck function B(λ,T), the ratio

$$\frac{B(\lambda_1, T)}{B(\lambda_2, T)} = \frac{\lambda_1^5 \left(e^{hc/kT\lambda_2} - 1\right)}{\lambda_2^5 \left(e^{hc/kT\lambda_1} - 1\right)}. \tag{56}$$

is a unique function of T. (Astronomers would call this ratio the *color* of a star.) However, because there will be some noise in the measurements, it may be necessary to make measurements at three wavelengths (red, blue, and ultraviolet) and use some kind of curve fitting to arrive at the temperature that best fits the three measurements. This is the method astronomers have used since the nineteenth century for measuring the temperatures of stars, and this method is still fundamental.

Therefore, if we know that we are observing something that is a black body, we can measure its temperature by measuring its color.

2.6.1 Effects of changes in emissivity

A second problem arises in the field of microwave remote sensing of the ocean surface. To measure the sea surface temperature (SST), we might build a radiometer that measures the intensity at a single microwave frequency somewhere between 1 and 85 GHz. Suppose that this radiometer has been placed in a satellite and it looks down at the surface of the ocean.

I introduced the concept of brightness temperature in §2.1. For a real material with an emissivity e_p, where p is the polarization, either v or h, the brightness temperature would be

$$T_b = e_p T_s, \tag{57}$$

where T_s is the physical temperature of that material and we assume that there is no intervening gas to absorb or emit radiation. Since $0 \le e_p \le 1$,

$$T_b \le T_s. \tag{58}$$

The emissivity of the ocean surface depends on the wind speed at the surface, because wind roughens the surface and produces foam. The size of this effect changes with frequency. Let w be the wind speed and $e_p(w)$, the emissivity for polarization p. Then the intensity we observe will be

$$T_b(w) = T_s e_p(w). \tag{59}$$

The brightness temperatures are different, in general, for the vertical and horizontal polarizations. If we knew the functional form of $e_p(w)$, we could infer the wind speed from

$$e_p(w) = \frac{T_b(w)}{T_s}. \tag{60}$$

Similarly, if we knew the wind speed, we would know the value of $e_p(w)$ and could infer the temperature of the surface from

$$T_s = \frac{T_b}{e_p(w)}. \tag{61}$$

But what happens if there is some uncertainty in the emissivity? If the uncertainty in the emissivity is δe_p, the corresponding uncertainty in the surface temperature is

$$\delta T_s = -\frac{T_b}{e_p^2}\delta e + \frac{dT_b}{e_p},, \tag{62}$$

where δT_b is the error in the measurement of T_b.

The perceptive reader will already have grasped the problem. Since, in real life, the wind speed and the sea surface temperature (SST) are both variable, we cannot infer either the wind speed or the SST from a single radiometric measurement. Therefore, we might try the following stratagem. We can make measurements simultaneously at two different frequencies and infer both the wind speed and SST from that pair of measurements. However, to do this, we would have to know (quite accurately) how the emissivity at one frequency is related to the emissivity at the second frequency.

To make some progress, let us assume that the emissivity of the ocean surface depends linearly on wind speed:

$$e_v(w) = e_{0,v} + \alpha_v w, \tag{63}$$

where v is the frequency and α_v is a coefficient that depends only on frequency.[35] Let T_1 and T_2 be the brightness temperatures measured at frequencies v_1 and v_2; let α_1 and α_2 be the corresponding wind-speed coefficients. Then

$$\begin{pmatrix} T_1 \\ T_2 \end{pmatrix} = T_s \begin{pmatrix} e_{1,0} + \alpha_1 w \\ e_{2,0} + \alpha_2 w \end{pmatrix}. \tag{64}$$

So we can solve this 2×2 problem for these two quantities simultaneously. If we could ignore the noise in the measurements of T_1 and T_2, we could solve for the wind speed by taking the ratio

$$\frac{T_1}{T_2} = \frac{e_{1,0} + \alpha_1 w}{e_{2,0} + \alpha_2 w}; \tag{65}$$

so

$$w = \frac{e_{2,0}T_1 - e_{1,0}T_2}{\alpha_1 T_2 - \alpha_2 T_1}. \tag{66}$$

---

[35] For the moment, we ignore the polarization p.

A similar algebraic manipulation will lead us to a solution for T_s. But what happens if there is some noise in the measurements of T_1 and T_2? Especially if $\alpha_1 T_2 - \alpha_2 T_1$ is small? What if $e_p(\omega)$ is not linear? I shall take this up in Chapter 7.

References

Born, M., and E. Wolf, *Principles of Optics*, Sixth Edition, Pergamon Press, Oxford (1980).

Chandrasekhar, S., *Radiative Transfer*, Dover Publications, New York (1960).

Cowley, C. L., *The Theory of Stellar Spectra*, Gordon and Breach, New York, 260 pp (1970).

Gaut, N. E., 'Studies of atmospheric water vapor by means of passive microwave techniques,' Massachusetts Institute of Technology, Research Laboratory of Electronics, Technical Report 467 (1968).

Jackson, J. D., *Classical Electrodynamics*, John Wiley & Sons, New York, 641 pp. (1967).

Liou, K. N., *Radiation and Cloud Processes in the Atmosphere*, Oxford University Press, New York, 487 pp (1992).

Liou, K.-N., *An Introduction to Atmospheric Radiation*, Academic Press, New York (1980).

Lipson, S. G., and H. Lipson, *Optical Physics*, second edition, Cambridge University Press, London (1969).

Mihalas, D., *Stellar Atmospheres*, W. H. Freeman and Co., San Francisco, 463 pp (1970), pp 81-85.

Pais, A., *'Subtle is the Lord...' The Science and the Life of Albert Einstein*, Clarendon Press, Oxford, 552 pp (1982).

Peach, G., 'The width of spectral lines,' *Contemporary Physics*, **16**, 17-34 (1975).

Ulaby, F. T., R. K. Moore, and A. K. Fung, *Microwave Remote Sensing Active and Passive*, Vol. 1, Addison-Wesley Publishing Co., Reading, MA (1981).

3. Instruments and Noise

If your experiment needs statistics, you ought to
have done a better experiment.

Ernest Rutherford (1871-1937)

In this chapter, I shall discuss a few topics relevant to the effects that instruments have on remote-sensing data. Basically, these instruments are all radiometers: they measure the intensity of the radiation incident from a certain direction within a certain range of frequencies. The properties that interest us here are:

1. Instrument noise, which is usually due to thermal noise or shot noise;
2. Calibration; and
3. Angular resolution.

Since there are many books on the instruments used in remote sensing, I shall not try to treat that subject here. However, since this is a book about deriving information from noisy data, and the noise arises primarily in the measuring instruments, I need to say just a few words about them. The discussion will be mostly generic, treating properties that are common to all radiometers. However, because I want to provide a motivation for the problem treated in Chapter 12, I shall discuss some physical principles that apply to radio antennas. With some modification, they would also apply to visible or infrared radiometers, but there is no need to pursue those ideas here.

For the reader who is interested in this subject, I recommend Kraus' *Radio Astronomy* (1986). Although the subject might, on the face of it, seem to be too specialized for most readers, the material in his Chapter 3 will actually serve as a good general introduction to the subject. Ulaby *et al.*, Chapter 4, is another good introduction.

3.1 Radiometers

All of the discussion in this book is based on the notion that we can measure the amount of light emitted (or absorbed) by an object. The general name for a device that measures the intensity of light is a *radiometer*.[1] The design and fabrication of radiometers is in itself an important area of engineering, but it is

---

[1] A device that measures the intensity of visible or infrared light is also called a *photometer*.

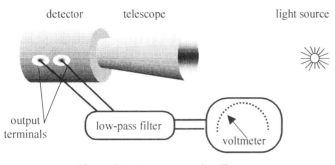

Figure 3-1. A conceptual radiometer.

far beyond the scope of this book. However, the reader should have some idea of what a radiometer is so that he can appreciate how noise arises in measurement of radiant intensity. I shall describe here what might be called a *conceptual* radiometer, in that it does not describe any particular physical device, only a concept. Avoiding the complexity of a real instrument will allow us to focus our attention on some essential aspects of its operation. Although the engineering principles that govern the specific devices that we use to measure radiation depend on the wavelength involved—microwave radiometers are entirely different from infrared or visible radiometers—we can ignore those differences here.

A radiometer has an input and an output. The input is the light we wish to measure; the output is the voltage produced between two terminals (on the outside of the radiometer, perhaps) as shown in Figure 3-1. A source of light is on the right. The radiometer consists of a device that may contain a combination of lenses, antennas, and mirrors that concentrate light. I shall use the term *telescope* as a generic term for the light-gathering and -focusing part of the instrument, regardless of its precise nature.

The second part of the radiometer is the detector, which converts the radiant energy focused on it by the telescope into an electric signal. An electromagnetic wave has an instantaneous voltage $E(t)$; the power is proportional to $E^2(t)$. The most common kind of detector is a *square-law detector*; *i.e.*, one that produces an output *voltage* proportional to the incident power. In our example, the voltage appears between two terminals on the outside of the detector housing.

Next, there is a low-pass filter that removes high-frequency fluctuations—and noise—from the output. This may be a filter or it may be an integrate-and-dump circuit. It can be described either by the cutoff frequency ν_c, which is the highest frequency that can pass through the filter, or by an integration time τ, which represents the sample time. They are related by

$$\tau = \frac{1}{2\nu_c};$$
(1)

the factor of ½ arising because the filter passes both positive and negative frequencies, as long as $|\nu| \leq \nu_c$.

The last part of the radiometer is a device that measures the output voltage and records it; we generally use some kind of voltmeter. The voltmeter might have a dial, requiring that someone write down the result of each measurement; it could be attached to some analog recording device like a chart recorder; or the signal might be digitized and recorded by a computer (clearly the best alternative). Whatever the details might be, the output of this voltmeter represents the fundamental measurement that we can make with this instrument.

Although the output voltage produced by a radiometer in response to a certain amount of light is arbitrary, since it is determined by the design of the electronic circuits and whatnot, our physical theories are related to absolute intensities in units of erg cm^{-2} ster^{-1} Hz^{-1}, or similar units. A calibration procedure is needed to convert these arbitrary voltages into radiances that can be expressed in known units. There are many ways that this calibration might be effected; I shall describe only one here. If we had a microwave or infrared radiometer, we would call it the "hot load, cold load" method. This method works well for a linear radiometer; *i.e.*, one where the change in the output voltage is proportional to a change in the incident radiant power.

The term *load* refers to a black body that can be placed in front of the radiometer and is used as a calibration source. In order to use this method, we must know both the emissivity e of the loads (which will be the same for both loads and very close to 1) and the temperatures of the loads (which will be quite different). Let V_h, V_c, and V_s be the voltages that we measure when we look at the hot load, the cold load, and the subject we want to measure. Let T_h and T_c be the temperatures of the hot and cold loads, respectively. Then the intensity that we measure can be expressed, if the radiometer is linear, using

$$\frac{eB(T_h) - eB(T_c)}{V_h - V_c} = \frac{B(T_s) - eB(T_c)}{V_s - V_c} \tag{2}$$

so the unknown brightness temperature is $I_s = B(T_s)$, or

$$I_s = \frac{1}{e}\left\{ \frac{V_s - V_c}{V_h - V_c}\left[B(T_h) - B(T_c)\right] + B(T_c)\right\}. \tag{3}$$

Here, $B(T)$ is the Planck black-body function at the appropriate frequency ν.

This general calibration method works at any frequency. At microwave frequencies, where the Planck function at frequency ν and temperature T is approximately $B_\nu(T) = 2kT\nu^2 / c^2 = 2kT / \lambda^2$, the Rayleigh-Jeans approximation (see §2.2.3). We usually express the intensity in units of an equivalent black-body temperature; this makes it easy to express the measured intensity in terms of the hot and cold loads.

As a practical matter, the calibration can be done with two pieces of microwave absorber, one at room temperature and the other cooled with liquid nitrogen, which has a temperature of 77 K. First one, and then the other, is placed in front of the antenna feed horn (the antenna itself being far too large) and the output voltage is measured each time.

Exercise 3-1: Why do I use a piece of microwave absorber? Why don't I buy a piece of microwave emitter instead? Or do I?

A radiometer is sensitive to radiation only in a certain range of frequencies. Let ν_{min} and ν_{max} be the minimum and maximum frequency; the difference $\Delta\nu = \nu_{max} - \nu_{min}$ is the *bandwidth* of the radiometer. Usually, $\Delta\nu \ll \nu_{min}$. Just as the power emitted from a black body is proportional to the bandwidth $\Delta\nu$, so is the power received by a radiometer (see §2.2.1). We shall see presently that this is an important consideration in determining the sensitivity of the radiometer. It is often convenient to consider the power *per unit bandwidth*, as I do in equation (4) below.

A word about nomenclature. In general, a satellite remote-sensing instrument will consist of several similar radiometers, each having a different range of frequencies. Sometimes, for visible or infrared measurements, there is a single radiometer and different filters are used sequentially to make measurements at different frequencies. These measurements are often referred to as different *channels* in the data stream, referring to different wavelengths used to make the measurements.

3.2 Noise

If real radiometers could produce an output voltage that depended only on the incident radiant power, we would not need this book. In fact, the output from any radiometer contains noise, and we should inquire as to what this noise is and where it comes from. I shall treat two types of noise here: thermal noise and shot noise.

3.2.1 Thermal noise

Thermal noise arises in any electrical circuit that has a resistance. Consider a metallic resistor with a resistance R at a temperature T. It will produce an electrical power per unit bandwidth

$$P_{th} = 2kTR, \tag{4}$$

where k is Boltzmann's constant; T, the temperature in kelvins; and R, the resistance in ohms. The spectrum of P_{th} is

$$S_{th}(\nu) = 2kTR\frac{\nu/\nu_0}{e^{\nu/\nu_0} - 1}, \tag{5}$$

where[2]

$$\nu_0 = 2.1 \times 10^{10} \, T \text{ Hz.} \tag{6}$$

At room temperature ($T = 300$), $\nu_0 = 6300$ GHz. Therefore, two measurements of the electric field are independent provided that they are at least $1/\nu_0 = 1.7 \times 10^{-11}$ apart. To see this, use the Fourier-transform relation between the spectrum and the autocorrelation function (see §9.1.4): the autocorrelation function of thermal noise is

$$C(\tau) = \int_{-\infty}^{\infty} S(\nu) e^{-2\pi i \nu \tau} \, d\nu = \int_{-\nu_0}^{\nu_0} e^{-2\pi i \nu \tau} d\nu = 2\nu_0 \, \text{sinc}\left(\frac{\tau}{2\nu_0}\right). \tag{7}$$

The width of $C(\tau)$ is $1/\nu_0 \approx 1.7 \times 10^{-11}$ s, so samples separated by more than this are uncorrelated.

Since $e^x - 1 \approx x$ for $x \ll 1$, the spectrum S_{th} is constant up to 1000 GHz or so. Therefore, for microwave radiometers, thermal noise has a spectrum proportional to T and independent of frequency. Since we generally express microwave intensities in terms of a black-body temperature, it is natural to express the noise also as a temperature. Note that the power available from a resistor has the same form as the Rayleigh-Jeans approximation to the Planck black-body function and, in fact, the power produced by a resistor is equivalent to thermal emission from a black body.

Most receivers have bandwidths considerably smaller than 6300 GHz. In general, the signal is low-pass filtered to remove as much noise as possible, with frequencies typically smaller than 1000 Hz. If $\Delta\nu$ is the bandwidth of the radiometer, *i.e.*, the range of frequencies the radiometer responds to, the noise in each *independent* measurement is

$$\Delta T = \frac{k T_{\text{sys}}}{\sqrt{\tau \cdot \Delta\nu}}, \tag{8}$$

where k is a number between 1 and 3 (or so), depending on the design of the radiometer, and T_{sys}, the *system temperature* of the radiometer, is a number that describes the noise output of the receiver part of the radiometer. It is generally 1 to 100 times the physical temperature of the hardware.[3] Both T_{sys} and ΔT are expressed in kelvins.

As a heuristic explanation of equation (8), we should note that the number of independent measurements of a signal with bandwidth $\Delta\nu$ and an integration time τ is $N = \tau \cdot \Delta\nu$. Low-pass filtering is equivalent to averaging N sam-

---

[2] See Gardner (1990, §10.67); Kraus (1986, §3-18); or Middleton (1960) for a derivation. This relationship was first derived by Nyquist (1928).

[3] See Kraus, 1986, Chapter 7; or Crane and Napier, 1986.

ples of the signal. As we know, the variance of a sum of N terms is proportional to N^{-1} (the standard deviation is proportional to $N^{-\frac{1}{2}}$). The system temperature, by definition, is the equivalent noise temperature for a one-hertz bandwidth and a one-second integration time.

Because the noise is given in terms of an equivalent black-body temperature, it is often called the NEΔT, or *noise-equivalent* ΔT.

Thermal radiation is called a *white-noise* process, since its spectrum is constant. Such a process has a Gaussian distribution,[4] which is why I have treated the noise as being a Gaussian random variable.

Throughout this book, I assume that the noise in a measurement is independent of the value of the quantity being measured. This is usually a very good assumption, although the system noise depends somewhat on the intensity being measured. For most earth-sensing applications, the variability of the intensity over a scene is usually small enough that we can ignore this effect.

3.2.2 Shot noise

The preceding discussion of thermal noise used a wave picture of electromagnetic radiation. This works well at microwave frequencies. At visible and infrared frequencies, the particle nature of light is more important in discussions of the noise that arises in visible and infrared radiometers.

The predominant source of noise in a visible or infrared radiometer is called *shot noise*. In this case, the output of a radiometer is proportional to the number of photons received during the integration time τ. They arrive randomly, so the number of photons varies from one interval to the next, even if the illumination remains (macroscopically) constant. The noise produced is similar—statistically—to the sound produced when a stream of lead shot is dropped into a bowl; hence the name. The spectrum of shot noise has two components, one at zero frequency and one that is independent of frequency ν for $\nu > 0$. This is also a white-noise process, and hence, Gaussian.

3.2.3 Why does a resistor emit radiation?

It is not obvious why a resistor should emit radiation. After all, doesn't it *absorb* power from an electric circuit? We all know that a resistor gets hot when too much current passes through it. The answer lies in the Second Law of thermodynamics.

Consider an antenna with a resistor connected to it (see Figure 3-2). If there is radiant energy incident on the antenna, it will produce a current in the resistor; this current will dissipate in the resistor and produce heat. Now sup-

---

[4] See Gardner, 1990, §10.7.1.

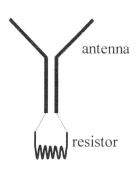

Figure 3-2. An antenna connected to a resistor absorbs radiation, so it must emit radiation also.

pose that the resistor only absorbed, but did not emit, radiation. Place the whole thing in a black-body enclosure at a temperature T. Since it absorbs but does not emit radiation, the resistor will get hotter at the expense of its environment. Eventually its temperature will be greater than that of the enclosure, which we know would violate the Second Law of thermodynamics. Therefore, the resistor must emit radiation also. This is precisely the same argument that gives rise to Kirchoff's law.

An interesting complication arises if we compare the discussion of the noise equivalent power of a resistor in the three references cited previously: Gardner gives equation (4) above.[5] Ulaby et al.[6] give the power as $P = 2kT \cdot \Delta v$; Kraus[7] gives $P = 2kT$ per unit bandwidth (which amounts to the same thing). They all quote Nyquist[8] as their source. But where did the R go in equation (4)? It seems clear that the resistance R should come in somewhere, since the power $P \to 0$ as $R \to 0$.

The answer seems to be that, while Gardner is writing about statistics in general, Kraus and Ulaby et al. are thinking of the problem only in relation to antennas (which is what their books are discussing where these equations arise). An antenna produces a current I when radiance of a certain intensity is incident on it; using Ohm's law, the power is $P = I^2 R_r$, where R_r is called the radiation resistance. Then we must have $R = R_r$ in a resistor that we are comparing with the antenna, so that the current, as well as the power, are the same for a resistor at temperature T and an antenna contained in a black-body enclosure at temperature T. We can look at the situation in the following way. Suppose we have two black-body enclosures, each at the same temperature T. One contains an antenna (assume that it has no resistive losses) while the other one contains a resistor. Each one produces a current that we can measure. The antenna has a certain radiation resistance R_r, which relates the current to the received power. Suppose the resistor has the same resistance. Then the current coming from each enclosure—as well as the power—will be the same and will vary according to the temperature T. If we did not know which enclosure contained the antenna and which one contained the resistor, we could not tell just from looking at the output.

---

[5] Gardner, (1990, p 235), although he implies that it is the total power, not per unit bandwidth.

[6] Ulaby et al. (1981) p 199.

[7] Kraus (1986) pp 3-39.

[8] Nyquist (1928).

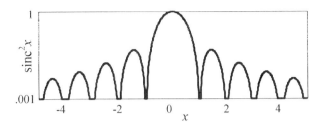

Figure 3-3. Antenna pattern of a uniformly illuminated aperture, plotted on a logarithmic scale. The nulls of the pattern at $x = \pm 1$, 2, ...; the side lobes are maximum at $x = \pm 3/2$, 5/2, ...

This gives rise to the notion of *antenna temperature*. Suppose I have antenna that receives a certain power P_{rec}. I could have a resistor that produced the same power; its temperature would be $T_r = P_{rec} / 2k \cdot \Delta v$. The received power is $P_{rec} = B\lambda^2 \cdot \Delta v$, where λ is the electromagnetic wavelength. We can describe the radiation incident on the antenna by defining the quantity

$$T_a = \frac{B\lambda^2 \cdot \Delta v}{2k \cdot \Delta v} = \frac{2kT\lambda^2 c^2 \cdot \Delta v}{v^2 \, 2k \cdot \Delta v} = T \tag{9}$$

where T is the temperature of the black body that would produce the given intensity.[9] It is also the temperature of the equivalent resistor. Because of the way we actually calibrate microwave radiometers, or radio telescopes, it is often very convenient to express measurements in units of antenna temperature T_a. (Note that the antenna temperature has nothing to do whatsoever with the physical temperature of the antenna structure, at least for ideal, lossless antennas).

3.3 Telescopes

The telescope was the first remote-sensing instrument: with it, Galileo observed mountains on the Moon, the rings of Saturn, and the phases of Venus. These observations would, in time, change forever our conception of the universe.

Most (but certainly not all) radiometers have a directional sensitivity: they measure the amount of light received from within a certain solid angle $\delta\Omega$ around the nominal direction that the radiometer is pointing (see Figure 3-3). The combination of antennas, mirrors, or lenses that focus the light on the radiometer that I am calling a telescope could work either as a receiver or as a transmitter. In either case, it will have the same properties. It seems to be

---

[9] For a lossless antenna, the received power is $B\lambda^2$ per unit area. See Kraus (1986) or Ulaby *et al.* (1981).

easier to think about the pattern of transmitted radiation, rather than of received radiation: our language seems to deal with the former concept better. So we can say that the telescope (as a generic term) transmits a beam of light—electromagnetic radiation—in some direction, called the boresight, and the beam spreads into a certain solid angle $\delta\Omega$ around that direction. We can plot the relative amount of radiation transmitted in each direction θ with respect to this direction (assuming azimuthal symmetry), as shown in Figure 3-4. We call this the *power pattern* of the telescope. For a transmitter, it is the relative amount of power emitted in the direction that makes an angle θ with respect to the boresight; for a receiver, it is the relative sensitivity to radiation coming from that direction.

The radiometer might have a single detecting element, or it could have an array of detecting elements. In the latter case, a fixed radiometer can produce a 1- or 2-dimensional image without having to scan it. In the former case, we can only form an image by pointing the radiometer at each part of a scene in turn, recording the output at each position. Ra-diometers that have multiple sensing elements can make an image without moving anything, but this does not affect the mathematical problems that are our major interest, that of infer-ring the brightness of a scene from the meas-ured image. If (x_i,y_i) is the coordinate of the ith position in the scene, we can form an image with $T(x_i,y_i)$ being the measured intensity—or antenna temperature—at that position, or in that direction.[10] The amount of detail in the im-age is determined by the *resolution* of the radi-ometer, or its ability to distinguish two very close objects.[11] The smaller the solid angle $\delta\Omega$ the radiation comes from, the better the resolu-tion is.

The resolution of a telescope is determined by the wavelength of the electromagnetic radia-tion being used and the size of the device that focuses that radiation. Because of diffraction, the presence of an aperture affects the shape of the beam: a smaller aperture produces a wider beam, and *vice versa*. A telescope, where the

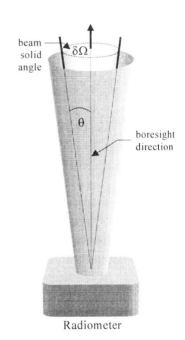

beam solid angle

$\delta\Omega$

θ

boresight direction

Radiometer

Figure 3-4. A radiometer looking up-ward, showing the nominal pointing direction and the solid angle of the beam.

---

[10] Note that (x,y) is a pair of *angular* coördinates if the point being imaged is changed by turning the radiometer in different directions, the way an astronomer points a radio telescope in different directions. We will assume that the approximation $\sin x = x$ is always valid, so the difference between angular and linear coördinates would be only formal.

[11] I will give a more precise definition later.

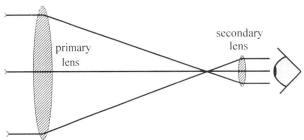

A schematic drawing of a refracting telescope (*i.e.*, a telescope with a primary *lens*).

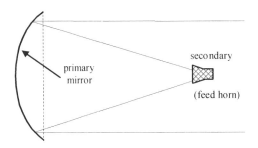

A reflecting telescope with a parabolic primary mirror and (secondary) feed horn. The dashed line shows the size of the aperture.

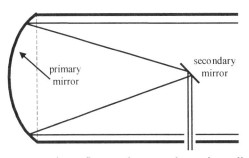

Newtonian reflector: the secondary mirror directs the light laterally out of the telescope to the viewer.

Figure 3-5. Telescopes.

primary lens or mirror (see Figure 3-5) is the largest and most expensive component, is usually designed so that the overall resolving power is limited by the size of the primary lens or mirror. When the primary element is a mirror, the aperture is the edge of the mirror; for a round mirror, the size of the aperture is denoted by its diameter (*e.g.*, a telescope with a 1-meter primary mirror has an aperture that is 1 meter in diameter). This acts like an aperture because radiation missing the mirror does not enter the radiometer. Let λ be the wavelength and d the diameter of the aperture;[12] the resolution is proportional to λ/d. If the wavelength is fixed by the physics of a remote-sensing problem— as is usually the case— then the resolution can only be improved by making the instrument bigger. However, bigger usually means more expensive— sometimes much, much more expensive. It would be very useful if we could improve the resolution somehow by processing the data in a certain way; especially when satellites are involved, computers are cheaper than large instruments.

Figure 3-6 illustrates a more formal definition of resolution. Suppose that there are two point sources—*i.e.*, sources with infinitesimal angular extent— separated by the angle δx. Figure 3-6a shows the image with a small aperture (compared with the wavelength λ); it looks like only one broad source is pre-

---

[12] *I.e.*, the diameter of the opening through which the radiation enters the radiometer.

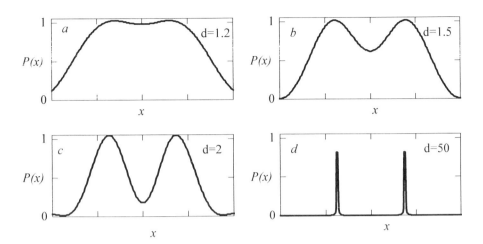

Figure 3-6. A pair of point sources imaged by antennas with the relative diameters shown.

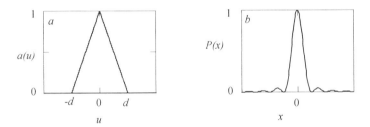

Figure 3-7. *a*) Autocorrelation of the electric field. *b*) Power pattern (linear plot).

sent. Make the aperture somewhat larger, as in Figure 3-6*b*, and the two peaks start to appear. In Figure 3-6*c*, we can see that there are indeed two sources; an aperture 25 times larger than the original one shows two very narrow sources widely separated (Figure 3-6*d*).

The image that results when we scan a microwave radiometer over an area of the sky is the convolution of the power pattern of the telescope and the brightness distribution (or intensity) incident on the sensor. The intensity measured by the sensor is[13]

$$T(x) = \int_{-\infty}^{\infty} B(u)\, P(x-u)\, d\,u, \tag{10}$$

where $B(x)$ is the brightness distribution; $x = \sin\phi$, where ϕ is the angular co-ordinate; and $P(x)$ is the power pattern of the antenna. (Throughout this dis-

---

[13] For the purposes of this chapter, I shall treat only a one-dimensional problem, but we can treat two- or higher-dimensional problems in the same way.

cussion, I shall assume that the total field-of-view is small enough that we can use the approximation $\sin x = x$. This simplifies the mathematics somewhat.)

Brightness is a radio astronomer's term: it means about the same thing as intensity.[14] It should be clear from equation (10) how the resolution is determined by $P(x)$. The broader $P(x)$ is, the wider the response will be to a point target, and the harder it will be to distinguish two targets that are close together.

We commonly use a source of infinitesimal extent, a point source, to measure the power pattern of an antenna. Substitute a δ-function into equation (10), and

$$T(x) = \int_{-\infty}^{\infty} \delta(u)P(x-u)du = P(x). \tag{11}$$

So when we scan across a point source, $T(x) = P(x)$.

3.3.1 The power pattern

Here, I want to make a digression.[15] As we shall see, an important property of the power pattern of any telescope is that it has a Fourier transform that is identically zero outside a certain interval $[-d/2, d/2]$, where d is usually the diameter of the primary mirror or lens. Because of this, the power pattern itself extends to $\pm\infty$; *i.e.*, there is no finite region $[-a, a]$ such that the power pattern itself is identically zero outside that region. This single property affects any attempt we might make to improve on the effect of the power pattern by manipulating the data. I provide a brief physical explanation of why this happens for those that might be interested.

Electromagnetic radiation is characterized by a time-varying electric field. Let $E(u,t)$ be the electric field in the aperture of the telescope; it depends on both space u and time t. Let $a(u)$ be the autocorrelation function of $E(u,t)$:

$$a(u) = \int_{-\infty}^{\infty} E(t)E^*(u+t)dt. \tag{12}$$

I shall only consider here the case where $E(u)$ is given by

$$E(u) = 1 \qquad |u| \le \frac{1}{2}\frac{d}{\lambda}$$

---

[14] To a radio astronomer. Workers in different wavelength bands use different terms for the quantities that arise in radiometry and the study of light. This lends a certain air of confusion and excitement to discussions across different fields (or, rather, different wavelength bands).

[15] This section and the next one use material on Fourier transforms that is introduced in Chapter 8.

$$E(u) = 0 \qquad |u| > \tfrac{1}{2}\frac{d}{\lambda}, \qquad (13)$$

in which case we say that the antenna is *uniformly illuminated*. The autocorrelation function of $E(u)$ is

$$a(u) = 1 - \frac{\lambda|u|}{d} \qquad |u| \leq \frac{d}{\lambda}$$

$$a(u) = 0 \qquad\qquad |u| \geq \frac{d}{\lambda}. \qquad (14)$$

This is illustrated in Figure 3-7. Here, the unit of length is λ, the electromagnetic wavelength. For a given diameter d, the resolution will be better if we use a smaller wavelength, and *vice versa*.

Since the diameter of the aperture is d, the electric field is zero for $|u| > d/2\lambda$; therefore the autocorrelation function is zero for $|u| > d$. Since the autocorrelation function is zero outside the finite interval $(-d, d)$, the power pattern, its Fourier transform, cannot also be zero everywhere outside any finite interval $(-b,b)$.[16]

The power pattern P is the Fourier transform of $a(u)$:

$$P(x) = \int_{-\infty}^{\infty} a(u)\,e^{-2\pi i u x}\,\mathrm{d}u = \int_{-d}^{d} a(u)\,e^{-2\pi i u x}\,\mathrm{d}u. \qquad (15)$$

The interested reader should consult *Antennas* or *Radio Astronomy* by Kraus. The assumption that $E(u)$ is constant within the interval $(-d/2\lambda, d/2\lambda)$ is not necessary: in the real world, engineers can produce a wide variety of antenna patterns by changing the design of the antenna, but describing this would take us too far afield. Figure 3-7 shows the autocorrelation function of a uniform electric field in the aperture, $a(u)$, and the resulting power pattern, $P(x)$. For an aperture with diameter d and a wavelength λ, the power pattern is

$$P(x) = \int_{-\infty}^{\infty} a(u)\,e^{-2\pi i u x}\,\mathrm{d}u = \int_{-1}^{1}\left(1 - \frac{\lambda|u|}{d}\right)e^{-2\pi i u x}\,\mathrm{d}u = \frac{d}{\lambda}\,\mathrm{sinc}^2\frac{xd}{\lambda}, \qquad (16)$$

where, by definition,[17]

---

[16] This is proved by Papoulis (1965). It is this feature, that there is no finite interval centered on 0 such $P(x) = 0$ outside that interval, that makes the problem so interesting.

[17] See Bracewell (1965, p 62).

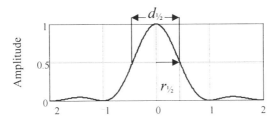

Figure 3-8. $\text{Sinc}^2 r$ power pattern. The half-width at half-maximum is denoted by $r_{1/2}$; the full-width at half-maximum, which is usually called just the *beamwidth*, is $d_{1/2} = 2r_{1/2}$.

$$\text{sinc}\,x \;=\; \frac{\sin \pi x}{\pi x}. \tag{17}$$

Note that

$$\underset{x \to 0}{Lim}\ \text{sinc}\,x \;=\; 1. \tag{18}$$

The power pattern is normalized so that

$$\int_{-\infty}^{\infty} P(x)\,dx \;=\; a(0) \;=\; 1. \tag{19}$$

I shall dwell on the uniformly-illuminated antenna because it is easy to characterize analytically. In practice, such a pattern has side lobes that are undesirably high. Antenna engineers often taper the illumination, making it decrease toward the edge of the antenna. This produces a more compact power pattern. Figure 3-7 shows the pattern of a uniformly illuminated antenna, which is $\text{sinc}\,x$. There is a main lobe at the center, where the response is greatest, but there is also a response from the peaks (lobes) on either side of the main lobe—reasonably enough, these are called *side lobes*.

3.3.2 Power patterns

We define the width of the power pattern $P(x)$ by the values of x where $P(x) = \frac{1}{2}P(0)$, assuming that $P(0)$ is the maximum value of $P(x)$. Let $x_a < 0$ and $x_b > 0$ be the points where $P(x_a) = P(x_b) = \frac{1}{2}P(0)$. Then x_a and x_b are called the half-power points, and $d_{1/2} = x_b - x_a$ is the width,[18] or 3-dB width, of the pattern (see

---

[18] The full width at half-maximum.

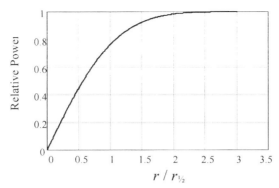

Figure 3-9. The relative amount of power contained within a distance r of the center of a Gaussian beam, where r is measured in units of $r_{1/2}$, as defined in the text.

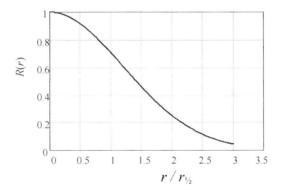

Figure 3-10. Correlation coefficient of two measurements separated by a distance r.

Figure 3-8; note that $\log_{10}\frac{1}{2} \approx 0.3$, or 3 decibels.) I shall also use the half-width $r_{1/2} = \frac{1}{2}d_{1/2}$ to describe certain properties of the patterns.[19]

I want to examine two possible power patterns briefly. One is a Gaussian pattern, which is useful because of its analytical simplicity. Since the Fourier transform of a Gaussian is another Gaussian, it is not, therefore, a bandwidth-limited function. (Since any real radiometer has a finite aperture, its power pattern must be bandwidth-limited.) However, a Gaussian is close enough to a possible antenna pattern that it is often an acceptable approximation. I shall deal with a more realistic pattern, $\text{sinc}^2 r$, presently.

Consider the two-dimensional Gaussian power pattern

$$G(x,y) = \frac{1}{2\pi w^2} e^{-\frac{1}{2}\left[(x-x_0)^2 + (y-y_0)^2\right]/w^2}, \qquad (20)$$

---

[19] Confusingly enough, we sometimes use the term "half-width" to mean $d_{1/2}$; the "one-half" refers to the "one-half" power points: it is more correctly called the "full width at half-maximum", or FWHM.

where w is the beamwidth; and $x = \sin\theta$ and $y = \sin\phi$ are two angular coördinates; and x_0 and y_0 are the nominal direction the radiometer is pointing in. For $x_0 = y_0 = 0$, this pattern is radially symmetric: letting $r^2 = x^2 + y^2$,

$$G(r) = \frac{1}{\sqrt{2\pi}w} e^{-\frac{1}{2}r^2/w^2} . \tag{21}$$

It is instructive to look at a plot of $G(r)$ (see Figure 3-9, which shows $G(r)$ for a $\mathrm{sinc}^2 r$ power pattern). We shall adopt the 3-db half-width $r_{\frac{1}{2}}$ as the unit of distance. The full width at half-maximum, denoted by $d_{\frac{1}{2}}$ in Figure 3-8, is commonly taken to be the resolution of the radiometer. We can see from equation (21) that $G(r_{\frac{1}{2}}) = \frac{1}{2}G(0)$ when $r_{\frac{1}{2}}^2 = 2w^2\ln(2)$. The 3-db half-width is $2r_{\frac{1}{2}}$. Figure 3-9 is a plot of the relative power contained within a radius r of the center, where r is measured in units of $r_{\frac{1}{2}}$; $r = 1$ corresponds to the 3-db radius of the pattern (which is one-half of the 3-db beamwidth). We can see that only 76% of the power is contained within this contour: almost one-quarter of the power comes from *outside* the 3-db contour.

3.3.3 Overlap

If we have two measurements from a satellite radiometer that are separated by a distance r, the measurements will not be independent if the two beams overlap. The overlap of two Gaussian beams separated by a distance r is

$$R(r) = \frac{1}{\sqrt{2\pi}w} \int_{-\infty}^{\infty} e^{-\frac{1}{2}\left[t^2 + (r-t)^2\right]/w^2} dt = e^{-\frac{1}{4}r^2/w^2} . \tag{22}$$

Since this is a correlation coefficient, it is normalized so that $R(0) = 1$. Suppose the scene being imaged has an infinitesimal correlation length. Then the correlation coefficient for two measurements separated by r is just $R(r)$. I show a plot in Figure 3-10. We can see that the correlation coefficient $R(2) \approx 0.255$. (This corresponds to one beam *diameter*, or $d_{\frac{1}{2}}$, which is the usual measure of beamwidth). Therefore, measurements spaced one beamwidth apart are correlated; they are uncorrelated if the measurements are spaced more than 1.5 beamwidths or so apart.

Exercise 3-2: As long as we are discussing power patterns, what is the Nyquist frequency corresponding to a Gaussian power pattern? a $\mathrm{sinc}^2 r$ pattern?

A graphical solution shows that the half-power point of the $\mathrm{sinc}^2 r$ power pattern is at $r_{\frac{1}{2}} \approx 0.443$ (see Figure 3-10). It shows the correlation coefficient of two scenes imaged with a $\mathrm{sinc}^2 r$ power pattern a distance r apart, where r here is measured in units of the half-power radius of the $\mathrm{sinc}^2 r$ pattern. The formula for the correlation of two overlapping patterns is somewhat more complicated analytically than for the Gaussian pattern, but the results will be about the same.

Consider the Nyquist sampling interval. I will show in §8.5 that a band-width-limited function, such as $\text{sinc}^2 r$, should be sampled at a minimum interval δr to preserve all of the information. Since the Fourier transform of $\text{sinc}^2 r$ is $\Lambda(s)$, which is identically zero outside the interval $[-1,1]$, the Nyquist sampling interval is $\frac{1}{2}$. To retain all of the information possible in the measurements, we need to sample this frequently; otherwise, information will be lost. Since $d_{1/2} = 2r_{1/2} \approx 0.9$, we generally say that we need to sample twice per beamwidth (which means full-width at half-maximum) for a uniformly illuminated aperture. This is usually an adequate rule for sampling radiometric data. I shall treat the question of what, if anything, we can do to improve the resolution of an image by processing the data in Chapter 12.

References

Bracewell, R., *The Fourier Transform and Its Application*, McGraw-Hill Book Co., New York (1965).

Crane, P. C., and P. J. Napier, 'Sensitivity', in *Synthesis Imaging*, R. A. Perley, F. R. Schwab, and A. H. Bridle, eds, National Radio Astronomy Observatory Course Notes, Workshop 13, NRAO, Green Bank, WV (1986).

Gardner, W. A., *Introduction to Random Processes with Applications to Signals & Systems*, 2nd Edition, McGraw-Hill Publishing Co., New York (1990).

Kraus, J. D., *Antennas*, 2nd Edition, McGraw-Hill Book Co, New York (1988).

Kraus, J. D., *Radio Astronomy*, 2nd Edition, Cygnus-Quasar Books, Powell, OH (1986).

Middleton, D., *An Introduction to Statistical Communication Theory*, McGraw-Hill Publishing Co., New York (1960).

Nyquist, H., 'Thermal agitation of electric charge in conductors,' *Physical Review*, **32**, 110-113 (1928).

Papoulis, A., *Probability, Random Variables, and Stochastic Processes*, McGraw-Hill Publishing Co., New York (1965).

Ulaby, F. T., R. K. Moore, and A. K. Fung, *Microwave Remote Sensing Active and Passive*, Vol. 1, Addison-Wesley Publishing Co., Reading, MA (1981).

4. Radiative Transfer

Perhaps the most surprising thing about mathematics is that it is so surprising. The rules which we make up at the beginning seem ordinary and inevitable, but it is impossible to foresee their consequences. These have only been found out by long study, extending over many centuries. Much of our knowledge is due to a comparatively few great mathematicians such as Newton, Euler, Gauss, or Riemann; few careers can have been more satisfying than theirs. They have contributed something to human thought even more lasting than great literature, since it is independent of language.

It can be of no practical use to know that Pi is irrational, but if we can know it, surely would be intolerable not to know.

E. C. Titchmarsh (1899-1963)[1]

The discussion in Chapter 2 of the physics of absorption and emission of light illustrates, in broad strokes, some of the tools needed to model the radiation in the atmosphere and how it is affected by changes in variables like temperature and pressure. The temperature is especially important because it affects the absorption coefficients of the various absorbers and also affects the intensity of radiation because of the nature of the Planck function $B(\lambda, T)$. We still need to calculate the radiation field in the atmosphere in order to determine what the relationship is between the physical variables and the radiance that a satellite sensor will see. Without this information, we would not be able to interpret the radiances we measure. In other words, we use this information to find the coefficients of the matrix $\mathbf{A}$ in the equation $\mathbf{y} = \mathbf{Ax}$ that is the focus of this book.

4.1 Derivation

I treated the equation of radiative transfer in §2.4 for a single isothermal layer of gas. Here, I extend the discussion to the calculation of the radiation emerging from a nonhomogeneous atmosphere. Assume that there is a plane-parallel atmosphere bounded at the bottom ($z = 0$) by the surface of the earth; z is positive in the upward direction. Assume that the atmosphere has a finite limit,

---

[1] Appointed Savilian professor of geometry at the University of Oxford in 1931. "He devoted his early research to the theory of Fourier integrals and series and added new findings to the study of Fourier transforms. He further contributed to the theory of conjugate functions and general integral transforms and thus formed a major part of his *Introduction to the Theory of Fourier Integrals* (1937)." (Encyclopædia Britannica CD98®)

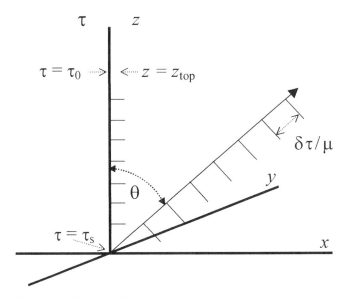

Figure 4-1. The coördinate system.

which I denote z_{top}. Consider a straight path that makes an angle θ with the z-axis (see Figure 4-1); define $\mu = \cos\theta$. We can calculate the radiance emitted upward from the atmosphere along this ray by dividing the atmosphere into layers. Consider a layer bounded by z and $z + \delta z$ with radiation with intensity $I_{0,\lambda}$ incident on it from the bottom. We need to calculate the amount of radiation absorbed between the positions s and $s + ds$ along a ray pointing in a given direction, and also the amount of radiation emitted in that direction. In order to do this, we will find it advantageous to use a coördinate that is related to the amount of radiation absorbed per unit distance. We shall replace the distance along a ray, $s = z/\cos\theta$, with the *optical depth*, τ.

Define the differential optical depth to be

$$d\tau_\lambda(z) = -\kappa_\lambda \rho(z) dz, \tag{1}$$

where κ_λ is the absorption coefficient at wavelength λ. The minus sign arises because, while z is measured as positive upwards, $\tau_\lambda = 0$ at the top of the atmosphere and $\tau_\lambda = \tau_{s,\lambda} > 0$ at the surface $z = 0$. It will be convenient to make optical depth the independent variable if the absorption coefficient does not change much over the bandwidth of interest. This establishes a new coordinate system, with τ_λ as the independent variable. Using Kirchoff's law (see §2.4), we see that the amount of radiation emitted along this path is

$$\delta I_\lambda(\tau_\lambda)_{emit} = -\kappa_\lambda B[T(z)]\rho(z)\delta z/\mu = -\varepsilon_\lambda \rho(z)\delta z/\mu, \tag{2}$$

where ε_λ is the emission coefficient. The optical depth is τ_λ, which depends on z, and the minus sign is present because

$$\delta I_\lambda = I_\lambda(z + \delta z) - I_\lambda(z) = I_\lambda(\tau_\lambda) - I_\lambda(\tau_\lambda + \delta \tau_\lambda). \tag{3}$$

If $T(z)$ is the temperature profile and κ_λ does not depend on temperature, then $T(\tau_\lambda) = T(z/\rho\kappa_\lambda)$. Note that, as a vertical coördinate,

$$\tau_\lambda(z) = \int_{z_{top}}^{z} \kappa_\lambda(t)\rho(t)\,dt, \tag{4}$$

while the optical depth along the ray that makes an angle θ with the vertical is

$$\tau_\lambda = \frac{1}{\mu} \int_{\tau = 0}^{\tau(z)} \kappa_\lambda(t)\rho(t)\,dt. \tag{5}$$

The amount of radiation that is absorbed is

$$\delta I_\lambda(\tau_\lambda)_{abs} = I_{0,\lambda} \kappa_\lambda \rho(z)\delta z/\mu. \tag{6}$$

The change in the upward-going intensity is (combine the last four equations)

$$\mu \frac{d I_\lambda(\tau_\lambda)}{d\tau_\lambda} = I_\lambda(\tau_\lambda) - B[\lambda, T(\tau_\lambda)]. \tag{7}$$

The right-hand side of equation (7) is $I - B$ because we are taking the top layer to have $\tau = 0$, with τ increasing downwards, but I_λ is intensity in the *upward* direction. For layer n, the intensity emitted by a ray that makes an angle θ with the vertical is

$$I_{\lambda,n} = B(T_n)\left(1 - e^{-\tau_n/\mu}\right). \tag{8}$$

Note again that we have used Kirchoff's law to relate the amount of radiation emitted to the amount absorbed; this allows us to write $B[\lambda, T(\tau_\lambda)]$ in place of $\varepsilon_\lambda/\kappa_\lambda$.

4.2 Formal solution

Ignore, for the moment, the radiation that is emitted from or reflected by the surface. We can find the formal solution of equation (7) by expressing I_λ as a function of τ_λ. Then use the relationship

$$\frac{d}{d\tau_\lambda}\left[e^{-\tau_\lambda/\mu} I_\lambda(\tau_\lambda)\right] = e^{-\tau_\lambda/\mu} \frac{d I_\lambda(\tau_\lambda)}{d\tau_\lambda} - \frac{1}{\mu} e^{-\tau_\lambda/\mu} I_\lambda(\tau_\lambda)$$

$$= \frac{e^{-\tau_\lambda/\mu}}{\mu}\left[\mu\frac{dI_\lambda(\tau_\lambda)}{d\tau_\lambda} - I_\lambda(\tau_\lambda)\right]. \tag{9}$$

Using equation (7),

$$\mu\frac{d}{d\tau_\lambda}\left[e^{-\tau_\lambda/\mu}I_\lambda(\tau_\lambda)\right] = -e^{-\tau_\lambda/\mu}B\left[\lambda, T(\tau_\lambda)\right]. \tag{10}$$

Integrating both sides,

$$\mu I_\lambda(\tau_\lambda) = \mu e^{\tau_\lambda/\mu}I_\lambda(0) - \int_0^{\tau_\lambda} e^{-(\tau_\lambda'-\tau_\lambda)/\mu}B\left[\lambda, T(\tau_\lambda')\right]d\tau_\lambda'. \tag{11}$$

In the special case where we want to calculate the intensity emitted from the top of the atmosphere (defined by $\tau_\lambda = 0$), we can assume that there is a layer deep enough in the atmosphere such that the upwelling intensity from that layer is negligible, and we can take $I_\lambda(\tau_\lambda) = 0$. In this limit,

$$I_\lambda(0) = \frac{1}{\mu}\int_0^{\tau_\lambda} e^{-\tau_\lambda'/\mu}B\left[\lambda, T(\tau_\lambda')\right]d\tau_\lambda'. \tag{12}$$

If the temperature is a constant T, we get the expected result that

$$I_\lambda(0) = B[\lambda, T]\left(1 - e^{-\tau_\lambda/\mu}\right). \tag{13}$$

4.3 Isothermal atmosphere

We can devise a simplified model of the radiative transfer in the earth's atmosphere as follows. Neglect the curvature of the earth: assume that there is a flat surface, with emissivity $\varepsilon_\lambda(\theta)^2$ and an isothermal atmosphere with temperature T_{atm}. We wish to calculate the radiation emitted from the top of the atmosphere. We must consider three components: the radiation emitted by the earth's surface; the downward radiation reflected from the surface; and the upward radiation emitted by the atmosphere.

   Suppose that the earth's surface is diffuse: radiation that is incident in one direction θ is reflected equally in all directions θ'; such a surface is said to be *Lambertian*. The downwelling radiation is found from equation (8) with a temperature T_{atm} and optical depth $\tau_{s,\lambda}$. A fraction $[1 - e_\lambda(\theta)] = r_\lambda(\theta)$ is reflected from the surface. The solid angle illuminated by the downwelling ray

---

[2] I use ε_λ for the emissivity of the atmosphere, and e_λ for the emissivity of a surface to avoid confusion.

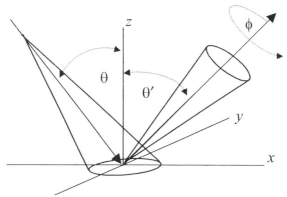

Figure 4-2. Incident and reflected rays; θ is the incidence angle; ϕ, the azimuthal angle around the direction of propagation.

with the incidence angle θ is roughly $d\Omega = \sin\theta\, d\theta\, d\phi$, where θ and ϕ are the incidence angle and the azimuthal angle around the direction of propagation (see Figure 4-2; we assume that the intensity is independent of ϕ). The factor of $\sin\theta$ arises because, as viewed by the observer, the solid angle that contributes to the beam increases as θ increases; there is also a factor of $\cos\theta$ because of the projection of the area (shown as a circle in Figure 4-2) along the line of sight. To an adequate approximation, the reflected radiation can be approximated by assuming that $\theta = \theta'$ and that the amount of reflected light is proportional to $\sin\theta\cos\theta$. Then the average downwelling intensity is[3]

$$< I_{dn,\lambda} > \ = \ 2\pi\, B(\lambda, T_{atm}) \int_0^{\pi/2} \sin\theta\cos\theta\Big[1 - e^{-\tau_{\lambda,s}/\cos\theta}\Big]d\theta\,. \tag{14}$$

Here, $\tau_{\lambda,s}$ is the optical depth at the surface (at wavelength λ). We can simplify this equation by letting $\mu = \cos\theta$ as before. Then, since $d(\cos\theta)/d\theta = -\sin\theta$, $d\mu = -\sin\theta\, d\theta$, and

$$< I_{dn,\lambda} > \ = \ 2\pi B(\lambda, T_{atm})\int_0^1 \Big[\mu e^{-\tau_{\lambda,s}/\mu} - 1\Big]d\mu. \tag{15}$$

Let $t = 1/\mu$: then

$$< I_{dn,\lambda} > \ = \ 2\pi\, B(\lambda, T_{atm}) \int_1^\infty \frac{1}{t^3}\Big(t - e^{-\tau_\lambda/t}\Big)dt. \tag{16}$$

We can simplify further this expression by using a family of functions called *exponential integrals* defined by

---

[3] See Liou, §4.3 (1980).

$$E_n(z) = \int_1^\infty \frac{1}{t^n} e^{-zt} dt \qquad (17)$$

for $n = 0, 1, 2, \ldots$. They obey a recurrence relation, as one can establish by integrating by parts,[4]

$$n E_{n+1}(z) = e^{-z} - z E_n(z) \qquad (18)$$

for $n = 1, 2, 3, \ldots$. Note that $E_n(z) \to e^{-z}/z$ as $z \to \infty$. Using these functions, we can write

$$\langle I_{dn,\lambda} \rangle = B(\lambda, T_{atm}) \left[E_3(\tau_{\lambda,s}) - \tfrac{1}{2} \right]. \qquad (19)$$

Note that when $\tau_{\lambda,s} \to 0$, $I_{dn} \to 0$, as it clearly must (a vacuum does not emit radiation). See Press $et\ al.$[5] and Chapter 8 on a method for calculating exponential integrals from this recursion relation. I show how to calculate the value of E_1 in §14.11.

Let T_s be the temperature of the surface: the intensity emitted from the surface is $B(\lambda, T_s) e_\lambda(\theta)$. The combined radiation reflected by and emitted from the surface is attenuated by a factor $\exp\{-\tau_{\lambda,s}\}$ on the way up. The upwelling radiation emitted from the atmosphere is also found from equation (7). Finally,

$$I_\lambda(\theta)_{up} =$$

$$\left\{ \langle I_{dn,\lambda} \rangle \left[1 - e_\lambda(\theta) \right] + B(T_s) e_\lambda(\theta) \right\} e^{-\tau_{\lambda,s}/\mu} + B(T_{atm}) \left[1 - e^{-\tau_{\lambda,s}/\mu} \right]$$

reflected	emitted	atmo –	emitted	(20)
down –	from	spheric	by atmo –	
welling	surface	absorpt.	sphere.	

The three terms on the r.h.s. of this equation represent the downwelling intensity reflected from the surface; the intensity emitted by the surface; and the intensity emitted by the atmosphere. Equation (20) shows how the radiation reflected from the surface, emitted by the surface, and emitted by the atmosphere all contribute to the radiance above the atmosphere.

---

[4] Abramowitz and Stegun , p 228.
[5] 1992, pp. 172 ff.

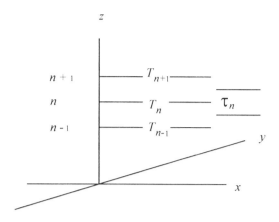

Figure 4-3. Temperatures and optical depths in the model atmosphere.

4.4 Inhomogeneous atmosphere

Now, if we allow the temperature in the atmosphere to change with height above the surface, we can solve the equation of radiative transfer in a manner similar to the one outlined above. However, when the temperature and pressure change with altitude, we can approximate the temperature behavior of the atmosphere by dividing it into many homogeneous, horizontal layers; we then perform the computation layer by layer. Figure 4-3 shows an atmosphere with several layers: the temperature of layer n is T_n, and its optical thickness τ_n. (Where the context makes it unnecessary, I will drop the subscript λ that indicates the wavelength dependence.) In order to calculate the reflected radiation, start at the topmost layer, and calculate the downwelling radiation at each level:

$$I_n^-(\mu) \,=\, I_{n+1}^-(\mu)\,e^{-\tau_n/\mu} \,+\, B(T_n)\!\left[1 \,-\, e^{-\tau_n/\mu}\right] \tag{21}$$

where $I_n^-(\mu)$ is the intensity emitted downward from the bottom of the nth layer in the direction μ. Average these intensities over angle to find the mean downwelling intensity; multiply by $[1 - e(\theta)]$ to get the reflected component. In practice, for a diffuse surface, we can calculate the downwelling radiation at an angle of $45°$, which is where $\sin\theta\cos\theta$ is maximum, and use that value. Let I_{dn} be the average downwelling radiance at $45°$: the radiance upward incident at the bottom layer is

$$I_{up} \,=\, \left[1 - e\!\left(45°\right)\right]I_{dn} + e(\theta)\,T_s. \tag{22}$$

Use this value to start the upward calculation. Finally, layer by layer, calculate

$$I_n^+(\mu) = I_{n-1}^+(\mu)e^{-\tau_n/\mu} + B(T_n)\left[1 - e^{-\tau_n/\mu}\right] \tag{23}$$

where $I_n^+(\mu)$ is the intensity emitted upward from the top of the nth layer. The thickness of the layers will determine the accuracy of the calculation. Note that this calculation is fairly tedious, even with a computer.

Exercise 4-1: Suppose that the absorption coefficient κ depends strongly on temperature. If the absorption coefficient is large enough, changes in κ will affect the rate of heating (or cooling) of the atmosphere, and therefore change the temperature in each layer, which changes the absorption coefficient, ... and so forth. How would we solve this problem? What happens if $d\kappa/dT > 0$? What happens if $d\kappa/dT < 0$? Does it make an important qualitative difference?

Exercise 4-2: It might be reasonable to try to solve this problem by iteration. Will such a method converge?

4.5 Weighting functions

Meteorologists often use the pressure at distance z above the surface instead of z as the independent variable representing vertical distance. Making this transformation has several advantages that will become clear as we proceed.

There is little vertical motion in the atmosphere: to a good approximation, it is in hydrostatic equilibrium, in that the pressure at each point in the atmosphere just balances the weight of the air above it. Let g be the acceleration of gravity (we ignore the change in g with height); hydrostatic equilibrium obtains when

$$g\rho(z) = -\frac{dp(z)}{dz}, \tag{24}$$

where $\rho(z)$ and $p(z)$ are the density and pressure at the height z.

Consider the change in intensity along an upward ray that makes an angle θ with respect to the vertical. Let τ be the independent variable.[6] The optical depth is $\tau = \tau_s$ at the surface and $\tau = 0$ at the top of the atmosphere (see Figure 4-4). The intensity emitted upward within the layer with optical depth $\delta\tau$ in the direction θ is [compare this with equation (8)]

---

[6] The optical depth still depends on wavelength and angle, but I do not denote this explicitly in this section.

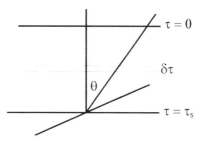

Figure 4-4. The optical depth is τ_s at the surface and 0 at the top of the atmosphere.

$$\delta I(\mu) = B[T(\tau)](1 - e^{-\delta\tau/\mu}) \cong \frac{\delta\tau}{\mu} B[T(\tau)].$$ (25)

Part of this is absorbed by the intervening atmosphere; the intensity leaving the top of the atmosphere from this layer is

$$\delta I(\mu) = B[T(\tau)]\frac{\delta\tau}{\mu} e^{-\tau/\mu}.$$ (26)

Integrating this, we get

$$I(\mu) = \frac{1}{\mu}\int_0^{\tau_s} B[T(\tau)]e^{-\tau/\mu}\,d\tau.$$ (27)

Now using the equation of hydrostatic equilibrium, $dp(z) = -g\rho(z)dz$, and the differential optical depth is $d\tau = \kappa(z)\rho(z)dz$, so

$$d\tau = -\frac{\kappa(z)dp(z)}{g},$$ (28)

where τ is the optical depth in the vertical direction. Substitute this into equation (4); the intensity at an angle $\theta = \cos^{-1}\mu$ is

$$I(\mu) = \int_0^{p_s} \frac{\kappa(p)}{\mu g} B[T(p)]\,e^{-\tau(p)/\mu}\,dp,$$ (29)

where p_s is the pressure at the surface. Define

$$w_\mu(p) = e^{-\frac{1}{\mu g}\int_0^p \kappa(p')dp'};$$ (30)

then

$$\frac{dw_\mu(p)}{dp} = -\frac{\kappa(p)}{\mu g} w_\mu(p). \tag{31}$$

Then dw/dp is called a *weighting function*; the intensity is now

$$I(\mu) = \frac{1}{\mu} \int_0^{p_s} B[T(p)] \frac{dw_\mu(p)}{dp} dp. \tag{32}$$

The largest contribution to $I(\mu)$ comes from the layer p where dw_μ/dp is largest.

Unless the atmosphere is optically thick, we also have to add the intensity emitted and reflected from the surface. Consequently, the upwelling intensity is

$$I_{up}(\mu) = \frac{1}{\mu} \int_0^{p_s} B[T(p)] \frac{dw_\mu(p)}{dp} dp + \left(r(\theta)\langle I_{dn}\rangle + T_s e(\theta)\right)e^{-\tau_s}. \tag{33}$$

where $\langle I_{dn}(\mu)\rangle$ is the average downwelling intensity; r and e are the reflection coefficients of the surface. When temperature sounding[7] is performed at wavelengths where the atmosphere is optically thick, equation (33), without the last term, adequately represents the intensity.

The advantage of writing the equation of radiative transfer in this form is that layers where dw_μ/dp are small contribute little to the upwelling intensity, while layers where it is large contribute most. Furthermore, as a practical matter, dw_μ/dp has a single maximum, the graph near this maximum has a certain width. The width of the peak determines the resolution that is possible for a particular channel (*i.e.*, band of wavelengths).

Note that equations (32) and (33) are integral equations; we can write them in the form

$$I(\mu) = \int_a^b K(\mu, p)T(p)dp, \tag{34}$$

where $K(\mu,p)$ is the *kernel* of the integral equation, and a and b some constants. Note carefully what this equation says: it relates the temperature at the level where the pressure is p to the intensity emerging from the top of the atmosphere in the direction $\mu = \cos\theta$. Figure 4-5 shows this schematically. At

---

[7] Originally, temperature profiles were measured with radiosondes, instruments that are carried aloft by balloons. The term for such a measurement was a sounding, presumably by analogy with measurements of water depth. The term sounding is also used today to refer to measurement of temperature profiles with satellite data.

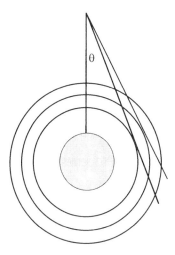

Figure 4-5. Observations of the earth's atmosphere from different incidence angles $\mu = \cos\theta$.

normal incidence, we can see furthest into the atmosphere; at larger incidence angles, we cannot see as far. In solar astronomy, where we can assume that the properties of the solar atmosphere depend only on the distance from the center of the sun, this is the customary method for determining how the temperature of the solar atmosphere changes with depth. When we look at the sun, we see a bright disk; at each radial distance from the center of this disk, we are looking into the solar atmosphere at different incidence angles θ. If we determine how the brightness changes with distance from the limb—the edge of the solar disk—we can invert equation (34) and infer the vertical temperature structure of the sun near the surface. It turns out that when we look at the center of the solar disk, we see down to the layer that has the minimum temperature; the temperature increases outwards and therefore, as we look toward the limb, we are seeing hotter material. Because of this, the brightness of the sun changes very little from the center to the edge.

Exercise 4-3: Why can't we use this method for finding temperature profiles in stars other than the Sun?

Exercise 4-4: Suppose that the temperature of the solar atmosphere were to decrease in the upward direction. How would that affect the relative brightness of the center and the edge of the solar disk?

There is a different way we can look at this situation: remember that the intensity depends on wavelength. If the incidence angle μ is fixed, but we

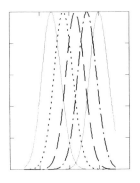 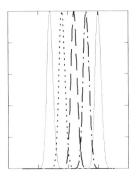

Wide Weighting Functions   Narrow Weighting Functions

Figure 4-6. Hypothetical weighting functions. The
horizontal axis represents depth into the atmosphere.
Curves correspond to different wavelengths used to
measure the radiance above the atmosphere. Wide
weighting functions, left, have more overlap and
poorer spatial resolution.

make observations at different wavelengths (which is what is usually done in
remote sensing), we can write equation (34) in the general form

$$I(\lambda) = \int_a^b K'(\lambda, p)T(p)\mathrm{d}p,\qquad(35)$$

where the kernel in this equation is different from the one in the previous
equation. Look at equation (32); the weighting function depends both on
wavelength and on the incidence angle. Again, if we measure the intensity at
several wavelengths, keeping μ fixed, we can find the temperature profile
from

$$I(\lambda) = \frac{1}{\mu} \int_0^{p_s} B[T(p)]\, \frac{\mathrm{d}w(\lambda, p)}{\mathrm{d}p}\, \mathrm{d}p.\qquad(36)$$

I discuss integral equations in Chapter 10.

4.6 Resolution

Some hypothetical weighting functions are shown in Figure 4-6. Each panel
corresponds to a different hypothetical instrument; the different curves within
a panel, to different channels in the instrument. The weighting functions on
the left are substantially wider than those on the right. In general, as we saw in
the section on thermal noise, the weighting functions will usually be narrower
if the bandwidth used for each channel is smaller. But using smaller band-

widths increases the thermal noise, reducing the sensitivity of the measurements and reducing their usefulness.

In Figure 4-6, the horizontal axis represents depth into the atmosphere. Having a wide weighting function means that the layer that contributes to the radiance in a given channel is relatively thick; a narrower weighting function indicates that the layer is thinner. Therefore, we can see that the spatial resolution of the measurement depends, ultimately, on the bandwidths of the channels being used (as well as other factors).

References

Abramowitz, M., and I. Stegun, *Handbook of Mathematical Functions*, Dover, New York, (1968).

Liou, Kuo-Nan, *An Introduction to Atmospheric Radiation,* Academic Press, New York, (1980).

Press, W. H., S. A. Teukolsky, W. T. Vettering, and B. R. Flannery, *Numerical Recipes in Fortran, the Art of Scientific Computing*, Second Edition, Cambridge University Press, Cambridge, England (1992).

5. Covariance Matrices

To avoide the tediouse repetition of these woordes: is equalle to: I will settle as I doe often in woorke use, a paire of paralleles, or gemowe [twin] lines of one lengthe: =, bicause noe .2. thynges, can be moare equalle.

Robert Recorde (1557)[1]

Mark all mathematical heads which be wholly and only bent on these sciences, how solitary they be themselves, how unfit to live with others, how unapt to serve the world.

Roger Ascham (1515-1568)[2]

As I said earlier, we cannot devise a sensible retrieval algorithm if we do not know the statistics of the situation at hand. We use covariance matrices to describe the statistics of the variables in question. Here, I review some general properties of matrices that we need to use, as well as some special properties that pertain particularly to remote sensing. In particular, I discuss the rotation of matrices, the properties of symmetric real matrices, and the properties of eigenvalues of positive-definite matrices. Some the material in this chapter will already be familiar to many readers, but some of it will be new, and it is important for understanding Chapter 7, where I discuss methods for solving the equations that arise in remote sensing. Covariance matrices are necessarily symmetric—we shall see presently why this is important.

This chapter also can serve as a reminder of some important theorems as they apply to the subject at hand. In this sense, it may be frustrating to look for a certain theorem in a math book because mathematicians usually want to prove the most general theorem possible, while we are interested only in a particular application. For example: what are the necessary and sufficient conditions for a square matrix to be diagonable (*i.e.*, equivalent to a diagonal matrix)? The most general answer is that the matrix must be *normal*: it must

---

[1] 'Perhaps his most noted work was *The Whetstone of Witte* (1557), in which he first proposed the use of the equals sign (=). Recorde died in prison. The reason for his imprisonment is not known' (Encyclopædia Britannica CD98®).

[2] Ascham wrote on archery as well as mathematics. 'Ascham's *Toxophilus* ("Lover of the Bow"), written in the form of a dialogue, was published in 1545 and was the first book on archery in English. In the preface Ascham showed the growing patriotic zeal of the humanists by stating that he was writing "Englishe matter in the Englishe tongue for Englishe men."' (Encyclopædia Britannica CD98®).

commute with its Hermitian adjoint. If we don't remember much about complex matrices, this isn't much help. It's more useful for our purposes to know that any real, symmetric matrix is equivalent to a diagonal matrix.

As I shall show presently, how we choose to solve an inverse problem of the kind we encounter in remote sensing depends on the statistics of the independent variables x_i and on the amount of noise in each measured quantity. Most other treatments tacitly assume that these variables are independent [$\text{cov}(x_i, x_j) = 0$] and that they have the same variance [$\text{var}(x_i) = \sigma^2$ for $i = 1, \ldots, m$]. However, nature isn't always that simple: the statistics of the independent variables and the measurement noise will influence the form of the algorithm we derive. Because this is an important aspect of the problem, I have taken pains to include these statistics in the formalism. We use a matrix, called a *covariance matrix*, to describe the joint statistics of a collection of several variables. This chapter treats those aspects of matrix algebra that are germane to our inversion problem, and, in particular, the properties of covariance matrices.

Sometimes it is convenient to use a particularly simple form of a matrix equation to examine a certain aspect of a problem. For instance, we shall see that, if we deal only with diagonal matrices, we can calculate the inverse of a matrix in our heads, while otherwise calculating the inverse of even a small matrix is very tedious. But, if we simplify the form of the equations we want to solve, will we retain the generality we need to apply the result to all of the problems we shall encounter? I shall answer that question at the end of this chapter, and show that there is indeed a canonical form of a matrix equation that is particularly simple and lucid.

Throughout this book, I will bold uppercase letters for matrices, and bold lowercase letters for vectors. As in other chapters, the variables **x** and **y** are related by $\mathbf{y} = \mathbf{Ax}$. In some cases, the matrix **A** is replaced with a matrix **R** in discussions of properties that apply to all matrices of a certain class.

5.1 Probability and moments

Consider several scalar random variables $x, y, \ldots$, and so forth. When I say that x is a random variable, I mean that it represents some quantity whose value changes in an unpredictable way every time I measure it. Suppose that there are M possible outcomes to a random experiment (like tossing dice). When there is a discrete set of possible outcomes, we let p_i represent the probability that the ith outcome will occur. If n_i represents the number of times the ith outcome occurs out of a total of N trials, the probability $p_i = n_i / N$ in the limit $N \rightarrow \infty$.

Whenever we perform an experiment, *something* has to happen: therefore, if there are M possible mutually exclusive outcomes, it is necessary that[3]

$$\sum_{i=1}^{M} p_i = 1. \tag{1}$$

For continuous variables, we use a *probability density distribution* $p(x)$; the probability that the value of the random variable x lies in the interval $(x, x+dx)$ is $p(x)dx$. The probability that $x' \le x$ is

$$\Pr(x' \le x) = \int_{-\infty}^{x} p(t)dt. \tag{2}$$

As with the discrete case, the probability density must be normalized to

$$\int_{-\infty}^{\infty} p(x)dx = 1. \tag{3}$$

In the case of the measurements that will concern us here, the random variables represent real physical quantities and have some maximum and minimum values that are imposed by the physical situation encountered. For example, the temperature at the airport in Ann Arbor, Michigan will certainly lie in the interval (-60, 60) degrees Celsius, the extremes representing bitter cold and intense heat. Higher or lower temperatures could possibly occur, but they have an extremely low probability (fortunately, for the residents of Ann Arbor). Since there is no practical difference between the statements *they cannot happen* and *they only happen very, very infrequently*, we can assume that temperature is a random variable that can have any value between -∞ and ∞. This formal assumption, that their limits are ±∞ has no physical significance, but it will often simplify the mathematical analysis considerably; otherwise, we would need to evaluate the effects of having finite limits in every situation we encounter. We will not worry here whether the extreme values are impossible, or only very infrequent.

5.1.1 Mean and variance

Suppose that there are N realizations of the random variable x. Denote them by x_k, $k = 1, ..., N$. The population mean and variance of x are

---

[3] We implicitly assume here that only *one* result can happen for each random experiment: results i and j cannot both happen at the same time if $i \ne j$.

$$<x> \; = \; \frac{1}{N} \sum_{k=1}^{N} x_k \tag{4}$$

and

$$\mathrm{var}(x) \; = \; \frac{1}{N} \sum_{k=1}^{N} \left(x_k - \langle x \rangle \right)^2 \tag{5}$$

in the limit as $N \to \infty$. I have already discussed this in Chapter 1. The variance of x is denoted either by $\mathrm{var}(x)$ or σ_x^2.

In terms of the probability distributions, the mean and variance are given by

$$\langle x \rangle \; = \; \int_{-\infty}^{\infty} x \, p(x) \, dx \tag{6}$$

and

$$\mathrm{var}(x) \; = \; \int_{-\infty}^{\infty} \left(x - \langle x \rangle \right)^2 p(x) \, dx \, . \tag{7}$$

These are the first and second moments of the probability distribution $p(x)$.

Higher moments are also defined. The nth moment is defined by

$$\mathrm{E}\!\left[x^n \right] \; = \; \int_{-\infty}^{\infty} x^n \, p(x) \, d x \, . \tag{8}$$

The notation $\mathrm{E}[f(x)]$ means the *expected value* of $f(x)$. The notations $\mathrm{E}[x]$ and $<x>$ denote the expected and mean values of x, respectively, and are, for our purposes, synonymous.

I shall assume that the random variables of interest to us have probability distributions that are *Gaussian; i.e.,*

$$p(x) \; = \; \frac{1}{\sqrt{2\pi} \, \sigma_x} e^{\left(x - \langle x \rangle \right)^2 / 2\sigma_x^2} \, . \tag{9}$$

Gaussian distributions are important for at least three reasons. First, it can be proved that, when there are many different independent variables with different distributions, the sum of those variables approaches a Gaussian distribution. This is called the *central limit theorem*; it is discussed in textbooks on probability and I discuss it further in Chapter 14. Therefore, many physical processes are well represented with Gaussian distributions. In addition, the form of the Gaussian is analytically simple, so integrals that are hard or impossible to do with other distributions are easily evaluated for Gaussian distributions.[4] Finally, given the mean and variance—$<x^2>$ and σ_x^2—of a Gaussian random variable, it is possible to evaluate all of the higher moments.

Exercise 5-1: Evaluate the first and second moments of the Gaussian distribution. Also, show that, if we know these moments of a Gaussian distribution, we can calculate all of the higher moments.

5.1.2 Covariance

We define the covariance between two variables x and y by

$$\text{cov}(x,y) \;=\; \frac{1}{N}\sum_{k=1}^{N}\big(x_k - \langle x\rangle\big)\big(y_k - \langle y\rangle\big) \;=\; \frac{1}{N}\sum_{k=1}^{N}x_k y_k - \langle x\rangle\langle y\rangle \tag{10}$$

where x_k and y_k are the values of the kth realization of x and y. If we assume that the mean values of x and y are zero,

$$\text{cov}(x,y) \;=\; \frac{1}{N}\sum_{k=1}^{N}x_k y_k. \tag{11}$$

I denote the covariance of x and y by $\text{cov}(x,y)$ or σ_{xy}^2. Note that $\text{cov}(x,x) = \text{var}(x)$.

Making the appropriate substitutions into equation (10), it is easy to show that

$$\text{cov}(ax,by) \;=\; ab\,\text{cov}(x,y) \;=\; ab\sigma_{xy}^2 \tag{12}$$

for any scalars a and b. The variance of the sum of two random variables is

$$\text{var}(x \pm y) \;=\; \sigma_x^2 \pm 2\sigma_{xy}^2 + \sigma_y^2. \tag{13}$$

We can show this directly from the equation for the variance.

Two variables are *independent* if knowing the value of one of them provides no information about the other one. The covariance of two variables is a measure of how similar they are; if x and y are independent, then $\text{cov}(x,y)=0$. The converse is not necessarily true: even if x and y are not independent, $\text{cov}(x,y)$ may still equal 0. If x and y are independent, $\sigma_{xy}^2 = 0$, and

$$\text{var}(x \pm y) \;=\; \sigma_x^2 + \sigma_y^2. \tag{14}$$

---

[4] In fact, Gaussian distributions may be too simple, in that investigators may tend to assume that everything is Gaussian-distributed, even when they know otherwise.

Exercise 5-2: Suppose $x(t) = \cos t$ and $y(t) = \sin t$. What is the average value of x? of y? of xy? Show that x and y are uncorrelated. Are they independent?

An important example of independent variables is the value of a geophysical variable that we want to measure and the *noise*—or uncertainty in the measurement—that arises in a measuring device. Two such variables that represent independent physical situations must be independent in the statistical sense. Let y be a quantity we want to measure, and n be the noise in the measurement of y. Normally, the source of the noise n is physically independent of the physical processes that affect y. For instance, the cloud patterns on the earth are determined by the balance of many meteorological processes. Suppose we make a picture with a radiometer in a satellite. The data will contain some noise that is generated in the electronics of the radiometer. However, the two processes are independent: the radiation does not affect the noise,[5] and the noise certainly does not change the cloud patterns. In other words, we assume that the measurements we are considering do not affect the things we are measuring. When the noise in the measured value of y is independent of the true value of y, the noise is uncorrelated with y.

5.2 Vectors and matrices

I shall introduce a few ideas about matrices and vector spaces in this section. A vector $\mathbf{v}$ that has n components is an n-dimensional vector and is an element of an n-dimensional vector space. The most familiar vector space is E^n, the n-dimensional Euclidian vector space. The basis vectors (see §5.2.2) are n mutually orthogonal vectors. E^3 is epitomized by the usual x,y,z axes so commonly used in physics. The basis vectors of E^3 are commonly denoted by the unit vectors $\hat{\mathbf{i}} = (1,0,0)$, $\hat{\mathbf{j}} = (0,1,0)$, and $\hat{\mathbf{k}} = (0,0,1)$.

Formally, a vector space $\mathcal{V}$ consists of a collection of objects—vectors—along with a field of real or complex numbers. Let α and β be any two scalars; let $\mathbf{u}$ and $\mathbf{v}$ be two vectors. The following operations must be defined:

1. Commutative vector addition: $\mathbf{u} + \mathbf{v} = \mathbf{v} + \mathbf{u}$.
2. Scalar multiplication: $\alpha\mathbf{u}$.
3. Distributive property for scalar multiplication: $(\alpha + \beta)\mathbf{u} = \alpha\mathbf{u} + \beta\mathbf{u}$ and $\alpha(\mathbf{u} + \mathbf{v}) = \alpha\mathbf{u} + \alpha\mathbf{v}$.

A vector space is *closed* under addition and scalar multiplication: given a vector space $\mathcal{V}$, if $\mathbf{u} \in \mathcal{V}$ and $\mathbf{v} \in \mathcal{V}$, then, for any scalars α and β, $\alpha\mathbf{u} + \beta\mathbf{v} \in \mathcal{V}$.[6]

---

[5] Except insofar as the overall level of the radiance of the scene being viewed will affect the amount of noise—the noise *variance*—in the radiometer. This is usually a small, constant effect.

[6] The symbol $\in$ means "is an element of."

5.2.1 Vector and matrix transpose

Throughout this book, I write vectors as *column vectors*; *i.e.*, they are represented by columns of numbers. I will often use the notation

$$\text{col}(x_1, \cdots x_n) = \begin{pmatrix} x_1 \\ \vdots \\ x_n \end{pmatrix} \tag{15}$$

to save space.

The *transpose* of the matrix

$$\mathbf{A} = \begin{pmatrix} a_{11} & a_{12} & a_{13} \\ a_{21} & a_{22} & a_{23} \\ a_{31} & a_{32} & a_{33} \end{pmatrix} \tag{16}$$

is

$$\mathbf{A}^t = \begin{pmatrix} a_{11} & a_{21} & a_{31} \\ a_{12} & a_{22} & a_{32} \\ a_{13} & a_{23} & a_{33} \end{pmatrix}. \tag{17}$$

That is, the matrix is reflected about its main diagonal. When a matrix is equal to its transpose, it is said to be *symmetric*. Many important properties of the matrices that we will deal with stem from this simple property.

For a vector, the notation $\mathbf{x}^t$ denotes the transpose; if $\mathbf{x}$ is a column vector, then $\mathbf{x}^t$ is a row vector:

$$\mathbf{x} = \begin{pmatrix} x_1 \\ \vdots \\ x_n \end{pmatrix}, \quad \mathbf{x}^t = \begin{pmatrix} x_1 & \cdots & x_n \end{pmatrix}. \tag{18}$$

Vectors and matrices are added component-by-component. For example,

$$\begin{pmatrix} x_1 \\ \vdots \\ x_n \end{pmatrix} + \begin{pmatrix} y_1 \\ \vdots \\ y_n \end{pmatrix} = \begin{pmatrix} x_1 + y_1 \\ \vdots \\ x_n + y_n \end{pmatrix}; \tag{19}$$

matrices are added similarly.

There are three possible ways to define the product of two vectors: the inner (or dot) product, the cross product[7], and the outer product, which pro-

---

[7] We will not need the cross product in this book.

duce a scalar, a vector, and a matrix, respectively. The dot product is the most familiar:

$$\mathbf{x} \cdot \mathbf{y} = \mathbf{x}^t \mathbf{y} = x_1 y_1 + x_2 y_2 + \cdots + x_n y_n, \tag{20}$$

where $\mathbf{x}$ and $\mathbf{y}$ are vectors of length n. Note that $\mathbf{x} \cdot \mathbf{y}$ is a scalar. We will encounter the outer product in §5.3.

The product of two matrices $\mathbf{A}$ and $\mathbf{B}$ is defined according to their components a_{ij} and b_{jk}, respectively. Let $\mathbf{A}$ be an $n \times m$ matrix and $\mathbf{B}$ be an $m \times p$ matrix.[8] If $\mathbf{C} = \mathbf{AB}$, then $\mathbf{C}$ is the $n \times p$ matrix whose components are

$$c_{ik} = \sum_{j=1}^{m} a_{ij} b_{jk}. \tag{21}$$

The transpose of $\mathbf{C}$ is

$$\mathbf{C}^t = \mathbf{B}^t \mathbf{A}^t. \tag{22}$$

Note that the order of $\mathbf{B}$ and $\mathbf{A}$ has been reversed. Unlike scalar multiplication, matrix multiplication is *not* commutative: in general, $\mathbf{AB} \neq \mathbf{BA}$. For scalars, if $ab = 0$, then either $a = 0$ or $b = 0$. For matrices, this is not necessarily true. It is possible that $\mathbf{AB} = \mathbf{0}$ even if $\mathbf{A} \neq \mathbf{0}$ and $\mathbf{B} \neq \mathbf{0}$. Here, $\mathbf{0}$ is the zero matrix. Finally, the product of a matrix and a vector is another vector: we write it in the form $\mathbf{y} = \mathbf{Ax}$. In component notation,

$$y_i = \sum_j a_{ij} x_j. \tag{23}$$

If we are concerned with row vectors, then multiplication is defined from the right: $\mathbf{y}^t = \mathbf{x}^t \mathbf{A}^t$.

Exercise 5-3: Find some pairs of matrices $\mathbf{A}$ and $\mathbf{B}$, such that $\mathbf{A} \neq \mathbf{0}$ and $\mathbf{B} \neq \mathbf{0}$, but $\mathbf{AB} = \mathbf{0}$. One such pair is $\mathbf{A} = \begin{pmatrix} 1 & 0 \\ 0 & 0 \end{pmatrix}$, $\mathbf{B} = \begin{pmatrix} 0 & 0 \\ 0 & 1 \end{pmatrix}$.

5.2.2 Bases

Consider a vector space $\mathcal{V}$ and a set of n vectors $\mathbf{e}_1, \cdots, \mathbf{e}_n \in \mathcal{V}$. Let $\mathbf{v}$ be a linear sum of these vectors:

$$\mathbf{v} = a_1 \mathbf{e}_1 + \cdots + a_n \mathbf{e}_n, \tag{24}$$

---

[8] The product $\mathbf{AB}$ is not defined unless $\mathbf{A}$ has the same number of rows as $\mathbf{B}$ has columns.

where the a's are scalars. Then every such vector $\mathbf{v} \in \mathcal{V}$. The vectors $\mathbf{e}_1, \cdots, \mathbf{e}_n$ are *linearly independent* if it is not possible to write

$$\sum_i a_i \mathbf{e}_i = 0 \tag{25}$$

unless all of the coefficients a_i are zero. To say that a vector space has dimension n means that a set of n linearly independent vectors exists, but no set of $n + 1$ linearly independent vectors exists. A *basis* is any set of n linearly independent vectors; any vector $\mathbf{v} \in \mathcal{V}$ can be written as a linear sum of these basis vectors. If one basis has n vectors, then every basis has n vectors. The basis vectors are customarily normalized so that

$$\mathbf{e}_i \cdot \mathbf{e}_j = \delta_{ij}, \tag{26}$$

where

$$\delta_{ij} = \begin{cases} 1 & if\ \ i = j \\ 0 & if\ i \neq j. \end{cases} \tag{27}$$

Such a basis is called an *orthonormal* basis.

Let $\mathcal{V}$ and $\mathcal{W}$ be two vector spaces with dimension m and n, respectively. A linear transformation—or linear mapping—maps vectors in $\mathcal{V}$ to vectors in $\mathcal{W}$. Every linear transformation from $\mathcal{V}$ to $\mathcal{W}$ can be represented by an $m \times n$ matrix. Now if we change the set of basis vectors we use to represent vectors in $\mathcal{V}$ and $\mathcal{W}$, then the coefficients of the matrix change, although the linear transformation remains the same. As a simple example, suppose that $\mathcal{V} = \mathcal{W} = E^2$. If the basis vectors are $(1,0)$ and $(0,1)$, a matrix $\mathbf{A}$ represents a particular transformation. If we change the basis vectors to be $(1,1)$ and $(1,-1)$, the same transformation would be represented by a different matrix.

Exercise 5-4: Prove that, if $\{\mathbf{e}_i\}$ is a basis of $\mathcal{V}$ and $\mathbf{v} = a_1\mathbf{e}_1 + \cdots + a_n\mathbf{e}_n = b_1\mathbf{e}_1 + \cdots + b_n\mathbf{e}_n$, then $a_i = b_i$, $i = 1, \cdots, n$. (Remember that the notation $\{\mathbf{e}_i\}$ means the set that contains the vectors $\mathbf{e}_i$, $i = 1, \ldots, n$.)

As an aside, a vector is often defined as having a magnitude and a direction—*i.e.*, we can think of it as an arrow of some kind. In fact, the notion of a vector is more general than that: we can say that any element of a vector space is a vector. We shall see later that vectors can also be continuous functions, and that the corresponding vector space may have an infinite number of dimensions (see the discussion of Hilbert spaces in Chapter 14).

5.2.3 Rank

Consider the rows of an $n{\times}n$ matrix $\mathbf{M}$. It operates on a vector $\mathbf{v} \in E^n$ to produce another vector $\mathbf{u} \in E^n$. We can write this matrix as a column of row vectors:

$$\mathbf{M} = \begin{pmatrix} \mathbf{m}_1^t \\ \vdots \\ \mathbf{m}_n^t \end{pmatrix} \tag{28}$$

where each row vector $\mathbf{m}_i^t$ has n components. I shall reserve the notation $\mathbf{v}$ for a column vector, and use the notation $\mathbf{v}^t$ for a row vector. Therefore, we can write the product of $\mathbf{M}$ with a vector as

$$\mathbf{Mv} = \begin{pmatrix} \mathbf{m}_1 \cdot \mathbf{v} \\ \vdots \\ \mathbf{m}_n \cdot \mathbf{v} \end{pmatrix}. \tag{29}$$

The row rank of $\mathbf{M}$ is the maximum number of row vectors of $\mathbf{M}$ that are linearly independent. We can similarly write $\mathbf{M}$ as a row of m-dimensional column vectors

$$\mathbf{M} = \begin{pmatrix} \mathbf{m}_1 & \cdots & \mathbf{m}_m \end{pmatrix}. \tag{30}$$

The column rank of $\mathbf{M}$ is the maximum number of linearly independent column vectors. A theorem of linear algebra is that the row rank of any matrix equals its column rank.

The *rank* of a matrix is its row rank or its column rank. If $\mathbf{M}$ is an $m{\times}n$ matrix, its rank is less than or equal to the smaller of m and n.

5.2.4 Null space

Consider the vector space E^n. It may happen that, for some vector $\mathbf{x} \in E^n$, $\mathbf{x} \neq \mathbf{0}$, and $\mathbf{Ax} = \mathbf{0}$, where $\mathbf{0}$ denotes the zero vector. If there are two vectors $\mathbf{x}$ and $\mathbf{y}$, such that $\mathbf{Ax} = \mathbf{Ay} = \mathbf{0}$, then $\mathbf{A}(\alpha\mathbf{x} + \beta\mathbf{y}) = \alpha\mathbf{Ax} + \beta\mathbf{By} = \mathbf{0}$ for any scalars α and β. Therefore, all of the vectors that map to $\mathbf{0}$ form a vector space. The space that contains all of the vectors $\mathbf{x}$ that $\mathbf{A}$ maps to $\mathbf{0}$ is called the *null space* of $\mathbf{A}$, which I shall denote by $\mathcal{N}$. Let $\mathcal{K}'$ be the complement of $\mathcal{N}$ in $\mathcal{V}$; which means that $\mathcal{K}'$ is the set that contains all of the vectors that $\mathbf{A}$ does *not* map to $\mathbf{0}$. While $\mathcal{K}'$ is not itself a vector space, since it does not contain $\mathbf{0}$, define $\mathcal{K} = \mathcal{K}' \cup \{\mathbf{0}\}$. Then $\mathcal{K}$ is a vector space. Note that the null space $\mathcal{N}$ is a subspace of

E^n that is determined by the matrix **A**. Since **A** either maps a vector **x** to **0** or it maps **x** to some vector **y** ≠ **0**, for any non-zero vector **x**, either **x** ∈ 𝒩 or **x** ∈ 𝒦. Therefore, E^n = 𝒩 ∪ 𝒦. Let m and k be the dimensions of 𝒩 and 𝒦. Then $m + k = n$. Although the components of **A** depend on the basis used for the vector space, the dimensions of 𝒩 and 𝒦 are the same in any basis.

Here, as elsewhere, I use the notation {$x, y, ...$} to denote the set that contains $x, y, ...$ In particular here, I use the notation {**0**} to denote the set that contains the vector **0** and nothing else. The null space always contains **0**. As we shall see, whether or not it contains anything else is important.

Exercise 5-5: Prove that any vector **v** ∈ E^n is the sum of a vector **x** ∈ 𝒦 and a vector **y** ∈ 𝒩, and that this decomposition is unique.

Exercise 5-6: Prove that the rank of the matrix **A** is $k = n - m$. In other words, the rank of a square matrix is the dimension of the vector space minus the dimension of the null space.

Exercise 5-7: Why must a vector space contain **0**?

5.2.5 Matrix inverse

The *inverse* of a matrix **A**, if it exists, is the matrix $\mathbf{A}^{-1}$ that has the property

$$\mathbf{A}^{-1}\mathbf{A} = \mathbf{A}\mathbf{A}^{-1} = \mathbf{I}. \tag{31}$$

Here, **I** is the identity matrix:

$$\mathbf{I} = \begin{pmatrix} 1 & 0 & \cdots & 0 \\ 0 & 1 & & \\ \vdots & & \ddots & \vdots \\ 0 & & \cdots & 1 \end{pmatrix}. \tag{32}$$

If $\mathbf{A}^{-1}$ does not exist, the matrix is said to be *singular*. Otherwise, it is *nonsingular*.

A square matrix is singular if and only if the determinant of **A** ≠ 0; this is usually written det(**A**). The determinant of a 2×2 matrix is

$$\det\begin{pmatrix} a & b \\ c & d \end{pmatrix} = ad - bc. \tag{33}$$

The generalization to larger matrices is treated in any book on linear algebra.[9] If $\mathbf{A}$ is not singular, the determinant of its inverse is

$$\det(\mathbf{A}^{-1}) = \frac{1}{\det(\mathbf{A})}. \tag{34}$$

Exercise 5-8: Prove that a square matrix $\mathbf{A}$ is singular if and only if its null space is not $\{\mathbf{0}\}$.

The property in equation (31) that $\mathbf{A}$ and $\mathbf{A}^{-1}$ commute is not at all trivial: it ensures that, if it exists, $\mathbf{A}^{-1}$ is unique. Suppose that two distinct matrices $\mathbf{P}$ and $\mathbf{Q}$ are both inverses of $\mathbf{A}$. Then

$$\mathbf{PA} = \mathbf{AP} = \mathbf{I}$$

$$\mathbf{I} = \mathbf{QA} = \mathbf{AQ}. \tag{35}$$

Multiply the top equation on the left by $\mathbf{Q}$ and the bottom one by $\mathbf{P}$. If a matrix commutes with its inverse, then

$$\mathbf{Q}(\mathbf{AP}) = \mathbf{P}(\mathbf{AQ}), \tag{36}$$

and

$$(\mathbf{QA})\mathbf{P} = (\mathbf{PA})\mathbf{Q}, \tag{37}$$

so

$$\mathbf{P} = \mathbf{Q}. \tag{38}$$

Therefore, the inverse is unique.

Consider the equation $\mathbf{y} = \mathbf{Ax}$. If $\mathbf{w} \in \mathcal{N}$, then $\mathbf{y} = \mathbf{A}(\mathbf{x} + \mathbf{w}) = \mathbf{Ax} + \mathbf{Aw} = \mathbf{Ax} + \mathbf{0}$ also. Therefore, there is an infinite number of vectors $\mathbf{x}$ that $\mathbf{A}$ will map to $\mathbf{y}$ unless its null space $\mathcal{N} = \{\mathbf{0}\}$. Now consider the set of solutions of $\mathbf{y} = \mathbf{Ax}$, *i.e.,* the set of all vectors $\mathbf{x}$ that satisfy this equation for a given vector $\mathbf{y}$. If $\mathcal{N} = \{\mathbf{0}\}$, then there is a unique vector $\mathbf{x}$ that could have produced $\mathbf{y}$, so the notion of a solution $\hat{\mathbf{x}} = \mathbf{A}^{-1}\mathbf{y}$ makes sense. But if $\mathcal{N} \neq \{\mathbf{0}\}$, there is at least one nonzero vector $\mathbf{w} \neq \mathbf{0}$ such that $\mathbf{y} = \mathbf{A}(\mathbf{x} + \mathbf{w})$. Therefore there is no unique solution of this equation.

Exercise 5-9: Consider a 2×2 matrix again. The product of a 2×2 matrix with a 2-dimensional vector can be written generally as

---

[9] See, for example, Deif (1982) Chapter 2.

$$\mathbf{Ap} = \begin{pmatrix} a & b \\ c & d \end{pmatrix}\begin{pmatrix} p \\ q \end{pmatrix} = \begin{pmatrix} ap + bq \\ cp + dq \end{pmatrix}. \tag{39}$$

The null space has dimension 2 if and only if $a = b = c = d = 0$. Otherwise, $\mathcal{N}$ = $\{\mathbf{0}\}$ if and only if there is no solution to $ap + bq = 0$ and $cp + dq = 0$ for $p \neq 0$ and $q \neq 0$. Show that this is equivalent to the statement $\mathcal{N} = \{\mathbf{0}\}$ if and only if $\det(\mathbf{A}) \neq 0$.

5.2.6 Diagonal matrices

An $n \times m$ matrix represents a linear transformation from the vector space $\mathcal{V}$ to another vector space $\mathcal{U}$. A function $f(\mathbf{x})$ is linear if

$$f(a\mathbf{x} + b\mathbf{y}) = af(\mathbf{x}) + bf(\mathbf{y}), \tag{40}$$

where a and b are any scalars. However, while the above notation can be quite abstract, a matrix is a more concrete object. The corresponding equation for the matrix $\mathbf{M}$ that represents the transformation f would be

$$\mathbf{M}(a\mathbf{x} + b\mathbf{y}) = a\mathbf{Mx} + b\mathbf{My}. \tag{41}$$

Exercise 5-10: The equation $\mathbf{v} = \mathbf{Mx}$ is equivalent to

$$v_i = \sum_j m_{ij} x_j. \tag{42}$$

Use this to show that $\mathbf{M}$ is linear.

There are many possible orthonormal bases for any vector space of dimension greater than 1. The same linear function will be represented by different matrices if we use different sets of basis vectors. A given linear function will often be much simpler to understand if we use the right set of basis vectors. I shall show how we can transform any symmetric, real matrix into a diagonal matrix (defined below) by rotating the basis vectors. Suppose the set of vectors $\{\mathbf{e}_1, \cdots, \mathbf{e}_n\}$ is an orthonormal basis. Define a new set of vectors

$$\mathbf{b}_j = \mathbf{Ue}_j, \quad j = 1, \cdots, n, \tag{43}$$

where $\mathbf{U}$ is an *orthogonal* matrix. A matrix is called *orthogonal* if it has the property that

$$\mathbf{U}\mathbf{U}^t = \mathbf{U}^t\mathbf{U} = \mathbf{I}. \tag{44}$$

The inverse of an orthogonal matrix is equal to its transpose, so

$$\mathbf{e}_j = \mathbf{U}^t\mathbf{b}_j. \tag{45}$$

Exercise 5-11: Show that, if $\mathbf{U}$ is orthogonal, then $(\mathbf{U}\mathbf{e}_j)\cdot(\mathbf{U}\mathbf{e}_k) = \delta_{jk}$.

We can write an orthogonal matrix as a row of column vectors $\mathbf{u}_j$:

$$\mathbf{U} = \begin{pmatrix} \mathbf{u}_1 & \cdots & \mathbf{u}_n \end{pmatrix}. \tag{46}$$

A matrix is orthogonal if and only if

$$\mathbf{u}_i \cdot \mathbf{u}_j = \delta_{ij}. \tag{47}$$

If $\mathbf{b} = \mathbf{U}\mathbf{e}$, the components of $\mathbf{b}$ are $b_i = \mathbf{u}_i \cdot \mathbf{e}$.

Consider how an $n \times n$ matrix $\mathbf{A}$ transforms each of the vectors $\mathbf{b}_j$. Suppose that there are vectors $\mathbf{q}$ and $\mathbf{b}$ such that

$$\mathbf{q}_j = \mathbf{A}\mathbf{b}_j, \quad j = 1, \cdots, n. \tag{48}$$

If $\mathbf{b}_j = \mathbf{U}\mathbf{e}_j$, then $\mathbf{q}_j = \mathbf{A}\mathbf{U}\mathbf{e}_j$. Define the vector $\mathbf{r}_j$ by $\mathbf{q}_j = \mathbf{U}\mathbf{r}_j$, or $\mathbf{r}_j = \mathbf{U}^t\mathbf{q}_j$. Then

$$\mathbf{U}\mathbf{r}_j = \mathbf{A}\mathbf{U}\mathbf{e}_j. \tag{49}$$

Multiplying from the left by $\mathbf{U}^t$,

$$\mathbf{U}^t\mathbf{U}\mathbf{r}_j = \mathbf{r}_j = \mathbf{U}^t\mathbf{A}\mathbf{U}\mathbf{e}_j. \tag{50}$$

Now define the matrix $\mathbf{M} = \mathbf{U}^t\mathbf{A}\mathbf{U}$. Then $\mathbf{r} = \mathbf{M}\mathbf{e}$ and $\mathbf{q} = \mathbf{A}\mathbf{b}$ represent the same linear transformation, except that the basis vectors have been changed. Therefore the transformation $\mathbf{U}^t\mathbf{A}\mathbf{U}$ produces a new matrix $\mathbf{M}$ that represents the same linear transformation as $\mathbf{A}$ did, but with different basis vectors.

A transformation of the form

$$\mathbf{M} = \mathbf{U}^t\mathbf{A}\mathbf{U}, \tag{51}$$

where $\mathbf{M}$ and $\mathbf{A}$ are both $n \times n$ matrices and $\mathbf{U}$ is orthogonal, is called an *orthogonal transformation*. Physically, it represents a rotation in a vector space. It has the property that it preserves the lengths of all vectors.

If the matrices $\mathbf{M}$ and $\mathbf{A}$ are related by an orthogonal rotation, $\mathbf{M} = \mathbf{U}^t\mathbf{A}\mathbf{U}$, then $\mathbf{M}$ and $\mathbf{A}$ are said to be *equivalent* or *orthogonally equivalent*. Since the only difference between $\mathbf{M}$ and $\mathbf{A}$ is the choice of a set of basis vectors, $\mathbf{M}$ and $\mathbf{A}$ contain the same information. Given the equation $\mathbf{y} = \mathbf{A}\mathbf{x}$, it might be easier to find an equivalent matrix $\mathbf{M}$ and solve the related equation $\mathbf{y}' = \mathbf{M}\mathbf{x}'$ instead.

Exercise 5-12: Show that $\mathbf{M}$ and $\mathbf{A}$ have the same rank.

Exercise 5-13: Show that the determinant of an orthogonal matrix is ± 1.

A *diagonal* matrix has the form

$$\mathbf{D} = \begin{pmatrix} \lambda_1 & 0 & \cdots & 0 \\ 0 & \lambda_2 & & \\ \vdots & & \ddots & \vdots \\ 0 & & \cdots & \lambda_n \end{pmatrix}. \tag{52}$$

To conserve space, I shall often write $\mathbf{D} = \mathrm{diag}(\lambda_1, \ldots, \lambda_n)$. A diagonal matrix has zeroes everywhere except—reasonably enough—on the main diagonal.

Diagonal matrices are easy to work with because they are easy to invert. The reader should show for himself that

$$\mathbf{D}^{-1} = \mathrm{diag}(1/\lambda_1, \quad 1/\lambda_2, \quad \cdots, \quad 1/\lambda_n). \tag{53}$$

If $\mathbf{S}$ is any real, symmetric matrix, then there exists an orthogonal matrix $\mathbf{U}$ with the property that

$$\mathbf{D} = \mathbf{U}^t\mathbf{S}\mathbf{U} \tag{54}$$

is a diagonal matrix (the proof of this statement is beyond the scope of this book). We say that $\mathbf{U}$ *diagonalizes* $\mathbf{S}$. Using the orthogonality relationship in equation (44),

$$\mathbf{S} = \mathbf{U}\mathbf{U}^t\mathbf{S}\mathbf{U}\mathbf{U}^t = \mathbf{U}\mathbf{D}\mathbf{U}^t. \tag{55}$$

An $n \times n$ matrix is nonsingular if and only if it has rank n. This is particularly easy to see for diagonal matrices. The matrix $\mathbf{D} = \mathrm{diag}(d_1, \ldots, d_n)$ has rank n only if none of the d_i's is zero.

Note that the identity matrix $\mathbf{I}$ has the property that $\mathbf{U}\mathbf{I}\mathbf{U}^t = \mathbf{I}$ for any orthogonal matrix $\mathbf{U}$. If we think of $\mathbf{U}$ as inducing a rotation in the underlying vector space, then $\mathbf{I}$ is the same, no matter how the vector space is rotated. The only nonzero matrices that have this property can be written as $\alpha\mathbf{I}$, where α is any scalar.

Exercise 5-14: Show that the product of two orthogonal matrices is also orthogonal.

5.2.7 Determinants

I mentioned the determinant function a few paragraphs back. I should summarize the properties of determinants that are important for us. For any square matrices **A**, **B**, …,

1. Det(**A**) = 0 if and only if **A** is singular.

2. Det(**AB**) = Det(**A**) ·Det(**B**).

3. Det(**A**$^{-1}$) = 1 / Det(**A**) (if Det(**A**) ≠ 0).

4. If **D** = diag(λ_1, …, λ_n), then Det(**D**) = $\lambda_1 \cdot \lambda_2 \cdot \ldots \cdot \lambda_n$.

5. If **A** is a 3×3 matrix,

$$\text{Det} \begin{pmatrix} a & b & c \\ d & e & f \\ x & y & z \end{pmatrix} = aez + bfx + dyc - xec - dbz - ayf. \tag{56}$$

The pattern continues for larger square matrices. To see this, write the above equation on a piece of paper and draw arrows connecting the points as they appear in the products on the right. For example, connect *a* to *e* to *z, x* to *e* to *c*, and so forth. Notice that the downward-sloping lines have positive contributions; the upward-sloping, negative contributions.

Exercise 5-15: Show that, if **U** is an orthogonal matrix and **D** is diagonal, and **S** = **UDU**^{t}, that Det(**S**) = Det(**D**).

5.3 Covariance matrices

If we consider only two scalar variables x and y (which, remember, have zero mean), all the statistical information about x and y is contained in their variances σ_x^2 and σ_y^2 and their covariance σ_{xy}^2 if they are Gaussian distributed.[10] However, when there are many variables, we need to generalize this concept and consider the *covariance matrix*. Let **x** = $(x_1, \cdots, x_n)$ be a random vector with n components; let **x**$_k$ be the kth realization of **x**. Then the covariance matrix of **x** is defined to be

---

[10] For random variables that are not Gaussian distributed, a complete statistical description would necessitate knowing the values of all of the moments of x and y.

$$S_x \equiv \begin{pmatrix} \sigma^2_{x_1 x_1} & \sigma^2_{x_1 x_2} & & \sigma^2_{x_1 x_n} \\ \sigma^2_{x_2 x_1} & \sigma^2_{x_2 x_2} & & \vdots \\ \vdots & & \ddots & \vdots \\ \sigma^2_{x_n x_1} & \cdots & \cdots & \sigma^2_{x_n x_n} \end{pmatrix}. \tag{57}$$

In terms of the components,

$$s_{ij} = \text{cov}(x_i, x_j) = \sigma^2_{x_i x_j}. \tag{58}$$

Remember that we always assume that variables have zero mean (or we have transformed them from x to $x - <x>$). If we did not do this, we would have to write $s_{ij} = \text{cov}[(x_i - <x_i>)(x_j - <x_j>)]$.

If $\mathbf{S}$ is a covariance matrix, then $\mathbf{S} = \mathbf{S}^t$. To see this, just note that each component is defined in terms of a covariance: $s_{ij} = \text{cov}(x_i, x_j) = \text{cov}(x_j, x_i)$. Since we are measuring real quantities, covariance matrices are also real. Real symmetric matrices have important properties that we shall exploit in some of the following sections.

A common notation for the covariance matrix is

$$\mathbf{S} = \mathbf{x}\mathbf{x}^t. \tag{59}$$

We should interpret this as follows. The *outer product* of the vectors $\mathbf{x}$ and $\mathbf{y}$ is the matrix.

$$\mathbf{x}\mathbf{y}^t = \begin{pmatrix} x_1 y_1 & x_1 y_2 & \cdots & x_1 y_n \\ x_2 y_1 & x_2 y_2 & & \vdots \\ \vdots & & \ddots & \\ x_n y_1 & \cdots & & x_n y_n \end{pmatrix}. \tag{60}$$

The notation $\mathbf{S} = \mathbf{x}\mathbf{x}^t$ really means $\mathbf{S} = <\mathbf{x}\mathbf{x}^t>$; *i.e.*, the covariance matrix is the average value of the outer product of $\mathbf{x}$ with itself. Each sample of the vector $\mathbf{x}$ produces a matrix $\mathbf{x}\mathbf{x}^t$. Since the components of the average of a matrix are just the average of each component, this definition of the covariance matrix agrees with the first one. That is,

$$\left\langle \mathbf{x}\mathbf{x}^t \right\rangle = \begin{pmatrix} \langle x_1 x_1 \rangle & \langle x_1 x_2 \rangle & \cdots & \langle x_1 x_n \rangle \\ \langle x_2 x_1 \rangle & \langle x_2 x_2 \rangle & & \vdots \\ \vdots & & \ddots & \\ \langle x_n x_1 \rangle & \cdots & & \langle x_n x_n \rangle \end{pmatrix}. \tag{61}$$

This notation for the covariance matrix is too convenient for us not to use it, even though outer vector product is not as familiar as the other two. I shall always use this notation in the context of taking an average value.

There is still a third notation that I will use for a covariance matrix. Suppose that there is a quantity p that is a function of two variables.[11] For example, p might be a function of space x and time t. We measure the values of p at points x_j, $j = 1, n$, and times t_k, $k = 1, m$.[12] If we arrange these measurements into a rectangular array, we can define the $n \times m$ matrix $\mathbf{P}$ whose elements are

$$p_{jk} = p(x_j, t_k). \tag{62}$$

There are two covariance matrices that we could form from $\mathbf{P}$. Let

$$\mathbf{S}_x = \frac{1}{m}\mathbf{P}\mathbf{P}^t \tag{63}$$

Then the components of $\mathbf{S}_x$ are

$$s_{jk} = \frac{1}{m}\sum_{l=1}^{m} p_{jl}p_{kl} = \left\langle p(x_j), p(x_k) \right\rangle_t \tag{64}$$

which represent the covariance of p at one spatial point with p at another spatial point, averaged over time. Similarly, we could define

$$\mathbf{S}_t = \frac{1}{n}\mathbf{P}^t\mathbf{P}. \tag{65}$$

Then the components of $\mathbf{S}_t$ are

$$s_{jk} = \frac{1}{n}\sum_{l=1}^{m} p_{lj}p_{lk} = \left\langle p(t_j)p(t_k) \right\rangle_x \tag{66}$$

which represents a temporal covariance matrix. The notations $\langle \cdot \rangle_t$ and $\langle \cdot \rangle_x$ denote temporal and spatial averages, respectively.

We shall return to this in §5.5, where we will use Singular Value Decomposition to show that $\mathbf{S}_x$ and $\mathbf{S}_t$ have the same singular values (or eigenvalues; *cf* §5.4).

---

[11] This is easily extended to more dimensions.

[12] Naturally, the independent variables could be anything, but it is convenient to call them something concrete.

5.3.1 Trace

The *trace* of a matrix is the sum of its diagonal elements. For any square matrix **S**,

$$\text{tr}(\mathbf{S}) = \sum_{i=1}^{n} s_{ii} . \tag{67}$$

An orthogonal transformation leaves the trace of a square matrix unchanged. If **S** is the covariance matrix of **x**, and it is transformed to a diagonal matrix

$$\mathbf{D} = diag(\lambda_1, \cdots, \lambda_n), \tag{68}$$

the trace of **S** equals the trace of **D**, since $\mathbf{D} = \mathbf{U}^t\mathbf{S}\mathbf{U}$, and

$$\sum_{i=1}^{n} \text{var}(x_i) = \sum_{i=1}^{n} \lambda_i . \tag{69}$$

The trace of the covariance matrix **S** of $\mathbf{x} = \text{col}(x_1, \ldots, x_n)$ is the sum of the variances of the different components x_i. It is the same in any orthonormal basis, *i.e.*, whether or not **S** is diagonal. Therefore, we can define the *total variance* Σ_{tot} to

$$\Sigma_{\text{tot}} \equiv \sum_{i=1}^{n} \lambda_i , \tag{70}$$

which is independent of the basis vectors used to represent **x**. In other words, the total variance is invariant under orthogonal transformations. Because of this, it makes sense to define the total variance and to compare the covariance matrices belonging to the same physical qua $\sum_{i=1}^{n} \text{var}(x_i) = \sum_{i=1}^{n} \lambda_i$ tities $x_1, \ldots, x_n$ when the vectors **x** are defined in different coördinate systems. The variance of each quantity separately is partly a mathematical property of how we chose to represent these variables, but the total variance depends only on the physics of the situation in question.

5.3.2 Correlation matrix

The *correlation coefficient* of x_i and x_j is the quantity

$$\rho_{ij} = \frac{\langle x_i x_j \rangle}{\sqrt{\langle x_i^2 \rangle \langle x_j^2 \rangle}} . \tag{71}$$

The correlation matrix $\mathbf{C}$ has components ρ_{ij}. If we define the diagonal matrix $\mathbf{D}_x$ to be

$$\mathbf{D}_x = \text{diag}\left(\sqrt{\langle x_1^2 \rangle}, \ \cdots, \ \sqrt{\langle x_n^2 \rangle} \right), \tag{72}$$

then the covariance matrix $\mathbf{S}$ is related to the correlation matrix $\mathbf{C}$ by

$$\mathbf{S} = \mathbf{D}_x \mathbf{C} \mathbf{D}_x . \tag{73}$$

We will use this relationship in §5.7.

One important property of the correlation coefficient is that $|\rho_{ij}| \leq 1$. This follows immediately from the Cauchy-Schwarz inequality (see §14.5).

5.3.3 Rotating a covariance matrix

We have n physical variables $x_1, \cdots, x_n$ that I denote by the vector $\mathbf{x}$. They affect the m measured variables $y_1, \cdots, y_m$ (denoted by the vector $\mathbf{y}$). In other words, $\mathbf{x}$ is the independent variable; $\mathbf{y}$, the dependent variable. Assume that, from our analysis of the physics involved, we have determined that they obey the linear relationship

$$\mathbf{y} = \mathbf{A}\mathbf{x}, \tag{74}$$

where $\mathbf{A}$ is an $m \times n$ matrix whose coefficients are known. The covariance matrix of the variables, $\mathbf{S}_x$, has components given by

$$s_{ij} = \text{cov}(x_i, x_j) = \langle x_i x_j \rangle \tag{75}$$

where, as always, $\langle x_i \rangle = 0$. The covariance matrix of the measured quantities, $\mathbf{S}_y$, has components

$$\sigma_{ij} = <y_i y_j>. \tag{76}$$

These two covariance matrices are related by equation (74). Write

$$y_i = \sum_{j=1}^{n} a_{ij} x_j . \tag{77}$$

Substitute this into the previous equation:

$$\sigma_{ij} = \sum_{k,l=1}^{n} a_{ik} a_{jl} \langle x_k x_l \rangle = \sum_{k,l=1}^{n} a_{ik} a_{jl} s_{kl} . \tag{78}$$

Note that, S_x is an $n \times n$ matrix and S_y, an $m \times m$ matrix. Since A is an $m \times n$ matrix, the dimensions of the matrices make sense. We can write this equation compactly in matrix form as

$$S_y = AS_xA^t. \tag{79}$$

There are n independent variables, so the rank of S_y cannot be greater than n. The transformation in equation (74) may cause some of the information about the variables to be lost, but it cannot create extra information. On the other hand, if $m < n$ (*i.e.*, if there are more variables than there are separate measurements) some information about the variables must be lost. In this case, the system of equations is underdetermined.

5.4 Eigenvalues and eigenvectors

Let U be the orthogonal matrix that diagonalizes a symmetric matrix S. That is, there is a diagonal matrix D such that $D = U^tSU$. Decomposing U into column vectors and using the diagonal property of D, it is not hard to show that

$$S u_i = \lambda_i u_i \tag{80}$$

for every column vector of U. That is, each of the vectors u_i has the property that, when it is multiplied by S, the same vector results, but it is multiplied by a scalar. These vectors are called the *eigenvectors*, or singular vectors, of S. The corresponding scalars λ_i are called the *eigenvalues*, or *singular* values. The use of λ for eigenvalues is customary; hence the use in the diagonal matrices above.

Obviously, if u_i is an eigenvector of S, then any scalar multiple of u_i is also an eigenvector. It is customary to define eigenvectors so that they have unit length; *i.e.*,

$$u_i \cdot u_i = 1. \tag{81}$$

For λ to be a solution to equation (80), it must satisfy

$$(S - \lambda I)u = 0 \tag{82}$$

for some vector u. However, this is only possible if

$$\det(S - \lambda I) = 0. \tag{83}$$

If S is an $n \times n$ matrix, this represents an nth-order polynomial in λ; equation (83) is called the *secular equation*. The eigenvalues are the n solutions of the secular equation. Therefore, every $n \times n$ matrix has n eigenvalues, although they might not all be distinct.

There are several important properties of eigenvalues that we shall use repeatedly. Let λ_k, $k = 1, \cdots, n$, be the eigenvalues of a symmetric matrix $\mathbf{S}$, and let $\mathbf{u}_k$ be the corresponding eigenvector:

1. The eigenvalues of a diagonal matrix are just the elements along the diagonal. The eigenvectors of the identity matrix are $\mathrm{col}(\lambda_1, 0, \ldots)$, $\mathrm{col}(0, \lambda_1, 0, \ldots)$, ...

2. The eigenvalues λ_k of a symmetric matrix are real.

3. If $\lambda_k \neq \lambda_j$, then $\mathbf{u}_k \cdot \mathbf{u}_j = 0$. That is, eigenvectors belonging to different eigenvalues are orthogonal. If two or more eigenvalues are equal, then we can find the corresponding number of eigenvectors that form an orthonormal set, but they are not unique.

4. The eigenvectors of a matrix form an orthonormal basis for the vector space. In this basis, the matrix is diagonal.

5. An orthogonal transformation of a matrix does not change its eigenvalues.

6. An $n \times n$ matrix is singular if and only if at least one of its eigenvalues $\lambda_k = 0$. The rank of such a matrix is equal to the number of its eigenvalues that are different from zero. It is easy to see this for a diagonal matrix.

7. The eigenvalues of the identity matrix are all $\lambda_k = 1$.

8. If the eigenvalues of $\mathbf{S}$ are $\{\lambda_k, k = 1, \ldots n\}$, then the eigenvalues of $\mathbf{S}^{-1}$ are $\{1/\lambda_k\}$; if one or more of the eigenvalues of $\mathbf{S}$ is zero, then the inverse of $\mathbf{S}$ does not exist.

We can demonstrate this last property easily; it is one that we will use often. Suppose $\mathbf{D} = \mathrm{diag}(\lambda_1, \ldots, \lambda_n)$ and $\mathbf{E} = \mathrm{diag}(\lambda_1^{-1}, \ldots, \lambda_n^{-1})$. (If some $\lambda_i = 0$, then $\mathbf{D}^{-1}$ does not exist.) Direct matrix multiplication shows that $\mathbf{DE} = \mathrm{diag}(\lambda_1 \lambda_1^{-1}, \ldots, \lambda_n \lambda_n^{-1}) = \mathbf{I}$. The rest of these properties are proved in any introductory book on linear algebra.

Exercise 5-16: What are the eigenvalues of the matrix $\begin{bmatrix} 1 & 0 \\ 0 & -1 \end{bmatrix}$? of $\begin{bmatrix} 1 & a \\ a & -1 \end{bmatrix}$?

Exercise 5-17: What are the eigenvalues of the matrix $\mathbf{A} = \begin{bmatrix} a & b \\ c & d \end{bmatrix}$? Write out the characteristic equation $\det(\mathbf{A} - \lambda\mathbf{I}) = 0$ in component form and solve the resulting quadratic equation. What are the constraints on $\mathbf{A}$ that guarantee that the eigenvalues will be ≥ 0?

5.4.1 Nonsymmetric matrices

Although the focus here is on covariance matrices, which are necessarily symmetric, I want to digress for a moment to consider more general real matrices. We shall see that some of the properties of symmetric matrices are not true in general. We can diagonalize any real symmetric matrix $\mathbf{A}$; if $\mathbf{A}$ is not symmetric, there is an orthogonal transformation—a transformation of the form

$$\mathbf{B} = \mathbf{U}^t\mathbf{A}\mathbf{U},\tag{84}$$

where $\mathbf{U}$ is an orthogonal matrix—that transforms into an upper triangular matrix:

$$\mathbf{U}^t\mathbf{A}\mathbf{U} = \begin{pmatrix} \lambda_1 & * & * & * & * & \cdots & * \\ 0 & \lambda_2 & * & * & * & \cdots & * \\ 0 & 0 & \lambda_3 & * & * & \cdots & * \\ 0 & 0 & 0 & \lambda_4 & * & \cdots & * \\ 0 & 0 & 0 & 0 & \lambda_5 & \cdots & \vdots \\ \vdots & \vdots & \vdots & \vdots & \vdots & \ddots & * \\ 0 & 0 & 0 & 0 & \cdots & 0 & \lambda_n \end{pmatrix};\tag{85}$$

i.e, all of the coefficients below the main diagonal are zero. (This is known as Schur's theorem.) Furthermore, the coefficients λ_k on the main diagonal are the eigenvalues of $\mathbf{A}$. Clearly, $\mathbf{U}^t\mathbf{A}\mathbf{U}$ is symmetric if and only if $\mathbf{A}$ is symmetric.

If $\mathbf{A}$ is symmetric, the eigenvectors that belong to different eigenvalues are orthogonal: *i.e.*, if $\lambda_j \neq \lambda_k$, then $\mathbf{u}_j\cdot\mathbf{u}_k = 0$. We can prove by writing

$$\begin{aligned} \mathbf{A}\mathbf{u}_j &= \lambda_j\mathbf{u}_j \qquad and \\ \mathbf{A}\mathbf{u}_k &= \lambda_k\mathbf{u}_k. \end{aligned}\tag{86}$$

Multiply the first equation by $\mathbf{u}_k$, the second, by $\mathbf{u}_j$. Since $\mathbf{A} = \mathbf{A}^t$, $\mathbf{u}_k\cdot\mathbf{A}\mathbf{u}_j = \mathbf{u}_j\cdot\mathbf{A}^t\mathbf{u}_k, = \mathbf{u}_j\cdot\mathbf{A}\mathbf{u}_k$. Therefore,

$$\mathbf{u}_k \cdot \mathbf{A}\mathbf{u}_j - \mathbf{u}_j \cdot \mathbf{A}\mathbf{u}_k = \left(\lambda_j - \lambda_k\right)\mathbf{u}_k \cdot \mathbf{u}_j = 0.\tag{87}$$

But since $\lambda_j \neq \lambda_k$, $\mathbf{u}_k\cdot\mathbf{u}_j = 0$.

If there is an eigenvalue with multiplicity $n > 1$, *i.e.*, there are $n > 1$ eigenvalues $\lambda_r, \ldots, \lambda_{r+n-1}$ that are equal, and $\mathbf{u}_r$ and $\mathbf{u}_p$ are two eigenvectors corresponding to this eigenvalue, then any linear combination of $\mathbf{u}_r$ and $\mathbf{u}_p$ is also an eigenvector. If the matrix $\mathbf{A}$ is symmetric, then there are n orthonormal eigenvectors that correspond to the eigenvalue λ_r that has multiplicity n. However, if **A** is not symmetric, there may be $m < n$ eigenvectors corresponding to that eigenvalue.

As an example, consider the matrix

$$\mathbf{A} = \begin{pmatrix} \lambda & c \\ 0 & \lambda \end{pmatrix}. \tag{88}$$

Exercise 5-18: Clearly the vector col(1,0) is an eigenvector. What is its eigenvalue? It is easy to show that there is no other eigenvalue. Show that there is no other eigenvector.

Because we deal mostly with symmetric matrices in this book, we may forget that the theorems we derive about these matrices do not necessarily apply to all matrices. Some of the properties of symmetric matrices are not properties of all matrices—like the property that an $n{\times}n$ symmetric matrix has n eigenvectors that span the vector space. As we have seen, this is not true in general.

The eigenvalues of a matrix are the roots of its secular equation. If $p(\lambda) = \det(\mathbf{A} - \lambda\mathbf{I})$ is the secular equation of the matrix **A**, then it is an nth order polynomial equation and has n solutions, although some of the solutions may have multiplicity greater than zero. The multiplicity of an eigenvalue λ is the number of times it occurs in a complete factorization of p; call this its *algebraic* multiplicity. Suppose that there are r *distinct* eigenvalues. Then there are at least r eigenvectors, but if $r < n$, there may be fewer than n eigenvectors. Given an eigenvalue with algebraic multiplicity $q > 1$, there is at least one corresponding eigenvector. The number of independent eigenvectors is called the *geometric* multiplicity; the geometric multiplicity is always $\leq$ the algebraic multiplicity.

If the algebraic and geometric multiplicities of a matrix **A** are the same for each eigenvalue, it is called *nondefective* (we shall encounter this term again in the discussion of integral equations). A matrix for which each eigenvalue has geometric multiplicity = 1 is called *nonderogatory*.[13]

I personally find this particular situation a bit unsatisfying. Consider the set of all $2{\times}2$ real matrices as an example. If **A** is a $2{\times}2$ matrix, then it has two orthogonal eigenvectors—unless it has the form shown in equation (88), when it has but one. Unless $c = 0$, in which case it has two again.

---

[13] See Horn and Johnson (1985, p 58).

Exercise 5-19: Consider the matrix $\mathbf{B} = \begin{pmatrix} \lambda & q & 0 & 0 \\ 0 & \lambda & r & 0 \\ 0 & 0 & \lambda & s \\ 0 & 0 & 0 & \lambda \end{pmatrix}$. How many eigen-

vectors does it have if q, r and s are all different from zero? What if $s = 0$? What if $r = s = 0$?

5.4.2 Eigenvector expansion

The eigenvectors of an $n \times n$ symmetric matrix can serve as a basis for the vector space, since there are always n orthogonal eigenvectors. Let $\mathbf{u}_1, \cdots, \mathbf{u}_n$ be these eigenvectors. Then we can write any vector $\mathbf{w}$ in the form

$$\mathbf{w} = \alpha_1 \mathbf{u}_1 + \cdots + \alpha_n \mathbf{u}_n \tag{89}$$

where the coefficients α_j are given by

$$\alpha_i = \mathbf{w} \cdot \mathbf{u}_i. \tag{90}$$

To see why this is so, we can write

$$\mathbf{w} \cdot \mathbf{u}_j = \sum_i \alpha_i \mathbf{u}_i \cdot \mathbf{u}_j = \sum_i \alpha_i \delta_{ij} = \alpha_j. \tag{91}$$

Since covariance matrices are symmetric, an $n \times n$ covariance matrix has n orthogonal eigenvectors that span the vector space. We will use this expansion later.

5.4.3 Shifting eigenvalues

Suppose, as before, that $\mathbf{A}$ is an $n \times n$ symmetric matrix whose eigenvalues are $\lambda_1, \dots, \lambda_n$ and whose eigenvectors are $\mathbf{e}_1, \cdots, \mathbf{e}_n$. Consider the matrix $\mathbf{B} = \mathbf{A} + \alpha \mathbf{I}$. For any $k \leq n$,

$$\mathbf{B}\mathbf{e}_k = \mathbf{A}\mathbf{e}_k + \alpha \mathbf{I}\mathbf{e}_k = (\alpha + \lambda_k)\mathbf{e}_k. \tag{92}$$

Therefore $\mathbf{B}$ has the same eigenvalues as $\mathbf{A}$, but they are shifted by an amount α; $\mathbf{B}$ has the same eigenvectors as $\mathbf{A}$.

5.5 Singular value decomposition

So far, I have not said anything about how to perform matrix calculations on a computer. Finding the inverse of a matrix, or finding its eigenvalues, requires computer code that is not easy to write. I strongly recommend the book by Press *et al.*, *Numerical Recipes in Fortran*, which provides Fortran programs that perform a wide variety of calculations. (There is also a similar volume that provides C programs that do the same things.) Their discussion of numerical methods is invaluable. Finding efficient numerical methods that are also numerically stable is a subject that is beyond the scope of this book, but its importance should not be minimized. Press *et al.* provide clear explanations of the methods for performing the different kinds of calculations and also listings of Fortran or C programs that do these calculations.

One subject that Press *et al.* describe in their book has an important application here: the *singular value decomposition* of a matrix. It is an important theorem for us because we can use it to show how to reduce a general matrix equation to one that involves a diagonal matrix. I shall not prove it here, but only state the theorem.

Singular Value Decomposition Theorem: Let $\mathbf{M}$ be any real $n \times m$ matrix, with $n \geq m$. This matrix can be factored in the following form:

$$\mathbf{M} = \mathbf{UDV}^t, \tag{93}$$

where $\mathbf{U}$ is an $n \times m$ matrix; $\mathbf{D}$ is an $m \times m$ matrix; and $\mathbf{V}$ is also $m \times m$. Each of these matrices has a special property.

1. The columns of $\mathbf{U}$ are orthogonal. We can denote the columns of $\mathbf{U}$ by the column vectors $\mathbf{u}_i$, $i = 1,...,m$, with each $\mathbf{u}_i$ having length n; *i.e.*,

$$\mathbf{U} = (\mathbf{u}_1, \ldots, \mathbf{u}_n). \tag{94}$$

Then

$$\mathbf{u}_i \cdot \mathbf{u}_j = \delta_{ij}, \tag{95}$$

or

$$\mathbf{U}^t \mathbf{U} = \mathbf{I}. \tag{96}$$

2. The columns of $\mathbf{V}$ are also orthogonal. Since $\mathbf{V}$ is square, its rows are orthogonal also. Therefore,

$$\mathbf{V}^t \mathbf{V} = \mathbf{V} \mathbf{V}^t = \mathbf{I}. \tag{97}$$

3. The matrix $\mathbf{D}$ has zeroes everywhere except on the main diagonal:

$$\mathbf{D} = \mathrm{diag}(\sigma_1, \quad \sigma_2, \quad \cdots, \quad \sigma_m), \tag{98}$$

where

$$\sigma_i \geq 0, \quad i = 1, \cdots m. \tag{99}$$

If the matrix $\mathbf{M}$ is square, the σ_i's are the eigenvalues, which are also called *singular values*, and hence the name of this factorization of $\mathbf{M}$. But singular values exist even if $\mathbf{M}$ is not square.

Note that, if $m < n$, then $\mathbf{M}^t$ has this decomposition.

Exercise 5-20: Write the singular value decomposition of $\mathbf{M}^t$ explicitly for the case $m < n$. Then find the corresponding decomposition of $\mathbf{M}$ by transposing it.

We can use this theorem to show that any matrix equation can be reduced to an equation that contains a diagonal matrix; this often simplifies the analysis of such equations. Consider the matrix equation

$$\mathbf{Mx} = \mathbf{y}, \tag{100}$$

with $\mathbf{M}$ again an $n \times m$ matrix with $n \geq m$; $\mathbf{x}$, a vector of length m; and $\mathbf{y}$, a vector of length n. Then, using the singular value decomposition,

$$\mathbf{UDV}^t\mathbf{x} = \mathbf{y}. \tag{101}$$

If none of the singular values of $\mathbf{D}$ is zero, then $\mathbf{D}$ is nonsingular and

$$\mathbf{D}^{-1} = \mathrm{diag}(1/\sigma_1, \quad 1/\sigma_2, \quad \cdots, \quad 1/\sigma_m). \tag{102}$$

Therefore, using the orthogonality properties of $\mathbf{U}$ and $\mathbf{V}$,

$$\mathbf{x} = \mathbf{VD}^{-1}\mathbf{U}^t\mathbf{y}. \tag{103}$$

This provides a simple method for finding the inverse of $\mathbf{M}$. The computational problem, of course, is to find the singular value decomposition of the given matrix $\mathbf{M}$. Press *et al.* show how to do this computationally.

But now suppose that $\mathbf{M}$ is singular. That means that at least one of the singular values of $\mathbf{D}$ is zero. Define the matrix[14]

$$\mathbf{D}^\dagger = \mathrm{diag}(\tau_1, \tau_2, \ldots, \tau_p, 0, \ldots, 0) \tag{104}$$

where $\tau_i = 1/\sigma_i$ if $\sigma_i > 0$, and $\tau_i = 0$ otherwise. $\mathbf{D}^\dagger$ has p values > 0, and $m - p$ values $= 0$. Then $\mathbf{D}^\dagger\mathbf{D}$ is a kind of identity matrix: it leaves some components of a vector unchanged, but maps the rest of them to zero.

---

[14] This is a *pseudoinverse*; I say more about the subject in Chapter 7.

We can rewrite equation (101) as

$$\mathbf{DV^t x} = \mathbf{U^t y}. \tag{105}$$

If we define

$$\mathbf{x'} = \mathbf{U^t x} \tag{106}$$

and

$$\mathbf{y'} = \mathbf{V^t y} \tag{107}$$

then equation (100) is just

$$\mathbf{Dx'} = \mathbf{y'} \tag{108}$$

with $\mathbf{D}$ being a diagonal matrix. The inverse of this equation is

$$\hat{\mathbf{x}}' = \mathbf{D}^\dagger \mathbf{y'}, \tag{109}$$

where the components of $\mathbf{x'}$ that were not mapped to zero are recovered from $\mathbf{y'}$, while the components of $\mathbf{x'}$ that *were* mapped to zero, and which are unrecoverable, are replaced with zeroes in $\hat{\mathbf{x}}'$.

Exercise 5-21: Suppose $n > m$. $\mathbf{U}$ has n rows and m columns, and the columns are orthogonal. Is it possible for the *rows* of $\mathbf{U}$ to be orthogonal also? What do we know about the $n \times n$ matrix $\mathbf{UU^t}$?

5.5.1 Application to covariance matrices

Often data fall naturally into a two-dimensional array, as they would if we made daily measurements of the temperature in, say, New York and San Francisco. Perhaps we have 10,000 days data and we want to study the statistical properties of these pairs of temperatures. Should we consider the data to consist of ten thousand realizations of a two-point random process, or two realizations of a 10,000-point random process? In the former case, we would calculate a 2×2 covariance matrix by averaging over 10,000 points; in the latter, a 10,000×10,000 covariance matrix averaged over two points. Is one better than the other? Certainly the latter possibility, which would result in a matrix with 10^8 entries, would be difficult to calculate and to use.

a) *column vectors*

$$locations \rightarrow$$

$$
\begin{matrix} t \\ i \\ m \downarrow \\ e \\ s \end{matrix}
\quad
\left(
\begin{pmatrix} p_{11} \\ p_{21} \\ \vdots \\ p_{n1} \end{pmatrix}
\begin{pmatrix} p_{12} \\ p_{22} \\ \vdots \\ p_{n2} \end{pmatrix}
\cdots
\begin{pmatrix} p_{1m} \\ p_{2m} \\ \vdots \\ p_{nm} \end{pmatrix}
\right)
= \left(\mathbf{p}_1, \mathbf{p}_2, \cdots, \mathbf{p}_m \right)
$$

b) *row vectors*

$$locations \rightarrow$$

$$
\begin{matrix} t \\ i \\ m \downarrow \\ e \\ s \end{matrix}
\quad
\left(
\begin{matrix}
\begin{pmatrix} p_{11} & p_{12} & p_{1m} \end{pmatrix} \\
\begin{pmatrix} p_{21} & p_{22} & p_{2m} \end{pmatrix} \\
\vdots \\
\begin{pmatrix} p_{n1} & & p_{nm} \end{pmatrix}
\end{matrix}
\right)
= \begin{pmatrix} \mathbf{p}^{t}_{1} \\ \mathbf{p}^{t}_{2} \\ \vdots \\ \mathbf{p}^{t}_{n} \end{pmatrix}
$$

Figure 5-1. Representation of space-time data as *a*) column vectors, or *b*) row vectors.

Suppose that we have nm measurements of a certain variable p[15] made at m different times and n different locations. We can arrange the data in an $n \times m$ matrix $\mathbf{P}$, with each column of $\mathbf{P}$ representing a different location, and each row of $\mathbf{P}$, a different time; see Figure 5-1. In other words, we can consider $\mathbf{P}$ to be a matrix composed of column vectors, each column representing values of p for one location, or composed of row vectors, each row showing values of p for one time.

As I showed in §5.3, there are two different covariance matrices we could form from these data. We could either form an $m \times m$ covariance matrix, taking the product of the values of p for two locations and averaging over the N locations, or else form an $n \times n$ covariance matrix, taking the product of two values for different times and averaging over m times. That is, let $\mathbf{S}$ be the covariance matrix of the data, and s_{ij} be an element of $\mathbf{S}$. Either

$$ s_{ij} = \frac{1}{n} \sum_{k=1}^{n} p_{ki} p_{kj} \quad or \quad s_{ij} = \frac{1}{m} \sum_{k=1}^{m} p_{ik} p_{jk} \, . \tag{110} $$

---

[15] The symbol p means here a physical property—*e.g.*, temperature—as a quantity that we measure. In the previous example, p would mean the temperature either in New York or San Francisco on any of the days in question.

Remember that here, as always, I have assumed that the mean value of p is zero. This is true separately for each column vector, or each row vector, as the case may be. In the first case, we get an $m \times m$ covariance matrix, treating $\mathbf{P}$ as eing composed of column vectors; in the second, an $n \times n$ covariance matrix, treating $\mathbf{P}$ as being composed of row vectors. The corresponding matrix equations are

$$\mathbf{S}_m = \frac{1}{n}\mathbf{P}^t\mathbf{P} \quad or \quad \mathbf{S}_n = \frac{1}{m}\mathbf{P}\mathbf{P}^t. \tag{111}$$

Which one should we use? Clearly, if $n \gg m$, we would prefer the first alternative, and *vice versa*. What difference does it make?

Suppose that $n \geq m$. If the reverse were true, we could just transpose $\mathbf{P}$ and interchange the words *time* and *location*; it would not change the mathematics any. Using the singular decomposition theorem, we can write $\mathbf{P}$ as

$$\mathbf{P} = \mathbf{U}\mathbf{D}\mathbf{V}^t, \tag{112}$$

where $\mathbf{U}$ is an $n \times m$ matrix whose columns are orthogonal; $\mathbf{V}$ is an $m \times m$ orthogonal matrix, and $\mathbf{D}$ is an $m \times m$ diagonal matrix whose diagonal elements are ≥ 0. In the first case,

$$\mathbf{S}_m = (\mathbf{U}\mathbf{D}\mathbf{V}^t)^t\,\mathbf{U}\mathbf{D}\mathbf{V}^t = \mathbf{V}\mathbf{D}\mathbf{U}^t\,\mathbf{U}\mathbf{D}\mathbf{V}^t. \tag{113}$$

Since the columns of $\mathbf{U}$ are orthogonal, $\mathbf{U}^t\mathbf{U} = \mathbf{I}$, so

$$\mathbf{S}_m = \mathbf{V}\mathbf{D}^2\mathbf{V}^t. \tag{114}$$

Since $\mathbf{D}$ is diagonal, so is $\mathbf{D}^2$. Explicitly, if $\mathbf{D} = \mathrm{diag}(d_1,...,d_m)$, then $\mathbf{D}^2 = \mathrm{diag}(d_1{}^2,...,d_m{}^2)$. Each element $d_k{}^2 \geq 0$. The rank of $\mathbf{S}_m$ is $\leq m$.

In the other case,

$$\mathbf{S}_n = \mathbf{U}\mathbf{D}\mathbf{V}^t(\mathbf{U}\mathbf{D}\mathbf{V}^t)^t = \mathbf{U}\mathbf{D}\mathbf{V}^t\,\mathbf{V}\mathbf{D}\mathbf{U}^t = \mathbf{U}\mathbf{D}^2\mathbf{U}^t \tag{115}$$

since $\mathbf{V}$ is orthogonal. Note, however, that $\mathbf{D}^2$ in this equation is still $m \times m$ and we assumed that $n \geq m$. The matrix $\mathbf{D}^2$ has a rank that is, at most, equal to m, but it contains all of the nonzero eigenvalues of $\mathbf{S}_n$, so it must also have a rank $\leq m$. Unless $n = m$, therefore, $\mathbf{S}_n$ is singular, since it has at least $n - m$ diagonal values that are equal to zero.

Now $\mathbf{S}_m$ has m eigenvectors (which are orthonormal). $\mathbf{S}_n$ has n eigenvectors, but at least $n - m$ of them correspond to eigenvalues that are equal to zero. That is, the variances of these eigenvector components are all zero. Finally, this means that there are at most m eigenvectors that have positive variance. Since the ranks of $\mathbf{S}_m$ and $\mathbf{S}_n$ are equal, and their eigenvalues are also equal, they must contain equivalent information about the behavior of the data in the matrix $\mathbf{P}$. We will make use of this in the next section.

5.6 Empirical orthogonal functions

Recall the space-time data from the last section. Suppose we take the column-vector approach, where we define the covariance matrix

$$\mathbf{S}_m = \frac{1}{n}\mathbf{P}'\mathbf{P} \tag{116}$$

That is, we have m different time series, one at each location. The averages in $\mathbf{S}_m$ are taken over time; each element of $\mathbf{S}_m$ corresponds to a different pair of locations. These locations could represent different pixels in an image; in this case, we would have images taken at n different times. From the last section, we know that the eigenvalues of $\mathbf{S}_m$ are $\lambda_1 = d_1^2, ..., \lambda_m = d_m^2$. Now take the data matrix $\mathbf{P}$ and multiply it by $\mathbf{E}$, which is the $(m \times m)$ matrix formed from the eigenvectors $\mathbf{e}_k$ of $\mathbf{S}_m$, $k = 1, ..., m$. Here, the eigenvectors $\mathbf{e}_k$ are column vectors. Since the definition of an eigenvector is

$$\mathbf{S}\mathbf{e}_k = \lambda_k\mathbf{e}_k, \tag{117}$$

we can also write

$$\mathbf{S}\mathbf{E} = \mathbf{E}\lambda, \tag{118}$$

where the matrix $\lambda = \text{diag}(\lambda_1, ..., \lambda_m)$. Since $\mathbf{E}$ is orthonormal,

$$\mathbf{E}'\mathbf{S}\mathbf{E} = \lambda. \tag{119}$$

Let

$$\mathbf{P}' = \mathbf{P}\mathbf{E}. \tag{120}$$

Since we considered the columns of $\mathbf{P}$ to be vectors corresponding to the m different locations, the columns of $\mathbf{P}'$ are linear combinations of the columns of $\mathbf{P}$, with the properties that
1. The different time series corresponding to $\mathbf{P}'$ vary independently; and
2. They have variances $\lambda_1, ..., \lambda_m$, respectively; *i.e.*,

$$\left\langle x_i'(t)x_j'(t)\right\rangle = \lambda_i\delta_{ij}. \tag{121}$$

In other words, at each time (corresponding to each row of $\mathbf{P}$), there are m measurements of x, one at each location, or we could say that we have an m-tuple of x's at each time. By rotating the data matrix $\mathbf{P}$, we get a new m-tuple

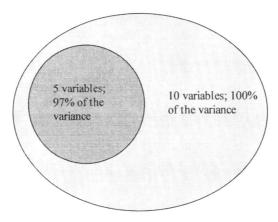

Figure 5-2. A hypothetical situation where a subset containing five variables contains 97% of the variance of the original set of ten variables.

of x's, but the different components of these rotated m-tuples vary independently, with their variances given by the eigenvalues of S_m. These eigenvectors e_k are called *empirical orthogonal functions*.[16] They are empirical, in the sense that they are determined by the statistics of the data. We can use them to represent the measurements of p more compactly than we could with the original data.

A small example is in order. Suppose we have $m = 2$ locations. We measure the variable x_1 at the first location and x_2 at the other one. So there are n measured pairs (x_1, x_2). Now if we find the 2×2 covariance matrix S_2 of variables x_1 and x_2, and the eigenvectors e_1 and e_2 of S_2, we can transform them into the new variables $x_1' = e_{11}x_1 + e_{12}x_2$ and $x_2' = e_{21}x_1 + e_{22}x_2$. These two new variables have variances λ_1 and λ_2, respectively, and vary independently [*i.e.*, $\text{cov}(x_1', x_2') = 0$]. Now it might turn out that x_1 and x_2 are highly correlated: in that case, the smaller eigenvalue—λ_2—will be close to zero. Then, with good accuracy, we could represent the same data *pairs* (x_1, x_2) with the single variable $x_1' = e_{11}x_1 + e_{12}x_2$. This could represent a significant savings in data analysis. Especially if $m \gg 2$, the potential savings in the data reduction could be very, very large. It is common in remote-sensing problems that the first few eigenvectors contain 95% or more of the total variance,[17] and in many cases the rest of the variance is not worth worrying about (see Figure 5-2). By judicious use of eigenvector analysis, we can sometimes reduce the number of independent measurements we need to make by a factor of two or more.

---

[16] See for example Alishouse *et al.* (1966).

[17] The *total variance* is just the sum of all of the eigenvalues, since each eigenvalue is the variance of the corresponding eigenvector, and they vary independently.

I defined the term *total variance* in §5.3.1; Σ_{tot} is the trace of the covariance matrix. Suppose we started out with m independent variables and the trace of the $m \times m$ covariance matrix is Σ_m. If we choose a subset $k < m$ of the variables, the trace of the smaller covariance matrix will be Σ_k. Clearly, $\Sigma_k \leq \Sigma_m$. We can judge how much of the information contained in the original set of m variables is also contained in the subset by calculating the fraction of the variance explained by the smaller number of variables:

$$f_{km} = \frac{\Sigma_k}{\Sigma_m}. \tag{122}$$

This concept is similar to the one described in §1.7. If there is a subset containing k variables that explain a fraction f_{km} of the variance of the larger set of m variables, there are $m - k$ variables that explain a fraction $1 - f_{km}$ of that variance. If we want to characterize measurements of m variables as succinctly as possible, we should choose the smallest set of $k \leq m$ variables that explain an acceptable fraction f_{km} of the total variance. By calculating the empirical orthogonal functions we can identify the component ϕ_1 with the largest variance λ_1; the component ϕ_2 with the next largest variance, λ_2; and so forth. Using the first k of these functions, $\phi_1, \phi_2, ..., \phi_k$ will give us the largest value f_{km} possible with k functions. This makes for the most compact representation of the data. The cost of doing this might be that it might be hard to understand what each compent means physically.

At the beginning of this section, I said that we would treat **P** as being composed of column vectors; *i.e.*, we have a different time series measured at each of m locations. But what would have happened if we had decided to treat **P** as being made up of row vectors?

The answer is, using the discussion of singular value decomposition in the last section, that we would have gotten the same eigenvalues, and therefore we would have been able to reduce the n sets of measurements—each corresponding to a different time—to the same number of sets as we would have gotten the other way. This shows that if we were given, for example, two time series, each with 1000 different times, we would be much better off forming a 2×2 covariance matrix by averaging over time, rather than a 1000×1000 covariance matrix by averaging over the two locations.

Exercise 5-22: Suppose that we have daily temperature measurements in the following cities: New York, NY; San Francisco, California; Chicago, Illinois; Paris, France; London, England; Melbourne, Florida; and Sydney, Australia. If we calculated the empirical orthogonal functions, what component would have the largest variance? The second-largest variance? Why? Would the answer change if the measurements were decadal average temperatures, rather than daily?

5.7 Non-negative definite matrices

In the following chapters, I will develop some principles for creating remote-sensing algorithms. In dealing with noise, we will see that, for certain matrices to be nonsingular, it will be necessary that their eigenvalues be of the form

$$\mu_i \equiv \lambda_i + \sigma^2, \tag{123}$$

where σ is a real number and the λ_i's are the eigenvalues of some covariance matrix. It will also be necessary that

$$\mu_i \geq \sigma^2. \tag{124}$$

This is only possible if we know that none of the λ_i's is negative. We therefore need to know that, quite generally, any covariance matrix (of real quantities) is non-negative definite, which means that no eigenvalue is < 0.

The basic theorem about non-negative definite matrices is as follows:[18] any real matrix $\mathbf{R}$ that can be written in the form

$$\mathbf{R} = \mathbf{P}^t \mathbf{P}, \tag{125}$$

is non-negative definite. In this case, $\mathbf{R}$ is symmetric. If $\mathbf{P}$ is nonsingular, then all of the eigenvalues of $\mathbf{R}$ are strictly greater than 0; if $\mathbf{P}$ is singular, some of the eigenvalues will equal 0.

Consider the case where $\mathbf{P}$ itself is symmetric. Then there is an orthogonal matrix $\mathbf{U}$ such that $\mathbf{D} = \mathbf{U}\mathbf{P}\mathbf{U}^t$ is a diagonal matrix. Then

$$\mathbf{R} = \left(\mathbf{U}^t\mathbf{D}\mathbf{U}\right)^t \mathbf{U}^t\mathbf{D}\mathbf{U} = \mathbf{U}^t\mathbf{D}\mathbf{U}\mathbf{U}^t\mathbf{D}\mathbf{U} = \mathbf{U}^t\mathbf{D}^2\mathbf{U}. \tag{126}$$

Since $\mathbf{D}$ is diagonal, $\mathbf{D}^2$ is also diagonal and each of its diagonal elements is ≥ 0. Since

$$\mathbf{D}^2 = \mathbf{U}\mathbf{R}\mathbf{U}^t \tag{127}$$

[multiply each side of equation (126) on the left by $\mathbf{U}$ and on the right by $\mathbf{U}^t$], the elements of $\mathbf{D}^2$ are the eigenvalues of $\mathbf{R}$: therefore, $\mathbf{R}$ is non-negative definite.

If $\mathbf{P}$ is not symmetric, we can use the singular value decomposition given in §5.5 to write $\mathbf{P} = \mathbf{U}\mathbf{D}\mathbf{V}^t$, where $\mathbf{V}$ is an orthogonal matrix and the columns of $\mathbf{U}$ are also orthogonal. Then $\mathbf{R} = \mathbf{P}^t\mathbf{P} = \mathbf{V}\mathbf{D}\mathbf{U}^t\mathbf{U}\mathbf{D}\mathbf{V}^t = \mathbf{V}\mathbf{D}^2\mathbf{V}^t$, so again $\mathbf{R}$ is non-negative definite.

Any covariance matrix $\mathbf{S}$ can be written in the form $\mathbf{S} = \mathbf{P}^t\mathbf{P}$ or $\mathbf{P}\mathbf{P}^t$, so it is symmetric and also non-negative definite.

---

[18] See, for example, Finkbeiner (1966) or Birkhoff and Mac Lane (1953).

5.8 Noise and the covariance matrix

Suppose we measure a vector quantity $\mathbf{y} = (y_1,\ldots,y_n)$ and there is noise in each measurement, denoted by the random vector $\mathbf{n}$. The result of this measurement is

$$\mathbf{y}' = \mathbf{x} + \mathbf{n}. \tag{128}$$

Let $\mathbf{S} = \langle\mathbf{y}\mathbf{y}^t\rangle$ be the covariance matrix of $\mathbf{y}$; $\mathbf{N} = \langle\mathbf{n}\mathbf{n}^t\rangle$ is the covariance matrix of the noise. The covariance matrix of $\mathbf{y}'$ is

$$\mathbf{S}'_y = \mathbf{y}'\mathbf{y}'^t = (\mathbf{y} + \mathbf{n})(\mathbf{y} + \mathbf{n})^t. \tag{129}$$

The terms proportional to $\langle\mathbf{y}\mathbf{n}^t\rangle$ are zero, since we assume that the noise is uncorrelated with the signal $\mathbf{y}$. In component notation, denoting the components of $\mathbf{S}_y'$ by s_{ij},

$$s_{ij} = \left\langle (y_i + n_i)(y_j + n_j) \right\rangle = \left\langle y_i y_j \right\rangle + \left\langle n_i^2 \right\rangle \delta_{ij}. \tag{130}$$

Therefore,

$$\mathbf{S}'_y = \mathbf{S}_y + \mathbf{N}. \tag{131}$$

The covariance matrix of any set of measurements that include uncorrelated noise will have the form of equation (131). As I said in the last section, we want to choose basis vectors so that the eigenvalues of $\mathbf{S}_y'$ are the sum of the eigenvalues of $\mathbf{S}_y$ and the eigenvalues of $\mathbf{N}$:

$$\lambda'_i = \lambda_i + \mu_i, \tag{132}$$

where λ_i and μ_i are the eigenvalues of $\mathbf{S}$ and $\mathbf{N}$ respectively, and λ_i', the eigenvalues of $\mathbf{S}_y'$. How can we accomplish this?

To find the eigenvalues of $\mathbf{S}_y'$, we need to find an orthogonal matrix $\mathbf{U}$ such that

$$\mathbf{U}(\mathbf{S} + \mathbf{N})\mathbf{U}^t = \mathbf{U}\mathbf{S}\mathbf{U}^t + \mathbf{U}\mathbf{N}\mathbf{U}^t = \mathbf{D}. \tag{133}$$

We would expect, however, that any rotation that diagonalized $\mathbf{S}_y$, would probably undiagonalize $\mathbf{N}$. Unless *both* $\mathbf{S}_y$ and $\mathbf{N}$ are diagonalized by the same transformation $\mathbf{U}$, equation (132) will not be satisfied.

We can guarantee that $\mathbf{S}_y$ and $\mathbf{N}$ can be diagonalized simultaneously if $\mathbf{N}$ is a scalar multiple of the identity matrix, in which case any rotation leaves $\mathbf{N}$ unchanged. We can always do this by renormalizing $\mathbf{y}$. If we normalize each component y_i by dividing by the square root of the noise variance of that component, then each of the transformed components will have unit noise variance.

Since $\mathbf{N}$ is positive definite, we can find matrices $\mathbf{N}^{\frac{1}{2}}$ and $\mathbf{N}^{-\frac{1}{2}}$ that satisfy

$$\mathbf{N}^{1/2}\mathbf{N}^{1/2} = \mathbf{N} \quad \text{and} \quad \mathbf{N}^{-1/2}\mathbf{N}^{1/2} = \mathbf{I}. \tag{134}$$

The parallel with the notation $2^{\frac{1}{2}} = \sqrt{2}$ is obvious, but not as commonly encountered with matrices. If we replace each vector $\mathbf{y}'$ with $\mathbf{y}'' = \mathbf{N}^{-\frac{1}{2}}\mathbf{y}'$, then

$$\mathbf{S}_y'' = \left\langle \mathbf{y}''\mathbf{y}''^t \right\rangle = \mathbf{N}^{-1/2}\left\langle (\mathbf{y}+\mathbf{n})(\mathbf{y}+\mathbf{n})^t \right\rangle \mathbf{N}^{-1/2} = \mathbf{N}^{-1/2}\mathbf{S}_y\mathbf{N}^{-1/2} + \mathbf{I}. \tag{135}$$

In other words, once we have transformed the data in this manner, the noise covariance matrix is the identity matrix. Any rotation of the basis vectors leaves the form of $\mathbf{S}_y''$ unchanged—it is always the sum of a (noise-free) covariance matrix and the identity matrix. This normalization is equivalent to changing the units that we use to measure each component of $\mathbf{y}$.

We could have used $a\mathbf{N}^{-\frac{1}{2}}$ instead of $\mathbf{N}^{-\frac{1}{2}}$ in this transformation, where a is any scalar. The form of $\mathbf{S}_y''$ would have remained unchanged.

It may be more convenient to let $a = \sigma_N^2$, where σ_N^2 is my usual symbol for the noise (variance) in each measurement, assuming that all of the components of $\mathbf{y}$ have about the same noise. Therefore, let us assume that either this is the case, or we have transformed the original measurements so that this has become so; then we can write the covariance matrix of $\mathbf{y}'$ as

$$\mathbf{S}_{y''} = \left\langle \mathbf{y}''\mathbf{y}''^t \right\rangle = \mathbf{S}_y + \sigma_N^2\mathbf{I}, \tag{136}$$

where $\mathbf{S}_y$ is the covariance matrix of $\mathbf{y}$ itself, without measurement noise. If $\mathbf{U}$ is the orthogonal matrix that diagonalizes $\mathbf{S}_y$, then

$$\mathbf{U}\mathbf{S}_{y''}\mathbf{U}^t = \mathbf{U}(\mathbf{S}_y + \sigma_N^2\mathbf{I})\mathbf{U}^t = \mathbf{D}_y + \sigma_N^2\mathbf{I} \tag{137}$$

where $\mathbf{D}_y = \mathrm{diag}(\lambda_1,\ldots,\lambda_n)$.

The important point here is that the covariance matrix of the measured quantities, $\mathbf{S}_{y''}$, is nonsingular and its smallest eigenvalue is not less than σ_N^2. The eigenvalues of $\mathbf{S}_y$ are the elements of $\mathbf{D}_y$, so the eigenvalues of $\mathbf{D}_y'' = \mathbf{D}_y + \sigma_N^2\mathbf{I}$ are

$$\lambda_i' = \lambda_i + \sigma_N^2. \tag{138}$$

Note that we have shifted the eigenvalues by σ_N^2. Since $\mathbf{S}_{y''}$ is a covariance matrix, $\lambda_j \geq 0$ for each i. Therefore, the smallest eigenvalue of $\mathbf{S}_{y''}$ is

$$\lambda_i' \geq \sigma_N^2. \tag{139}$$

While $\mathbf{D}_y$ might be singular (if one or more eigenvalues are zero), $\mathbf{D}_y''$ is not singular, and its inverse

$$\mathbf{D}_{y''}^{-1} = \text{diag}\left(\frac{1}{\lambda_1 + \sigma_N^2}, \cdots, \frac{1}{\lambda_n + \sigma_N^2}\right) \tag{140}$$

will be well behaved.

Recently, Roger[19] (1994) has published a similar analysis, which includes references to basic theorems cited in Graybill (1969).

5.9 Parallel component analysis

Suppose we have measured many realizations of the vector[20] $\mathbf{x}$ of length m and we have calculated the covariance matrix $\mathbf{S}_x$. Are all m measurements in each vector really necessary? How many independent pieces of information are there in the data, anyway? The preceding discussion shows that, in a system of m independent variables, there may be only $k < m$ linear combinations of the variables that explain almost all of the variance. Knowing these functions may not help us in reducing the number of variables that we measure, since each of them will, in general, depend on all m of the measured values. While the first component might be $x_1' = \frac{1}{2}x_1 + \frac{1}{4}x_2$, say, we need to measure both x_1 and x_2 to evaluate x_1'. If we want to reduce the number of independent variables, we have to eliminate one or more of the x's entirely. But which components should we delete from $\mathbf{x}$? One way is to perform the following analysis, which I call the *parallel-component* analysis.

We can arrange the independent variables so that the variances of x_i are nonincreasing as i increases; x_1 will have the largest variance. The parallel-component model works as follows:

For each measurement x_i, $i = 2, \cdots, n$, assume that it is composed of two parts: one is proportional to x_1, the other, uncorrelated with it. Write this as

$$x_i = a_i x_1 + x_i'. \tag{141}$$

Then the covariance of x_1 with x_i is, for $i > 1$,

$$\text{cov}(x_i, x_1) = \langle x_1(a_i x_1 + x_i')\rangle = a_i \text{var}(x_1) \tag{142}$$

and

---

[19] Roger (1994).

[20] This represents a slight change in the notation. The vector $\mathbf{x}$ may or may not contain measurement noise.

$$\mathrm{var}(x_i) = \left\langle \left(x_i' + a_i x_1\right)^2 \right\rangle = \mathrm{var}(x_i') + a_i^2 \, \mathrm{var}(x_1) \qquad (143)$$

since $\mathrm{cov}(x_1,x_i') = 0$, $i > 1$, by hypothesis. Substituting $a_i = \mathrm{cov}(x_i,x_1) \,/\, \mathrm{var}(x_1)$ into equation (143),

$$\mathrm{var}(x_i') = \mathrm{var}(x_i)\left[1 - \frac{\mathrm{cov}^2(x_1,x_i)}{\mathrm{var}(x_1)\,\mathrm{var}(x_i)}\right]. \qquad (144)$$

The fraction in equation (144) is the correlation coefficient of x_1 and x_i

$$\rho_{1i}^2 = \frac{\mathrm{cov}^2(x_1,x_i)}{\mathrm{var}(x_1)\,\mathrm{var}(x_i)}. \qquad (145)$$

Then

$$\mathrm{var}(x_i') = \mathrm{var}(x_i)\left(1 - \rho_{1i}^2\right). \qquad (146)$$

Similarly,

$$\mathrm{cov}(x_i,x_j) = \left\langle \left(a_i x_1 - x_i'\right)\left(a_j x_1 - x_j'\right)\right\rangle = a_i a_j \, \mathrm{var}(x_1) + \mathrm{cov}\left(x_i',x_j'\right), \quad (147)$$

so

$$\mathrm{cov}(x_i',x_j') = \mathrm{cov}(x_i,x_j) - \frac{\mathrm{cov}(x_1,x_i)\,\mathrm{cov}(x_1,x_j)}{\mathrm{var}(x_1)}$$

$$= \mathrm{cov}(x_i,x_j)\left(1 - \frac{\rho_{1i}\,\rho_{1j}}{\rho_{ij}}\right). \qquad (148)$$

The same information is contained in the set of variables $\{x_1, x_2', \ldots, x_n'\}$ as in the original set. However, since $\mathrm{var}(x_i') < \mathrm{var}(x_i)$, some of the new variables will have a variance very close to zero. We may not need to measure these variables, since they contain very little extra information. We can repeat this process as many times as appropriate, reducing the variance of the remaining x_i's at each step. This will allow us to minimize the number of measurements we need to make.

If there were no noise in the measurements, there would be no point in making redundant measurements. But when there is noise present, there may be some value in redundant measurements: if we measure something twice, we can average the two measurements. When we do this, we reduce the effect of the noise by a factor of $\sqrt{2}$. Depending on the circumstances, this may or may not be worth the cost of making extra measurements.

5.10 Finding eigenvectors

There is not room in a book like this for a discussion of the mathematical methods that are used for performing mundane chores like actually inverting a matrix or finding its eigenvalues. However, I do wish to treat one problem here because it will come up again in Chapter 11. That is, given a large matrix $\mathbf{A}$, how can we find some of its eigenvalues, if not all of them? Assume that $\mathbf{A}$ has the form $\mathbf{P^t P}$, so its eigenvalues are non-negative definite. Then there is clearly a largest eigenvalue λ_1, and successively smaller ones λ_2, λ_3, ..., with $\lambda_i \geq 0$. When $\mathbf{A}$ is very large, the canned routines for finding eigenvalues may not work, or may work only very slowly. Here is a method that will find the largest eigenvalues.

Given a real symmetric matrix $\mathbf{A}$, pick a vector $\mathbf{x}^{(0)}$ at random. Here, $\mathbf{x}^{(n)}$ means the nth vector in a series. Define recursively

$$\mathbf{x}^{(n+1)} = \frac{\mathbf{A}\mathbf{x}^{(n)}}{\left|\mathbf{A}\mathbf{x}^{(n)}\right|}, \tag{149}$$

$n = 0, 1, \ldots$. Let us examine the possible values of $\mathbf{x}^{(n)}$ as $n \to \infty$. Note that $\mathbf{x}^{(n)}$ has unit magnitude.

Let $\{\mathbf{e}_i\}$ be the set of eigenvectors of $\mathbf{A}$, and $\{\lambda_i\}$, the corresponding eigenvalues. As of yet, we do not know what their values are. However, there is a set of coefficients $\alpha_1^{(n)}$, $\alpha_2^{(n)}$, ..., such that

$$\mathbf{x}^{(n)} = \sum_i \alpha_i^{(n)} \mathbf{e}_i. \tag{150}$$

Now, whatever the values of $\mathbf{e}_i$ and λ_i might be, we know that, since $\mathbf{A}\mathbf{e}_i = \lambda_i \mathbf{e}_i$,

$$\mathbf{A}\mathbf{x}^{(n)} = \mathbf{A}\sum_i \alpha_i^{(n)} \mathbf{e}_i = \sum_i \alpha_i^{(n)} \mathbf{A}\mathbf{e}_i = \sum_i \alpha_i^{(n)} \lambda_i \mathbf{e}_i = \sum_i \alpha_i^{(n+1)} \mathbf{e}_i. \tag{151}$$

Therefore,

$$a_i^{(n+1)} = \lambda_i a_i^{(n)}. \tag{152}$$

Therefore, $\mathbf{A}^m \mathbf{x}^{(0)}$ is given by (here, $\mathbf{A}^m \mathbf{x}$ means $\mathbf{A}$ operating on $\mathbf{x}$ m times)

$$\mathbf{A}^m \mathbf{x}^{(0)} = \sum_i \alpha_i^{(0)} \lambda_i^m \mathbf{e}_i. \tag{153}$$

The renormalization in equation (149) serves to prevent the magnitude of $\mathbf{x}^{(n)}$ from becoming too large or too small; otherwise we could not actually evaluate equation (153) on a computer for large values of m without a serious loss of precision.

Since λ_1 is the largest eigenvalue, when n becomes large enough, the sum is dominated by $\lambda_1{}^n$. Therefore, $\mathbf{x}^{(n)} \to \mathbf{e}_1$ as $n \to \infty$. Therefore, we find $\mathbf{e}_1$ by repeated application of equation (149), and the value of λ_1 from $\mathbf{A}\mathbf{x}^{(n)} = \lambda_1\mathbf{x}_1$ in the limit as $n \to \infty$.

This gives us the largest eigenvalue and its corresponding eigenvector. To find the second one, note that for any vector $\mathbf{v}$,

$$\frac{\mathbf{v} - (\mathbf{v}\cdot\mathbf{e}_1)\mathbf{e}_1}{\left|\mathbf{v} - (\mathbf{v}\cdot\mathbf{e}_1)\mathbf{e}_1\right|} \equiv \mathbf{v}_{1,\perp} \tag{154}$$

is perpendicular to $\mathbf{e}_1$, so $\mathbf{e}_1\cdot\mathbf{v}_{1,\perp} = 0$. Therefore, $\mathbf{v}_{1,\perp}{}^{(n)}$ has the expansion

$$\mathbf{v}_{1,\perp}^{(n)} = \sum_{i=2} \alpha_i^{(n)}\mathbf{e}_i = \sum_{i=2} \alpha_i^{(0)}\lambda_i^n\mathbf{e}_i. \tag{155}$$

If we alternate applications of equation (149) with equation (154) (to keep $\mathbf{v}^{(n)}$ $\perp \mathbf{e}_1$), we will find the *second* eigenvalue and its associated eigenvector.

Exercise 5-23: Expand this argument to finding the 3rd, 4th, …, eigenvectors.

Note that, if the rank of $\mathbf{A}$ is large, we cannot find *all* of the eigenvectors. When we already have found many of them, finding the next one requires that we use an equation analogous to equation (154) to keep the vector perpendicular to all of the previous ones, but roundoff errors will prevent our doing this well enough for this scheme to work. However, when $\mathbf{A}$ is large and non-negative definite, finding the largest few eigenvalues and their associated eigenvectors may be all we need to do.

In the discussion of integral equations in Chapter 10, I mention in several places the similarities between matrix equations and integral equations. The procedure I outlined here has a direct analog in that context. It may will be useful in understanding the possible solutions when we try to solve integral equations by iteration.

5.11 A canonical form for matrix equations

As I said at the beginning of this chapter, if we want to invert the equation $\mathbf{y} = \mathbf{A}\mathbf{x}$, we need to know the statistics—*i.e.*, the covariance matrix—of $\mathbf{x}$. Most treatments of this subject tacitly assume that the covariance matrix of $\mathbf{x}$ is $\mathbf{S}_x = \sigma^2\mathbf{I}$. This means that the independent variables x_i are independent and have the same variance. While this makes it easier to write down the appropriate equations, it also obscures an important part of the problem.

To be specific, consider the statistical quantities involved. The covariance matrix of $\mathbf{x}$ is specified by nature—we generally have no control over it. If $\mathbf{y} = \mathbf{Ax}$, then the covariance matrix of $\mathbf{y}$ is $\mathbf{S_y} = \langle(\mathbf{Ax})(\mathbf{Ax})^t\rangle = \mathbf{A S_x A^t}$. How well we can retrieve $\mathbf{x}$ from measurements of $\mathbf{y}$ will depend, in large part, on the relative magnitudes of $\mathbf{S_y}$ and $\mathbf{N}$, the noise covariance matrix. Because, for remote sensing measurements, $\mathbf{N}$ usually represents independent noise in different radiometers, I am satisfied to assume that $\mathbf{N} = \sigma_N^2 \mathbf{I}$. If it is a more general covariance matrix, the formalism that I discuss in this book can easily be modified to accommodate this change. Let us consider a general linear retrieval problem, that is specified by a matrix $\mathbf{A}$, the covariance matrix $\mathbf{S_x}$ of the independent variables, and σ_N^2, the noise variance.

Exercise 5-24: What is the rank of $\mathbf{N}$?

Because it is often advantageous to consider only a very simple problem, I shall show that a problem specified by these three quantities can always be reduced to one where $\mathbf{A}$ is diagonal and both $\mathbf{x}$ and $\mathbf{y}$ have diagonal covariance matrices. To do so, I shall use the singular value decomposition (SVD) described in §5.5.

To begin with, consider the equation $\mathbf{y} = \mathbf{Ax}$, where $\mathbf{y}$ is a vector of length n; $\mathbf{x}$, a vector of length m; and $\mathbf{A}$ and $n \times m$ matrix. Assume that $n \geq m$, as otherwise the problem would be underdetermined. If we have n measurements, we cannot estimate the values of more than n independent variables.

If we ignore the statistical part of the problem for a moment, we can use the SVD theorem to write

$$\mathbf{A} = \mathbf{UQV} \qquad (156)$$

where $\mathbf{Q}$ is an $m \times m$ diagonal matrix; $\mathbf{V}$ is an $m \times m$ orthogonal matrix; and $\mathbf{U}$ is an $n \times m$ matrix whose columns are orthogonal (so $\mathbf{U^t U} = \mathbf{I}$). Without loss of generality, we can require that that $\mathbf{Q}$ be non-negative definite. Then

$$\mathbf{y} = \mathbf{Ax} = \mathbf{UQVx}, \qquad (157)$$

so

$$\mathbf{U^t y} = \mathbf{QVx}. \qquad (158)$$

Next, define $\mathbf{y}' = \mathbf{U^t y}$ and $\mathbf{x}' = \mathbf{Vx}$; then

$$\mathbf{y}' = \mathbf{Qx}' \qquad (159)$$

and if $\mathbf{Q} = \text{diag}(\lambda_1^2, \ldots, \lambda_m^2)$, this is just a set of uncoupled linear equations $y_i' = \lambda_i x_i'$. I showed how to solve this equation in Chapter 1.

But now we want to take the covariance matrix $\mathbf{S_x}$ into account explicitly. I shall show that it is always possible (with a small exception I shall mention presently) to write equation (156) in a way that the matrix $\mathbf{Q}$ is still diagonal, but the components of $\mathbf{x}$ all vary independently with variance σ_x^2 and the noise covariance matrix still has the form $\mathbf{N} = \sigma_N^2\mathbf{I}$. That is, each independent component of $\mathbf{x}$ (each eigenvector of $\mathbf{S}_x$) maps to a single component of $\mathbf{y}$. To do this, we need to assume that $\mathbf{S}_x$ is not singular. If it were, it would mean either that x_k has a vanishingly small variance for some value of k, or that two of the variables, say x_k and x_i, are highly correlated. However, if the variance of x_k is very small, we know that its value is always close to zero, so there is no need to retrieve its value. If two or more of the independent variables are highly correlated, we only need to measure one to know the value of the other. Therefore, a sensible inversion problem will always be characterized by a nonsingular covariance matrix $\mathbf{S}_x$.

If we define

$$\mathbf{x}' = \sigma_x\mathbf{S}_x^{-\frac{1}{2}}\mathbf{x}, \tag{160}$$

then the covariance matrix of $\mathbf{x}'$ is

$$\mathbf{S}_{x'} = \left\langle\mathbf{x}'\mathbf{x}'^{t}\right\rangle = \sigma_x^2\left\langle\left(\mathbf{S}_x^{-\frac{1}{2}}\mathbf{x}\right)\left(\mathbf{S}_x^{-\frac{1}{2}}\mathbf{x}\right)^{t}\right\rangle = \sigma_x^2\mathbf{S}_x^{-\frac{1}{2}}\left\langle\mathbf{xx}^{t}\right\rangle\mathbf{S}_x^{-\frac{1}{2}} = \sigma_x^2\mathbf{I}, \tag{161}$$

where σ_x^2 is chosen to be some suitable scalar. Then equation (157) becomes

$$\mathbf{y} = \sigma_x\left(\mathbf{AS}_x^{\frac{1}{2}}\right)\mathbf{S}_x^{-\frac{1}{2}}\mathbf{x} = \sigma_x\left(\mathbf{AS}_x^{\frac{1}{2}}\right)\mathbf{x}' = \sigma_x\mathbf{UPVx}', \tag{162}$$

where I have used the SVD theorem to write $\mathbf{AS}_x^{\frac{1}{2}} = \mathbf{UPV}$, where $\mathbf{P}$ is an $m \times m$ diagonal matrix and $\mathbf{U}$ and $\mathbf{V}$ have orthogonal columns as before [although they are different matrices from the ones in equation (157); I need to recycle some of my symbols]. If we let

$$\mathbf{x}'' = \mathbf{Vx}' \tag{163}$$

and

$$\mathbf{y}' = \mathbf{U}^t\mathbf{y}, \tag{164}$$

then we get the simple equation

$$\mathbf{y}' = \sigma_x\mathbf{Px}'', \tag{165}$$

where $\mathbf{P} = \mathrm{diag}(\lambda_1^2,\ldots,\lambda_m^2)$.[21]

Now assume that the noise covariance matrix of $\mathbf{y}$ is $\sigma_N^2\mathbf{I}$; the noise in each component of $\mathbf{y}'$ will be the same as the noise in each component of $\mathbf{y}$, since the noise covariance matrix is $\mathbf{N}' = \mathbf{U}^t\mathbf{NU} = \sigma_N^2\mathbf{U}^t\mathbf{U} = \sigma_N^2\mathbf{I}$. Therefore,

---

[21] Also, different λ's.

we have again an uncoupled system of equations whose solutions will be in the form

$$\hat{x}_i = r_i(y_i + n_i), \tag{166}$$

where each coefficient r_i is to be determined, and where the variance of n_i is σ_N^2. Note also that we can, if we want to, let $\sigma_x^2 = 1$ (so it drops out of the notation), as long as we interpret the noise variance to mean σ_N^2 / σ_x^2.

We should note that σ_x^2 and σ_N^2 do not enter this system of equations independently, only the ratio σ_x^2 / σ_N^2 is important. We could call equation (165) the *canonical form* of the matrix equation, since any linear problem can be reduced to this form. I shall make use of this form later, especially in Chapter 10, but I shall use it sparingly, since we need to keep the role of the statistics of the situation in mind as we examine different ways to solve inverse problems.

5.12 Rank of the covariance matrix

We should, at this point, stop and think about the covariance matrices we will use in inversion problems. Suppose again that $\mathbf{y}$ is an n-dimensional vector, $\mathbf{x}$ is m-dimensional, and $\mathbf{A}$ is an $n{\times}m$ matrix. We have seen that the covariance matrix

$$\mathbf{S}_{y'} \equiv \langle \mathbf{y}'\mathbf{y}'^t \rangle = \langle (\mathbf{y} + \mathbf{n})(\mathbf{y} + \mathbf{n})^t \rangle, \tag{167}$$

where $\mathbf{n}$ is random noise uncorrelated with $\mathbf{y}$, can be written in the form of equation (136), so none of the eigenvalues of $\mathbf{S}_{y'}$ is $< \sigma_N^2$. Now we may want to rotate the vectors $\mathbf{y}'$ into the coördinate system where the covariance matrix is diagonal. Let $\mathbf{U}$ be the matrix that diagonalizes $\mathbf{S}_{y'}$, so $\mathbf{S}_{y''} = \mathbf{U}\mathbf{S}_{y'}\mathbf{U}^t = \mathbf{D}_y + \sigma_N^2\mathbf{I}$. Then the basis vectors of the rotated vectors, $\mathbf{y}''$, are the eigenvectors of $\mathbf{S}_{y''}$. The elements of $\mathbf{y}''$ vary independently, while the variance of the ith component y_i'' is $\lambda_i' = \lambda_i + \sigma_N^2$. As a purely mathematical operation, this is fine.

But suppose that we want to interpret the eigenvectors of $\mathbf{S}_{y''}$ (call them $\mathbf{u}_i$) as being empirical orthogonal functions (see §5.6)? Since $\mathbf{S}_{y''}$ is an $n{\times}n$ matrix, there are n of them. How many of them are *physically* meaningful? Remember that the covariance matrix of $\mathbf{y}$ is related to the covariance matrix of $\mathbf{x}$ by $\mathbf{S}_y = \mathbf{A}\mathbf{S}_x\mathbf{A}^t$, so $\mathbf{S}_y$ is determined by two factors: the statistical properties of $\mathbf{x}$ and the relationship between $\mathbf{y}$ and $\mathbf{x}$ (*i.e.*, the matrices $\mathbf{S}_x$ and $\mathbf{A}$). It is natural to try to make inferences about the statistical properties of $\mathbf{x}$ from the statistical properties of $\mathbf{y}$.

Since the smallest eigenvalue of $S_{y''}$ is σ_N^2, we would be tempted to say that the rank of $S_{y''}$ is n. If $n = m$ (the number of measurements = number of independent variables), we might be tempted to say that S_x also has rank n and try to give each eigenvector of S_x a physical interpretation. However, even if $AS_xA^t = S_y$ has a rank $r < n$, because the noise covariance matrix has rank n, so will $S_{y''}$. So, if we look only at $S_{y''}$, we might try to provide a physical interpretation to eigenvectors of S_x that don't exist.

When we look at the statistical properties of our measurements, we need to remember how much of the variance is due to the variance of the independent variable, and how much is due to noise.

References

Alishouse, J. C., L. J. Crone, H. E. Fleming, F. L. Van Cleef, and D. Q. Wark, 'A discussion of empirical orthogonal functions and their application to vertical temperature profiles,' *Tellus*, **XIX,** 477-482 (1966)

Birkhoff, G. and S. Mac Lane, *A Survey of Modern Algebra*, Macmillan Co, New York, p 256 ff. (1953).

Deif, A. S., *Advanced Matrix Theory for Scientists and Engineers*, Abacus Press, Tunbridge Wells & London (1982).

Finkbeiner, D. T., II, *Introduction to Matrices and Linear Transformations*, W. H. Freeman & Co., San Francisco, p. 204 (1966).

Graybill, F. A., *Introduction to Matrices with Applications to Statistics*, Wadsworth Publishing Company, Inc., Belmont, California, (1969).

Horn, R. A., and C. R. Johnson, *Matrix Analysis*, Cambridge University Press, Cambridge (1985).

Press, W. H., S. A. Teukolsky, W. T. Vettering, and B. R. Flannery, *Numerical Recipes in Fortran, the Art of Scientific Computing*, Second Edition, Cambridge University Press, Cambridge, England, 963 pp (1992).

Roger, R. E., 'A faster way to compute the noise-adjusted principal components transform matrix,' *IEEE Tr. Geoscience & Remote Sensing,* **32,** 1194-1196 (1994).

6. Regression

He uses statistics as a drunken man uses lamp posts—for support rather than illumination.

Andrew Lang (1844-1912)

Throughout this book, I will generally assume that we know the relationship between the quantities we measure and the quantity we want to retrieve, based on our understanding of the physics of the situation. This is expressed by the matrix $\mathbf{A}$ in the equation

$$\mathbf{y} = \mathbf{A}\mathbf{x}. \tag{1}$$

But what if we do not know what $\mathbf{A}$ is? In many remote-sensing problems, the physics of the situation is not known well enough to calculate it. In this case, we can try to use regression to find in inverse to equation (1) without knowing anything about the physics involved at all. In order to do this, we need to make simultaneous measurements of $\mathbf{y}$ and $\mathbf{x}$; from a set of such measurements, we can find the coefficients of the matrix $\mathbf{R}$ so that we can estimate the values of $\mathbf{x}$ from

$$\hat{\mathbf{x}} = \mathbf{R}\mathbf{y}. \tag{2}$$

The real advantage of this method is that you do not have to know very much—or think very much—to use it. The main disadvantage is that it usually does not work very well. I would suggest that the reader stop, at this point, and think of reasons why it might not work. I shall return to them later.

There are several ways that we can use—or misuse—regression techniques to solve equation (1). My philosophy is that we should employ a method that makes use of the covariance matrix $\mathbf{S}_x$ of the independent variable $\mathbf{x}$ noise covariance matrix $\mathbf{N}$. Not everybody agrees with this philosophy, but it seems to me that it is fundamental. I shall explore these ideas in this short chapter.

A variation on this theme involves models, rather than data. In the last paragraph, I said that we would measure $\mathbf{x}$ and $\mathbf{y}$. But since measurements are often difficult and expensive to make, we could use calculations instead. Suppose we believe we have a model that gives $\mathbf{y}$ as a known function of $\mathbf{x}$:

$$\mathbf{y} = \mathbf{A}(\mathbf{x}), \tag{3}$$

where $\mathbf{A}$ is a known vector function which may be nonlinear. We could use a random number generator to produce $\mathbf{x}$'s; calculate the corresponding $\mathbf{y}$'s; and then perform the regression on the pairs $(\mathbf{x}_i, \mathbf{y}_i)$ to calculate $\mathbf{R}$. This method can be especially useful when the function that relates $\mathbf{y}$ to $\mathbf{x}$ is very complicated, so that we cannot find an analytical inverse.

There are many books on statistics that treat multiple regression. They are generally involved with finding the coefficients of the matrix $\mathbf{A}$ (or of $\mathbf{R}$). Here, we are after something different: a matrix that, when applied to a measurement $\mathbf{y}$, will produce a stable estimate of $\mathbf{x}$.

As in Chapter 5, let $\mathbf{S}$ be the $n \times n$ covariance matrix of the independent variable:

$$\mathbf{S} = \left\langle \mathbf{x}\mathbf{x}^t \right\rangle. \tag{4}$$

Usually, we do not know the components of $\mathbf{S}$ are because the extensive sets of measurements of temperature, wind speed, or whatever, are rarely available. I have stressed many times that, without such information, we cannot be assured that the retrieval algorithm that we have devised is close to being optimal—or even that it is close to being very good.

There are m components of $\mathbf{x}$ and n components of $\mathbf{y}$. Define the $n \times m$ cross-covariance matrix $\mathbf{Q}$ by

$$q_{ij} = \left\langle x_i y_j \right\rangle \tag{5}$$

or

$$\mathbf{Q} = \left\langle \mathbf{x}\mathbf{y}^t \right\rangle. \tag{6}$$

We determine $\mathbf{Q}$ either from a set of simultaneous measurements of $\mathbf{y}$ and $\mathbf{x}$ or from a set of calculations with our model. In many cases, this depends on having simultaneous and coincident[1] measurements of $\mathbf{x}$ and $\mathbf{y}$ at many times and locations. In the case of remote sensing, if the $\mathbf{y}$'s may be measured radiances above the earth's atmosphere, the corresponding $\mathbf{x}$'s, called *ground truth*,[2] are the corresponding values on the earth's surface or in its atmosphere. It is usually the case that the coincidence in time and space is only approximate, adding a further source of error in the retrieval when $\mathbf{Q}$ is calculated.

Finally, let $\mathbf{T}$ be the $m \times m$ covariance matrix

$$\mathbf{T} = \left\langle \mathbf{y}\mathbf{y}^t \right\rangle. \tag{7}$$

To find the coefficients for estimating the value of $\mathbf{x}$ in equation (2), write the error covariance matrix

---

[1] Coincident in space.

[2] With a good bit of wishful thinking.

$$\chi_{ij}^2 = \left\langle (\hat{x}_i - x_i)(\hat{x}_j - x_j) \right\rangle. \tag{8}$$

As before, we assume that all of the variables, both independent and dependent, have zero mean. Substitute equation (2) into this one. Then

$$\chi_{ij}^2 = \left\langle \left(x_i - \sum_{k=1}^{m} r_{ik} y_k \right) \left(x_j - \sum_{l=1}^{m} r_{jl} y_l \right) \right\rangle. \tag{9}$$

Let **N** be the noise covariance matrix and assume that the different noise components are uncorrelated and have equal variance σ_N^2. That is,

$$\mathbf{N} = \sigma_N^2 \mathbf{I}. \tag{10}$$

The noise in $\hat{\mathbf{x}}$ will be $\mathbf{RNR^t}$. Using the definitions of **S, T,** and **Q,**

$$\chi_{ij}^2 = s_{ij} - \sum r_{ik} q_{kj} - \sum r_{jl} q_{li} + \sum \sum r_{ik} r_{jl} \left(t_{kl} + \sigma_N^2 \delta_{kl} \right) \tag{11}$$

or

$$\chi^2 = \mathbf{S} - \mathbf{RQ} - \mathbf{QR^t} + \mathbf{R(T+N)R^t}. \tag{12}$$

Take the derivative of χ_{ij}^2 with respect to each coefficient r_{ik}:

$$\frac{\partial \chi_{ij}^2}{\partial r_{ik}} = \sum_l r_{jl} \left(t_{kl} + \sigma_N^2 \delta_{kl} \right) - q_{kj} = 0 \tag{13}$$

or

$$\mathbf{R(T+N)} = \mathbf{Q}. \tag{14}$$

Therefore,

$$\mathbf{R} = \mathbf{Q(T+N)}^{-1}. \tag{15}$$

Also, the error covariance matrix is

$$\chi^2 = \mathbf{S} - \mathbf{Q(T+N)}^{-1}\mathbf{Q}. \tag{16}$$

Since **T** is a covariance matrix, its eigenvalues are non-negative definite, so **(T+N)** is nonsingular. This was explained in Chapter 5. At least this solution will not be unstable. Note that equation (15) is not the equation we would have gotten by simply regressing the values of **x** against the (measured or calculated) values of **y**.

Four covariance matrices, **S**, **T**, **Q**, and **N**, are all involved in determining the retrieval matrix **R**. If these matrices are wrong, so will **R** be. In particular, suppose that we have a set of pairs of vectors $\mathcal{H} = \{\mathbf{y}_k, \mathbf{x}_k\}$; we have estimated the covariance matrices from a set of N of these pairs. (If we have used models instead of observations, we have simply *assumed* some form for the covariance matrix **S** and calculated the other two.) If the set $\mathcal{H}$ contains a truly representative sample of the things that *do* happen, the regression will work well. However, if there are vector pairs $\{\mathbf{y}_k, \mathbf{x}_k\}$ that occur in nature but are not represented in $\mathcal{H}$, the regression will work poorly.

We sometimes see the following kind of discussion from many scientists who should know better. One of them wants to develop an algorithm for determining wind speed over the ocean, perhaps, from microwave measurements made simultaneously at several different frequencies. "We have 168 cases of coïncident wind-speed measurements and microwave radiance measurements. We divided the data randomly into two equal sets, a *training set* and an *evaluation set*. We used the first 84 data pairs to determine the regression coefficients. Then we applied them to the other 84 samples. We found that the r.m.s. error in the wind speed was…."

What is going on here? This person should have used all 168 samples to determine the regression coefficients and then used equation (9) to calculate the error variance. Why didn't he? It is hard to tell. The use of "training" and "evaluation" sets cannot produce as accurate result as the one I have outlined above.

Exercise 6-1: Suppose that I have $2N$ data pairs $(\mathbf{x}_i, \mathbf{y}_i)$ and I divide them randomly into two equal sets. I find a regression matrix $\mathbf{R}_1$ that minimizes $<|\mathbf{x}_i - \mathbf{R}_1\mathbf{y}_i|^2>$ for set 1. Alternatively, we could use equation (15) to find another estimate; call it $\mathbf{R}_2$, also derived from set 1. Compare the errors that would result when we apply each one to the other set of data. What happens in each case when we apply $\mathbf{R}_1$ or $\mathbf{R}_2$ to new data that have different covariance properties (*i.e.*, the independent variables $\mathbf{x}_i$ have a different covariance matrix)?

7. Matrix Solution of Linear Equations

> *Mathematicians are like lovers. Grant a mathematician the least principle, and he will draw from it a consequence which you must also grant him, and from this consequence another.*
>
> Bernard Le Bovier Fontenelle (1657-1757)

S tarting with this chapter, I will take up the mathematics proper of remote sensing, in the sense of providing methods for inverting systems of equations. I shall start this chapter with a discussion of what is, I believe, the optimal method of solving these problems, and why it is optimal. Then I will take up some other possible methods.

Suppose that we are given the relationship

$$\mathbf{y} = \mathbf{A}\mathbf{x} \tag{1}$$

where $\mathbf{A}$ is an $m{\times}n$ matrix and the vectors $\mathbf{x} = \mathrm{col}(x_1,...,x_n)$ and $\mathbf{y} = \mathrm{col}(y_1,...,y_m)$ are $n{\times}1$ and $m{\times}1$ column vectors, respectively. For the sake of concreteness, we will suppose that $\mathbf{x} \in \mathrm{E}^n$ and $\mathbf{y} \in \mathrm{E}^m$. The matrix $\mathbf{A}$ is a mapping from E^n to E^m. The vector $\mathbf{x}$ represents the independent variables whose values we want to know. The vector $\mathbf{y}$ represents a set of variables that we measure.

When I say we "measure" $\mathbf{y}$, I mean that we use some instrument that lets us assign a numerical value to each component $\mathbf{y} = \mathrm{col}(y_1, \ldots, y_n)$. However, what we really want to know is the values of $\mathbf{x} = \mathrm{col}(x_1, \ldots, x_m)$. We could say that we want to "measure" $\mathbf{x}$, because that's what we're really interested in knowing. For example, $\mathbf{x}$ might represent the near-surface wind speed and surface temperature in a certain area of the ocean; we want to know these values for meteorological forecasting, perhaps. The vector $\mathbf{y}$ would represent the intensity of the radiation emitted from the sea surface in the microwave part of the spectrum. The values of $\mathbf{y}$ are of no particular interest to us, except insofar as we can use them to infer the values of $\mathbf{x}$. In this sense, I will try to be consistent throughout this book and use the word *measure* to indicate the physical process whereby we assign a value to $\mathbf{y}$, while we infer or *retrieve* the value of $\mathbf{x}$: we are doing all of this, after all, because we cannot measure $\mathbf{x}$ directly.

If we want an optimal solution to equation (1), we first need to decide what *optimal* means. As we shall see, there is more than one plausible definition, and there is no purely mathematical basis for choosing which one to use; we have to choose it on philosophical or practical grounds instead. Quite gen-

147

erally, we can derive an optimal algorithm by minimizing the average value of some quantity. It seems natural to me to minimize the mean error, given by $<(\hat{x} - x)^2>$, where $\hat{x}$ is the estimated value of x. We could, however, use other criteria, such as the smoothness of $\hat{x}$ or its variance. It is common to derive a solution that contains an arbitrary parameter—one that is chosen, perhaps, on aesthetic grounds—and therefore does not minimize any mathematical quantity, but I don't think that this is a sound way to proceed. So we shall look first at the method that minimizes the mean squared error in $\hat{x}$. I assume that we know, from physical analysis, what the components of $\mathbf{A}$ are. The problem at hand is to find a linear inverse of equation (1), an estimate of the form

$$\hat{x} = \mathbf{R}y, \tag{2}$$

where $\mathbf{R}$ is a matrix to be determined.[1] That is, given the measured value y, we want to estimate the value of x that must have given rise to it. It should satisfy $y = \mathbf{A}\hat{x}$ to within the error in the measurement of y. Although the problem looks simple, there is a lot of complexity hidden in this equation. We shall see that the choice of what matrix $\mathbf{R}$ we should use depends not only on the matrix $\mathbf{A}$, but also on two other factors: the noise in the measurements of the dependent variable y and the covariance matrix of the independent variable x.

Could we use $\mathbf{A}^{-1}$ to solve equation (1)? Probably not. The matrix $\mathbf{A}$ might be singular; even if it is nonsingular, it might be ill-behaved, in that some of its eigenvalues might be very close to zero. Furthermore, if $\mathbf{A}$ is not square, $\mathbf{A}^{-1}$ is not even defined. I showed in §5.2.5 that, if $\{\lambda_i\}$ is the set of all singular values of $\mathbf{A}$, then $\{1/\lambda_i\}$ is the set of all the singular values of $\mathbf{A}^{-1}$. Because of the noise in the measurements, the results will be unacceptable if some of the λ_i are too close to zero.

Again supposing that we know what the matrix $\mathbf{A}$ is, there are two sources of error in the inversion of equation (1): measurement noise and goodness-of-fit. I will discuss noise in a moment. By goodness-of-fit, I mean the difference between the true value of x and the value of $\hat{x}$ that would have been retrieved from noiseless data; we can measure it with the variable

$$\varepsilon_x^2 \equiv \left| \mathbf{R}y - x \right|^2. \tag{3}$$

Now one might naïvely think that the best solution would be the one that makes $\varepsilon_x^2 = 0$. If this were so, you would not be reading this book, since there would be little for me to say on the subject. However, it turns out that, in gen-

---

[1] Note that this is different from the covariance matrix in Chapter 5.

eral, we need to minimize the sum of ε_x^2 and another quantity. While the choice of what that other quantity should be is arbitrary, I think that the best choice by far is to minimize the sum of ε_x^2 and the error in $\hat{\mathbf{x}}$ due to the noise in the measurements. If we represent the noise by a random vector $\mathbf{n}$, each component of which has zero mean and variance σ_N^2, then the noise error is

$$\varepsilon_N^2 = \left\langle \left| \mathbf{Rn} \right|^2 \right\rangle. \tag{4}$$

(The factor $\alpha \equiv \varepsilon_N^2/\sigma_N^2$ is called the *noise amplification factor*.) We usually want to minimize the sum $\varepsilon_x^2 + \varepsilon_N^2$.

We will examine only linear problems in this chapter, except for a section on quadratic equations at the end. Throughout this book, I will assume that the mean value of any variable x is zero. If it were not, we could replace x with a new variable

$$x' = x - \langle x \rangle, \tag{5}$$

where $<x>$ denotes the mean value of x. Note, however, that while $<x>$ may equal 0, $<x^2> > 0$ (unless $x \equiv 0$), and we have to keep this in mind when we consider quadratic equations.

7.1 Noise

Any measurement of $\mathbf{y}$ will include some random error, which is denoted by a vector $\mathbf{n} = (n_1,...,n_m)$. I assume that each component of $\mathbf{n}$ is Gaussian distributed with mean 0 and variance $\sigma_{N,i}^2 = <n_i^2>$. In general, I will use a subscript N to denote a quantity associated with noise. The quantities that we measure always contain noise:

$$\mathbf{y}' = \mathbf{y} + \mathbf{n} = \mathbf{Ax} + \mathbf{n}. \tag{6}$$

We assume throughout this book that the source of noise is somewhere in the instrument that is used to measure $\mathbf{y}$ and that it is independent of the value of $\mathbf{x}$.

It is usually the case that the noise components are uncorrelated; *i.e.*, their covariance matrix has components

$$\sigma_{ij}^2 = \left\langle n_i n_j \right\rangle = \sigma_{N,i}^2 \delta_{ij}. \tag{7}$$

I cannot stress too strongly the importance of considering the amount of noise that is present in a measurement, because the amount of noise, along with the variances of the variables we want to measure, will help to determine what retrieval algorithm we will use.

There is, in general, no unique solution to equation (1). I discussed the null space of a matrix in §5.2.2. Since $\mathbf{A}$ is a mapping from E^n to E^m, the null space $\mathcal{N}$ of $\mathbf{A}$ is the set of all vectors $\mathbf{v} \in E^n$ such that $\mathbf{Av} = \mathbf{0}_m$, where $\mathbf{0}_m$ is the zero vector in E^m. If $m < n$, $\mathcal{N}$ cannot be empty. Also, if the rank of $\mathbf{A}$ is less than m, $\mathcal{N}$ is not empty. This means that, if $\mathbf{y} = \mathbf{Ax}$, then also $\mathbf{y} = \mathbf{A}(\mathbf{x} + \mathbf{v})$ for any $\mathbf{v} \in \mathcal{N}$. Therefore, if we measure $\mathbf{y}$, there are infinitely many vectors that could have produced that measurement.

Exercise 7-1: Why is $\mathcal{N}$ different from $\{\mathbf{0}\}$ when $m < n$?

Since the measurement of $\mathbf{y}$ contains noise [see equation (6)], the true value of the independent variable is $\mathbf{x}$ and the measured value $\mathbf{y}'$ may not satisfy $\mathbf{y}' = \mathbf{Ax}$ exactly. The residual vector $\mathbf{n} = \mathbf{y}' - \mathbf{Ax}$ is a random variable whose value is, of course, unknown (since we don't know what $\mathbf{x}$ is). A solution $\hat{\mathbf{x}}$ that satisfies $\mathbf{y}' = \mathbf{A}\hat{\mathbf{x}}$ exactly might be very far from the true value of $\mathbf{x}$. We shall see that, in general, choosing $\hat{\mathbf{x}}$ by minimizing $|\mathbf{y}' - \mathbf{A}\hat{\mathbf{x}}|$ may not produce good results.

There are two reasons, then, why the solution to $\mathbf{y}' = \mathbf{Ax}$ may not be unique. But we must choose one of the many possible methods of estimating $\mathbf{x}$. Which one is best? The definitive answer is, "that depends."

Let $\{\hat{\mathbf{x}}\}$ be the set of all $\mathbf{x}$'s that satisfy $\mathbf{y}' = \mathbf{Ax}$, in the sense that $|\mathbf{A}\hat{\mathbf{x}} - \mathbf{y}'| < \sigma_N$. In general, we shall pick some measure of how well each particular $\mathbf{x} \in \{\mathbf{x}\}$ fits our idea what "the best solution" means, and we will pick the $\mathbf{x}$ that minimizes that measure. For instance, we might choose the $\mathbf{x}$ that minimizes $|\mathbf{x}| = (\mathbf{x} \cdot \mathbf{x})^{1/2}$. The *least squares* solution picks the value $\hat{\mathbf{x}}$ that minimizes the error

$$\chi^2 = \left\langle (\mathbf{x} - \hat{\mathbf{x}})(\mathbf{x} - \hat{\mathbf{x}})^t \right\rangle, \tag{8}$$

where $\langle \cdot \rangle$ denotes an ensemble average value and $\mathbf{x}$ denotes the true value of the independent variable. This is a reasonable and useful method of choosing the best estimate $\hat{\mathbf{x}}$ but it is not the only one.

7.2 Retrieval

The derivation of the least squares solution is quite straightforward; it has appeared in many places. However, all of the derivations that I am aware of appear in various articles about atmospheric temperature sounding that are not intelligible to a reader who is not familiar with the physics involved. In addition, each author adopts his own notation, which is not always completely explained, and which would further confuse the uninitiated reader.

Generally, with remote sensing problems, the number of variables we want to retrieve (n) is not be the same as the number of variables that we measure (m); we have no assurance that $\mathbf{A}$ is a square matrix. The problem may be underdetermined (i.e., $m < n$), or overdetermined ($m > n$). In either case, $\mathbf{A}^{-1}$ is not defined. Even where the problem is well-determined ($n = m$), the matrix $\mathbf{A}$ may turn out to be singular. Unless the physics of the situation requires that some components of $\mathbf{A}$ be identically zero, $\mathbf{A}$ might not be singular in the strict mathematical sense, but, from the point of view of computing an inverse, $\mathbf{A}$ might just as well be singular, because the eigenvalues of $\mathbf{A}$ are small [see equation (102) in Chapter 5]. The small eigenvalues cause problems because, as we shall see, the inversion involves, essentially, multiplying different components[1] of $\mathbf{y}'$ by λ_i^{-1}, which means that there is a term of the form $\lambda_i^{-1} n$, where n is the random noise. Remember that the variance of ax, where a is a scalar, is given by $\mathrm{var}(ax) = a^2 \mathrm{var}(x)$. So, if the $\mathrm{var}(n) = \sigma_N^2$, the variance of $\mathrm{var}(\lambda_i^{-1} n) = \sigma_N^2 / \lambda_i^2$. When λ_i is small, this comes to dominate the retrieved value of $\mathbf{x}$.

Let me illustrate these points with the 2×2 matrix

$$\mathbf{A} = \begin{pmatrix} 1 & 1-\varepsilon \\ 1+\varepsilon & 1 \end{pmatrix}, \tag{9}$$

where ε is a small positive quantity. The determinant of $\mathbf{A}$ is

$$\det\begin{pmatrix} 1 & 1-\varepsilon \\ 1+\varepsilon & 1 \end{pmatrix} = 1 - (1-\varepsilon)(1+\varepsilon) = \varepsilon^2. \tag{10}$$

When $\varepsilon \to 0$, $\mathbf{A}$ approaches being singular.

But what happens for small values of ε? The inverse of this matrix is

$$\mathbf{A}^{-1} = \frac{1}{\varepsilon^2}\begin{pmatrix} 1 & \varepsilon-1 \\ -\varepsilon-1 & 1 \end{pmatrix}. \tag{11}$$

The reader can confirm this by multiplying the two matrices: $\mathbf{A}\mathbf{A}^{-1} = \mathbf{A}^{-1}\mathbf{A} = \mathbf{I}$.

Mathematically speaking, for $\varepsilon \neq 0$, $\mathbf{A}^{-1}$ is defined, although its components could be very large. However, for computational purposes, $\mathbf{A}^{-1}$ might just as well not exist when ε is small because of its effect on the random errors in the measurement of $\mathbf{y}$. I should hasten to point out that this problem does not arise from any computational limitation of our computer, and it cannot be eliminated by carrying out the computations with more precision. The problem, rather, is the effect that the very small eigenvalues of $\mathbf{A}$ (which correspond to very large eigenvalues of $\mathbf{A}^{-1}$) have on the measurement noise.

---

[1] In a suitable coördinate system; see §7.5.

In terms of our example, if we use $\mathbf{A}^{-1}$ to invert equation (1), we will get

$$\mathbf{x}' = \mathbf{A}^{-1}\mathbf{y}' = \mathbf{A}^{-1}(\mathbf{y} + \mathbf{n})$$

$$= \frac{1}{\varepsilon^2}\begin{pmatrix} y_1 - (1-\varepsilon)y_2 \\ y_2 - (1+\varepsilon)y_1 \end{pmatrix} + \frac{1}{\varepsilon^2}\begin{pmatrix} n_1 - (1-\varepsilon)n_2 \\ n_2 - (1+\varepsilon)n_1 \end{pmatrix}. \tag{12}$$

It is possible to argue that, as $\varepsilon \to 0$, $\mathbf{A}^{-1}\mathbf{y}$ remains finite, since, after all, we have been assuming that the relationship $\mathbf{y} = \mathbf{A}\mathbf{x}$ is exact (and that $\mathbf{x}$ is finite). However, $\mathbf{A}^{-1}\mathbf{n}$ is proportional to $1/\varepsilon^2$. Therefore, the noise variance is

$$\frac{1}{\varepsilon^2}\sigma_N^2 \to \infty \quad \text{as} \quad \varepsilon \to 0. \tag{13}$$

The property we have just encountered is called *noise amplification*, and this idea is central to this book. In most cases, the process of inverting a matrix equation amplifies the noise present in the measurement of $\mathbf{y}$. A common problem that the unwary will encounter is that, if we ignore the errors in our measurements, we may end up with results that depend more on the noise than the physical processes we are interested in measuring.

We will see that there are, in general, two conflicting requirements. On the one hand, we would like to minimize the error defined in equation (3); this measures how well our solutions fit the true values of the variables. On the other hand, since there is noise in the measurements, we also need to minimize the noise amplification, which is defined in equation (4). In general, we will minimize a weighted sum of these two terms. This leads to a unique solution. In the Backus-Gilbert approach (see Chapter 12), we will obtain a family of possible solutions using a different criterion, that of minimizing the *spread* of the retrieved values. We have to pick the criterion is the most suitable for a given problem.

7.3 Least-squares inverse

I shall develop a least-squares inverse of equation (1) in this section. I say "a least-squares inverse" because, as I said in the introduction to this chapter, there are many possible principles we might use to select the "best" from an infinite number of possible inverses. The term "least squares" simply means that we minimize some quadratic quantity. The principle that I will discuss here, and one that is appropriate in most situations, minimizes the sum of the errors due to goodness-of-fit ε_x^2 and noise in the measurements. This formulation handles both over- and under-determined systems of equations.

I shall seek a solution of the form

$$\hat{\mathbf{x}} = \mathbf{R}\mathbf{y}' = \mathbf{R}(\mathbf{y} + \mathbf{n}), \tag{14}$$

for some matrix $\mathbf{R}$ whose properties we shall derive now. The ij^{th} component of the error covariance matrix [see equation (8)] is

$$\chi_{ij}^2 = \left\langle (\hat{x}_i - x_i)(\hat{x}_j - x_j) \right\rangle \tag{15}$$

where $\hat{x}_i$ is the estimated value of x_i. That is,

$$\hat{x}_i = \sum_{k=1}^{m} r_{ik} y_k' = \sum_{k=1}^{m} r_{ik}(y_k + n_k). \tag{16}$$

Substitute equation (1) into this one:

$$\hat{x}_i = \sum_{k=1}^{m} r_{ik} \left(\sum_{j=1}^{n} a_{kj} x_j + n_k \right), \tag{17}$$

which is the component form of equation (14). Then, substituting equation (17) into (15),

$$\chi_{ij}^2 = \left\langle \left[x_i - \sum_k r_{ik}(y_k + n_k) \right] \left[x_j - \sum_{k'} r_{jk'}(y_{k'} + n_{k'}) \right] \right\rangle$$

$$= \left\langle \left[x_i - \sum_k r_{ik} \left(\sum_l a_{kl} x_l + n_k \right) \right] \left[x_j - \sum_{k'} r_{jk'} \left(\sum_{l'} a_{k'l'} x_{l'} + n_{k'} \right) \right] \right\rangle. \tag{18}$$

Now, expanding this expression and using the property that the noise $\mathbf{n}$ is uncorrelated with $\mathbf{x}$,

$$\chi_{ij}^2 = \left\langle x_i x_j \right\rangle - \left\langle x_j \sum_k r_{ik} \left(\sum_l a_{kl} x_l + n_k \right) \right\rangle - \left\langle x_i \sum_{k'} r_{jk'} \left(\sum_{l'} a_{k'l'} x_{l'} + n_{k'} \right) \right\rangle$$

$$+ \left\langle \sum_k \sum_{k'} r_{ik} r_{jk'} \left(\sum_l \sum_{l'} a_{kl} a_{k'l'} x_l x_{l'} + n_k n_{k'} \delta_{kk'} \right) \right\rangle, \tag{19}$$

where $\delta_{kk'} = 1$ if $k = k'$, and equals 0 otherwise. Note that, to get this result, I have used $\langle n_k x_j \rangle = 0$. Also, I will use the notation that

$$\left\langle x_i x_j \right\rangle = \sigma_{ij}^2 = \text{cov}(x_i, x_j) \tag{20}$$

(since $\langle x_i \rangle = 0$). Finally, we can reduce this to

$$\chi_{ij}^2 = \sigma_{ij}^2 - \sum_k r_{ik} \sum_l a_{kl} \sigma_{lj}^2 - \sum_{k'} r_{jk'} \sum_{l'} a_{k'l'} \sigma_{l'i}^2$$

$$+ \sum_k \sum_{k'} r_{ik} r_{jk'} \left(\sum_l \sum_{l'} a_{kl} a_{k'l'} \sigma_{ll'}^2 + \sigma_{N,k}^2 \delta_{kk'} \right). \tag{21}$$

In matrix notation,

$$\chi^2 = \mathbf{S}_x - \mathbf{RAS}_x - \mathbf{S}_x \mathbf{A^t R^t} + \mathbf{RAS}_x \mathbf{A^t R^t} + \mathbf{RNR^t}. \tag{22}$$

Here, $\mathbf{N} \equiv \langle \mathbf{nn^t} \rangle$ is the covariance matrix of the noise; since the different noise terms are uncorrelated, it is a diagonal matrix, and we can write

$$\mathbf{N} = \begin{pmatrix} \sigma_{N,1}^2 & 0 & \cdots \\ 0 & \ddots & \\ \vdots & & \sigma_{N,m}^2 \end{pmatrix}. \tag{23}$$

Here, $\sigma_{N,j} \equiv \langle n_j^2 \rangle$.

Remember that $\mathbf{S}_x$, the covariance matrix of $\mathbf{x}$, is an average property of the ensemble of $\mathbf{x}$'s and is usually determined from other sets of measurements. For instance, in remote sensing, we could measure $\mathbf{S}_x$ with conventional methods—with radiosondes, in the case of vertical temperature profiles. If all else fails, we can find $\mathbf{AS}_x \mathbf{A^t}$ directly from the measurements of the $\mathbf{y}$'s (see §7.7.2).

To find the coefficients r_{ik} that minimize the error, take the derivative of each component χ_{ij}^2 with respect to r_{ik} and set the result equal to zero.

$$\frac{\partial \chi_{ij}^2}{\partial r_{ik}} = - \sum_l a_{kl} \sigma_{lj}^2 + \sum_{k'} r_{jk'} \sum_l \sum_{l'} a_{kl} a_{k'l'} \sigma_{ll'}^2 + r_{jk} \sigma_{N,k}^2 = 0 \tag{24}$$

(if $i = j$, each term on the left is multiplied by 2, but this makes no difference). In matrix form, this equation is simply

$$\mathbf{R}\left(\mathbf{AS}_x \mathbf{A^t} + \mathbf{N} \right) = \mathbf{S}_x \mathbf{A^t}. \tag{25}$$

The matrix $\mathbf{R}$ is found by taking the inverse of $(\mathbf{AS}_x \mathbf{A^t} + \mathbf{N})$:

$$\mathbf{R} = \mathbf{S}_x \mathbf{A^t} (\mathbf{AS}_x \mathbf{A^t} + \mathbf{N})^{-1}. \tag{26}$$

Exercise 7-2: Use the method of §14.2 to derive equation (23) from equation (22). That is, symbolically take the derivative $d\chi^2/d\mathbf{R}$.

Exercise 7-3: Suppose that **A** is the 2×2 matrix $\begin{pmatrix} 1 & t \\ -t & 1 \end{pmatrix}$. Suppose that $\mathbf{S}_x$ is diagonal, so $\mathbf{S}_x = \sigma_x^2 \mathbf{I}$, and that $\mathbf{N} = \sigma_N^2 \mathbf{I}$. Calculate the error χ^2 for different values of σ_x^2 and σ_N^2. What are the eigenvalues and eigenvectors of **A**?

The reader who has followed all of this may now be wondering, "Why did we do all of this?" After all, I said that

$$\mathbf{R} = \mathbf{A}^{-1} \tag{27}$$

is unacceptable, mostly because $\mathbf{A}^{-1}$ is generally either not defined or unstable. Why is the inverse we have derived here any better? To answer this question, we should look at two properties of $(\mathbf{AS}_x\mathbf{A}^t + \mathbf{N})$:

1. **Dimension.** Even if **A** is not square, the matrix $(\mathbf{AS}_x\mathbf{A}^t + \mathbf{N})$ is square, since **A** is $m \times n$, $\mathbf{S}_x$ is $n \times n$, and $\mathbf{A}^t$ is $n \times m$. The noise covariance matrix **N** is $m \times m$. Therefore, $(\mathbf{AS}_x\mathbf{A}^t + \mathbf{N})$ is an $m \times m$ matrix, so its inverse is defined unless that matrix is singular.

2. $(\mathbf{AS}_x\mathbf{A}^t + \mathbf{N})$ **is nonsingular.** To see this, we need to use some properties of non-negative definite matrices that are discussed in Chapter 5; the proof was given in the previous chapter. Since[2]

$$\mathbf{AS}_x\mathbf{A}^t = \mathbf{AS}_x^{\frac{1}{2}}\mathbf{S}_x^{\frac{1}{2}}\mathbf{A}^t = \mathbf{AS}_x^{\frac{1}{2}}(\mathbf{AS}_x^{\frac{1}{2}})^t, \tag{28}$$

$\mathbf{AS}_x\mathbf{A}^t$ is non-negative definite. Therefore, the eigenvalues λ_i of $\mathbf{AS}_x\mathbf{A}^t + \mathbf{N}$ are large enough so that σ_N^2 / λ_i will not be too large.

Note that I mean here that $(\mathbf{AS}_x\mathbf{A}^t + \mathbf{N})$ is nonsingular in the computational sense, that its inverse can be computed and is well-behaved. As I discussed earlier, a matrix can be mathematically nonsingular, but still be so ill-conditioned that we cannot calculate its inverse, even with the best computers. On the other hand, $(\mathbf{AS}_x\mathbf{A}^t + \mathbf{N})^{-1}$ depends both on $\mathbf{S}_x$ and **N**, and it does not commute with **A** the way a true inverse should. Despite this, it is the best way to estimate **x** from a measurement of **y**.

The components of the error covariance matrix are

$$\chi_{ij}^2 = \left\langle (x_i - \hat{x}_i)(x_j - \hat{x}_j) \right\rangle = \sigma_{ij}^2 - \left\langle x_i\hat{x}_j \right\rangle - \left\langle x_j\hat{x}_i \right\rangle + \left\langle \hat{x}_i\hat{x}_j \right\rangle. \tag{29}$$

We can use equation (21) to evaluate the error covariance matrix χ^2. Writing equation (21) in matrix notation,

$$\chi^2 = \mathbf{S}_x - \mathbf{RAS}_x - \mathbf{S}_x\mathbf{A}^t\mathbf{R}^t + \mathbf{R}(\mathbf{AS}_x\mathbf{A}^t + \mathbf{N})\mathbf{R}^t \tag{30}$$

---

[2] Remember that $\mathbf{S}^{\frac{1}{2}}$ is defined by the relationship $\mathbf{S} = \mathbf{S}^{\frac{1}{2}}\,\mathbf{S}^{\frac{1}{2}}$.

Using equation (26), we can see that

$$\mathbf{R}\left(\mathbf{A}\mathbf{S}_x\mathbf{A}^t + \mathbf{N}\right)\mathbf{R}^t = \mathbf{S}_x\mathbf{A}^t\mathbf{R}^t = \mathbf{R}\mathbf{A}\mathbf{S}_x, \qquad (31)$$

(since $\mathbf{R}\mathbf{A}\mathbf{S}_x$ is symmetric), so

$$\chi = \mathbf{S}_x - \mathbf{R}\mathbf{A}\mathbf{S}_x = \mathbf{S}_x\left(\mathbf{I} - \mathbf{R}\mathbf{A}\right), \qquad (32)$$

where $\mathbf{I}$ is the $n \times n$ identity matrix. We can consider two limiting cases:

1. *Perfect retrieval.* If $\mathbf{R} = \mathbf{A}^{-1}$, then the error is proportional to $(\mathbf{I} - \mathbf{I}) = 0$, and there is no error.

2. *No retrieval.* Suppose that $\mathbf{R}$ is the zero matrix, $\mathbf{R} = \mathbf{0}$. This means that we ignore the data and use the estimate $\mathbf{x} = <\mathbf{x}>$ every time. In this case, $\chi^2 = \mathbf{S}_x$; *i.e.*, the error in x_i is just the variance of x_i. This no-information case should be the worst we can do. If our retrieval algorithm performs worse than this, we have done something very, very wrong.

Exercise 7-4: Show that the least squares solution is also the one that minimizes the average value of $|\hat{\mathbf{x}}|$.

Consider the special case where all of the independent variables have the same variance σ_x^2, and the noise in each measurement is σ_N^2. We also assume that the independent variables—as well as the noise—are uncorrelated. That is, assume that

$$\mathbf{S} = \sigma_x^2\mathbf{I} \quad and \quad \mathbf{N} = \sigma_N^2\mathbf{I}. \qquad (33)$$

Then equation (26) will take on the simple form

$$\mathbf{R} = \mathbf{A}^t\left(\mathbf{A}\mathbf{A}^t + \gamma\mathbf{I}\right)^{-1}, \qquad (34)$$

where $\gamma \equiv \sigma_N^2 / \sigma_x^2$. Furthermore, we can show that

$$\mathbf{R} = \left(\mathbf{A}^t\mathbf{A} + \gamma\mathbf{I}\right)^{-1}\mathbf{A}^t. \qquad (35)$$

Exercise 7-5: Prove this last statement. (Hint: use the result of §10.4.2.) Show that, in general,

$$\mathbf{R} = \mathbf{S}_x^{\frac{1}{2}}\left(\mathbf{S}_x^{\frac{1}{2}}\mathbf{A}^t\mathbf{A}\mathbf{S}_x^{\frac{1}{2}} + \gamma\mathbf{I}\right)\mathbf{S}_x^{\frac{1}{2}}\mathbf{A}^t. \qquad (36)$$

Furthermore, since the eigenvalues of $\mathbf{A}^t\mathbf{A}$ are > 0, so are the eigenvalues of $\mathbf{R}$. Why?

Exercise 7-6: Evaluate the covariance matrix of $\mathbf{R}(\mathbf{y} + \mathbf{n})$, where $\mathbf{y}$ is fixed and $\mathbf{n}$ has a covariance matrix $\sigma_N^2 \mathbf{I}$.

7.4 An example and discussion

I shall illustrate these ideas with, again, a 2×2 example. Take the case where $\mathbf{A}$ is diagonal:

$$\mathbf{A} = \begin{pmatrix} \lambda_1 & 0 \\ 0 & \lambda_2 \end{pmatrix}, \tag{37}$$

where λ_1 and λ_2 are non-negative, real numbers. By convention, we choose $\lambda_1 \geq \lambda_2$.

The inverse of $\mathbf{A}$ is

$$\mathbf{A}^{-1} = \begin{pmatrix} \lambda_1^{-1} & 0 \\ 0 & \lambda_2^{-1} \end{pmatrix}, \tag{38}$$

as the reader can prove by multiplying it by $\mathbf{A}$. Note that $\det(\mathbf{A}) = \lambda_1 \lambda_2 = 0$ if and only if $\lambda_2 = 0$ (if $\lambda_2 > 0$, then $\lambda_1 > 0$). If λ_2 is very small, however, $\mathbf{A}^{-1}$ is very close to the matrix

$$\mathbf{A}^{-1} = \begin{pmatrix} \lambda_1^{-1} & 0 \\ 0 & \infty \end{pmatrix}, \tag{39}$$

which is certainly not a very useful item.

However, let us look at the solution embodied in equation (25) as it applies to this case. Let σ_x^2 be the variance of each component of $\mathbf{x}$, and let σ_N^2 be the noise in the measurement of each component of $\mathbf{x}$. Then the covariance matrix of $\mathbf{x}$ is

$$\mathbf{S}_x = \begin{pmatrix} \sigma_x^2 & 0 \\ 0 & \sigma_x^2 \end{pmatrix} = \sigma_x^2 \begin{pmatrix} 1 & 0 \\ 0 & 1 \end{pmatrix}; \tag{40}$$

while

$$\mathbf{N} = \sigma_N^2 \begin{pmatrix} 1 & 0 \\ 0 & 1 \end{pmatrix}. \tag{41}$$

Then

$$\mathbf{A S_x A^t + N} = \sigma_x^2 \begin{pmatrix} \lambda_1^2 & 0 \\ 0 & \lambda_2^2 \end{pmatrix} + \sigma_N^2 \begin{pmatrix} 1 & 0 \\ 0 & 1 \end{pmatrix}$$

$$= \begin{pmatrix} \sigma_x^2 \lambda_1^2 + \sigma_N^2 & 0 \\ 0 & \sigma_x^2 \lambda_2^2 + \sigma_N^2 \end{pmatrix}. \tag{42}$$

The inverse given in equation (26) is, in this example

$$\mathbf{R} = \begin{pmatrix} \sigma_x^2 & 0 \\ 0 & \sigma_x^2 \end{pmatrix} \begin{pmatrix} \lambda_1 & 0 \\ 0 & \lambda_2 \end{pmatrix} \begin{pmatrix} \dfrac{1}{\sigma_x^2 \lambda_1^2 + \sigma_N^2} & 0 \\ 0 & \dfrac{1}{\sigma_x^2 \lambda_2^2 + \sigma_N^2} \end{pmatrix}. \tag{43}$$

Multiplying this out,

$$\mathbf{R} = \begin{pmatrix} \dfrac{\sigma_x^2 \lambda_1}{\sigma_x^2 \lambda_1^2 + \sigma_N^2} & 0 \\ 0 & \dfrac{\sigma_x^2 \lambda_2}{\sigma_x^2 \lambda_2^2 + \sigma_N^2} \end{pmatrix}. \tag{44}$$

Exercise 7-7: What happens when the noise becomes very small? When it becomes very large? What is the error in the retrieval? (Hint: evaluate **RA**.)

This is the central result of this book. The reader who has made the effort to get this far should think about it carefully and convince himself that he understands why the result has this form. Again, let me stress that this result depends on knowing both **S** and **N**. It cannot be achieved knowing only **A**.

A similar solution that we often find in the literature has the form

$$\mathbf{R} = \mathbf{A^t (A A^t + \alpha I)^{-1}}; \tag{45}$$

i.e., the covariance matrix **S** is taken to be the identity matrix. This form is called a *Tikhonov regularizer.*[3] Here, $\alpha > 0$ is a free parameter. Since the eigenvalues of $\mathbf{A A^t}$ are all ≥ 0, the matrix in parentheses cannot be singular. When the linear inverse problem is approached strictly from a mathematical point of view, as it was here, it makes sense to assume that all state vectors are equally likely and that the different variables are uncorrelated. Therefore it makes sense to assume that $\mathbf{S_x}$ (1986) = **I**. However, in most physical prob-

---

[3] See Beteró (1986, p 79); Groetsch (1984); or some of the articles in Sabatier (1987).

lems, some situations are more common than others, and we should make use of our knowledge of the statistics embodied in the covariance matrix $\mathbf{S}_x$.

I discussed shifting eigenvalues in Chapter 5. If the eigenvalues of $\mathbf{S}_x$ are $\lambda_1, \cdots, \lambda_n$, then the eigenvalues of $(\mathbf{S}_x + \alpha\mathbf{I})$ are $\lambda_1 + \alpha, \ldots, \lambda_n + \alpha$. So the inverse given by equation (45) has been achieved by shifting the eigenvalues of $\mathbf{S}$ by an amount sufficient to make $\mathbf{R}$ well behaved.

Exercise 7-8: Suppose that $\mathbf{S}_x = \sigma_x^2\mathbf{I}$ and $\mathbf{N} = \sigma_N^2\mathbf{I}$. Show that equation (26) gives the same value for $\mathbf{R}$ as equation (45), with $\alpha = \sigma_N^2 / \sigma_x^2$.

7.5 Singular value decomposition

I introduced singular value decomposition in §5.5. It is instructive to look at the singular value decomposition of equation (26). We can write the matrix $\mathbf{AS}_x^{1/2}$ in the form

$$\mathbf{AS}_x^{1/2} = \mathbf{UD}_x\mathbf{V}^t, \tag{46}$$

where the columns of $\mathbf{U}$ are orthogonal; $\mathbf{D}_x$ is diagonal; and $\mathbf{V}$ is orthogonal (both rows and columns, since $\mathbf{V}$ is square). With this definition, equation (26) becomes

$$\mathbf{R} = \mathbf{S}_x^{1/2}\mathbf{VD}_x\mathbf{U}^t\left(\mathbf{UD}_x\mathbf{V}^t \, \mathbf{VD}_x\mathbf{U}^t + \mathbf{N}\right)^{-1}. \tag{47}$$

Suppose that $\mathbf{N}$ is of the form $\alpha\mathbf{I}$. Substituting these into equation (47) and remembering that $\mathbf{V}^t\mathbf{V} = \mathbf{I}$,

$$\mathbf{R} = \mathbf{S}_x^{1/2}\mathbf{VD}_x\mathbf{U}^t\left(\mathbf{UD}_x^2\mathbf{U}^t + \alpha\mathbf{I}\right)^{-1}. \tag{48}$$

But $\mathbf{UD}_x\mathbf{D}_x\mathbf{U}^t + \alpha\mathbf{I} = \mathbf{U}(\mathbf{D}_x^2 + \alpha\mathbf{I})\mathbf{U}^t$, so

$$\mathbf{R} = \mathbf{S}_x^{1/2}\mathbf{VD}_x\mathbf{U}^t\left[\mathbf{U}\left(\mathbf{D}_x^2 + \alpha\mathbf{I}\right)\mathbf{U}^t\right]^{-1}. \tag{49}$$

Now note that, for any nonsingular matrices $\mathbf{P}$ and $\mathbf{Q}$, $(\mathbf{PQ})^{-1} = \mathbf{Q}^{-1}\mathbf{P}^{-1}$. Since we know that $\mathbf{U}^t\mathbf{U} = \mathbf{I}$, we can write

$$\mathbf{R} = \mathbf{S}_x^{1/2}\mathbf{VD}_x\left[\left(\mathbf{D}_x^2 + \alpha\mathbf{I}\right)\right]^{-1}\mathbf{U}^t. \tag{50}$$

Finally, we can let $\mathbf{D}_x = \mathrm{diag}(\lambda_1, \ldots, \lambda_n)$. Then $\mathbf{R}' \equiv \mathbf{V}^t\mathbf{S}^{-1/2}\mathbf{RU}$ has the simple form

$$\mathbf{R}' = \begin{pmatrix} \dfrac{\lambda_1}{\lambda_1^2 + \alpha} & & \\ & \ddots & \\ & & \dfrac{\lambda_n}{\lambda_n^2 + \alpha} \end{pmatrix}. \qquad (51)$$

Notice that, if none of the λ_i's is small, then $\lambda_i / (\lambda_i^2 + \alpha) \approx 1 / \lambda_i$, but when λ_i is small, the term α in the denominator keeps $\lambda_i / (\lambda_i^2 + \alpha)$ from being larger than $1 / \alpha$.

We can simplify things further if we define a new measurement variable $\mathbf{y}' \equiv \mathbf{U}^t\mathbf{y}$. Then

$$\mathbf{V}^t\mathbf{S}_x^{-1/2}\hat{\mathbf{x}} \; = \; \mathbf{V}^t\mathbf{S}_x^{-1/2}\mathbf{R}\mathbf{U}\mathbf{U}^t\mathbf{y} \; = \; \mathbf{R}'\mathbf{U}^t\mathbf{y} \; = \; \mathbf{R}'\mathbf{y}' \; = \; \hat{\mathbf{x}}'. \qquad (52)$$

This shows that the given problem can be reduced to one where the measurements are statistically independent and the variables x_i are estimated independently by the matrix $\mathbf{R}'$. This is accomplished by a combination of a rescaling of the independent variable $\mathbf{x}$ and a rotation of the basis vectors.

Therefore, it is possible to rewrite the problem in the following terms if $n = m$:

1. The components of $\mathbf{x}'$ vary independently; $\mathrm{var}(x_i') = \sigma_{x,i}'^2$.

2. The matrix $\mathbf{A} = \mathrm{diag}(\lambda_1, \dots, \lambda_n)$.

3. The components of $\mathbf{y}'$ also vary independently. Each component of $\mathbf{x}'$ is mapped to a single component of $\mathbf{y}'$.

4. If the noise covariance matrix is $\mathbf{N} = \sigma_N^2\mathbf{I}$, the system of equations has the simple form $y_i' = \lambda_i x_i' + n_i$, where n_i is a random variable and $\mathrm{var}(n_i) = \sigma_N^2$.

5. We can invert this system of equations by inspection:

$$\hat{x}_i \; = \; \frac{\sigma_{x,i}^2 \lambda_i y_i'}{\sigma_{x,i}^2 \lambda_i + \sigma_N^2}. \qquad (53)$$

7.6 A geometrical interpretation

Mathematical analysis is often unsatisfying because primates are, above all, visual beings: our brains are best at processing and interpreting information that is in the form of mental images. Therefore, it will be helpful to see if we can develop a geometrical interpretation of these concepts.

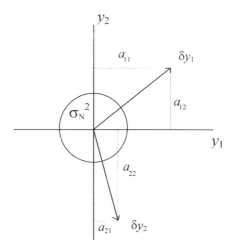

Figure 7-1. The vectors showing the change
in radiance due to a change in each variable.

Let us begin by re-stating the problem. There are two[4] variables, represented by the vector $\mathbf{x} = (x_1, x_2)$, and two measurements, denoted by $\mathbf{y} = (y_1, y_2)$. Assume that the two components of $\mathbf{x}$ vary independently. The vectors $\mathbf{x}$ and $\mathbf{y}$ are related by equation (1). One possible interpretation if this equation is that

$$a_{ij} = \frac{\partial y_i}{\partial x_j}. \tag{54}$$

Therefore, a change in variable x_1 produces changes proportional to a_{11} in y_1 and proportional to a_{21} in y_2. The effects of changing x_2 are similar. Geometrically, we can think of a change δx_1 producing a change in the *vector* $\mathbf{y}$. That is,

$$\delta y_i = \delta x_i \begin{pmatrix} a_{1i} \\ a_{2i} \end{pmatrix}. \tag{55}$$

This is illustrated in Figure 7-1.

Let the noise in the measurement of each component of $\mathbf{y}$ be σ_N^2. Each vector that we measure has a circle of radius σ_N^2 around it. This is also shown in Figure 7-1. (If the noise in the measurement of y_1 were different from that in y_2, this would be an ellipse.) Define δy_i to be the change in $\mathbf{y}$ due to a change in x_i. How well we can infer the value of $\mathbf{x}$ from the measurements depends on the magnitudes of δy_1 and δy_2; the angle between them; and the radius of the circle representing the noise. Qualitatively, the minimum detectable change $\delta \mathbf{x}$ that we can retrieve happens when the tips of δy_1 and δy_2 both lie on the circle

---

[4] Of course, in general there are more than two variables, but it won't help us at all to try to *visualize* a 5-dimensional problem, since we can't do it. At least, I can't.

with radius σ_N^2. Obviously, if they lie within that circle, we can say almost nothing about $\mathbf{x}$ from measurements of $\mathbf{y}$.

However, how well we can measure *both* components of $\mathbf{x}$ will depend on the angle between $\delta\mathbf{y}_1$ and $\delta\mathbf{y}_2$. If they are perpendicular, we can separate x_1 and x_2 well; if they are parallel, all we can measure is $x_1 + x_2$. This suggests the following idea.

Consider the pair of vectors $\delta\mathbf{y}_1$ and $\delta\mathbf{y}_2$, and

$$\delta\mathbf{y}_2' = \delta\mathbf{y}_2 - \frac{(\delta\mathbf{y}_1 \cdot \delta\mathbf{y}_2)}{|\delta\mathbf{y}_1|^2} \delta\mathbf{y}_1. \tag{56}$$

Exercise 7-9: Show that, when $\delta\mathbf{y}_1$ and $\delta\mathbf{y}_2$ are perpendicular, $\delta\mathbf{y}_2 = \delta\mathbf{y}_2'$ and when they are parallel, $\delta\mathbf{y}_2' = 0$. Therefore, when they are parallel, there is really but one piece of information present.

Another way to look at this is to realize that, if $\delta\mathbf{y}_1$ and $\delta\mathbf{y}_2$ are perpendicular, then the two columns of $\mathbf{A}$ are perpendicular, and $\mathbf{A}$ has rank 2 (the maximum for a 2×2 matrix). Similarly, if they are parallel, the matrix is singular (has rank 1) and we can retrieve only one quantity. The closer the angle between $\delta\mathbf{y}_1$ and $\delta\mathbf{y}_2$ is to 90°, the closer we get to being able to retrieve two independent components of $\mathbf{x}$.

Now consider this same problem in three dimensions. Suppose that we have the same two variables that we want to retrieve, x_1 and x_2, but there is also a third one (x_3) that we know exists, but that we cannot measure very well. In reality, when we considered the 2-dimensional problem, we were sweeping x_3 under the rug. Changes in x_3 will affect the values of $\mathbf{y}$, but we previously chose to ignore them. We had two equations in two unknowns: what could be easier to solve? Perhaps we were overly optimistic before. We should have made at least three independent measurements instead of just two. What happens when we make three measurements and include, at least implicitly, the effects of this third variable?

We now interpret the matrix $\mathbf{A}$ in equation (1) as being a 3×3 matrix and each vector as having three components. The basis vectors for $\mathbf{A}$ span a 3-dimensional space, and $\delta\mathbf{y}_1$ and $\delta\mathbf{y}_2$ are now 3-component vectors. As such, they might be perpendicular in 3-dimensional space, even though their projection onto the 2-dimensional subspace that we considered earlier is not. (Remember that *perpendicular* in this context means independent.) The retrieval of x_1 and x_2 from three measurements, even though we were taught that this is an overconstrained problem, will usually be better than the retrieval from only two measurements. In this case, we see that we might need to measure *three* variables y_i in order to retrieve two quantities, x_1 and x_2, well.

Another way we can look at this is to see that there will be situations where, although we measure three variables, we do not have enough information to retrieve more than one or two independent quantities. *In general, the belief that we can retrieve* n *independent variables from a set of* n *measurements is ill-founded. The number of independent variables we can retrieve is often much smaller, and may be equal to one.*

The unspecified component x_3 is a source of ambient noise, in the terminology of Chapter 1. It affects our measurements, but we do not take it into account explicitly. Usually, we are unsure of its precise nature, its mean value, or its degree of variability. Geometrically, changes in x_3 produce a vector $\delta\mathbf{y}_3$. The effect of changes in x_3 will be proportional to the cosine of the angle between $\delta\mathbf{y}_3$ and $\delta\mathbf{y}_i$, $i = 1$ or 2. If $\delta\mathbf{y}_3$ is perpendicular to the other two—we should be so lucky—the effect of changes in x_3 on the retrievals will be negligible. However, if $\delta\mathbf{y}_3$ has a substantial component parallel to $\delta\mathbf{y}_1$ or $\delta\mathbf{y}_2$, changes in y_3 will affect our estimates of x_1 and x_2.

7.7 Other matrix methods

So far, we have considered only one least-squares solution to equation (6), the one that we arrived at by minimizing the sum in equation (22). It depends on knowing the covariance matrix $\mathbf{S}$ and the noise in the measurements σ_N^2. There are other ways to find an inverse, however; I shall discuss some of them in this section. I would maintain, however, that the solution I have given in equation (26) is optimal provided that we know the relevant statistics.

If $\mathbf{A}$ is square, then the obvious solution might seem to be

$$\hat{\mathbf{x}} = \mathbf{A}^{-1}\mathbf{y}; \tag{57}$$

but I have already pointed out that the solution will be unstable if some of the eigenvalues of $\mathbf{A}$ are very small. If $\mathbf{A}$ is not square, the simplest solution is obtained by multiplying equation (1) by $\mathbf{A}^t$:

$$\mathbf{A}^t\mathbf{y} = \mathbf{A}^t\mathbf{A}\mathbf{x}. \tag{58}$$

$\mathbf{A}^t\mathbf{A}$ is a square matrix. Assume that it is nonsingular. Then we can estimate $\mathbf{x}$ from

$$\hat{\mathbf{x}} = \left(\mathbf{A}^t\mathbf{A}\right)^{-1}\mathbf{A}^t\mathbf{y}. \tag{59}$$

Note that both of these solutions ignore the effects of noise in the measurements. Now, even if $\mathbf{A}$ is square, we still could use equation (59) to estimate $\mathbf{x}$. Would that work better than using equation (57)? Twomey (1977) points out that it will generally give worse results, since the smallest eigenvalues of $\mathbf{A}^t\mathbf{A}$ will generally be smaller than the corresponding eigenvalues of $\mathbf{A}$, so the noise amplification will consequently be greater.

In Chapter 11, I describe an iterative method that will give the same result as the matrix method of equation (59). Neither one is stable in the presence of realistic amounts of noise, so we must look farther for stable solutions.

Exercise 7-10: Suppose that $\mathbf{A}$ is the 2×2 matrix $\begin{pmatrix} 1 & a \\ a & 1 \end{pmatrix}$. What are the eigenvalues of $\mathbf{A}^{-1}$ for $0 < a < 1$? What are the eigenvalues of $(\mathbf{A}^t\mathbf{A})^{-1}\mathbf{A}^t$?

Note that either equation (57) or equation (59) will yield a value of $\hat{\mathbf{x}}$ that satisfies equation (6); *i.e.*, it will satisfy

$$\mathbf{y} = \mathbf{A}\hat{\mathbf{x}} \qquad (60)$$

to within the accuracy of our computational capability.[5] However, to within the uncertainty in the measurement of $\mathbf{y}'$, there will be an infinite number of vectors $\mathbf{x}$ that will satisfy equation (60) about as well as the solution to equation (57) or equation (59) does. In other words, the solution will not be unique because of the effect of noise in the measurement $\mathbf{y}'$. In order to make the solution unique, we need to use a constraint of some kind. This constraint will be arbitrary; we cannot tell what the constraint is from the measurements themselves. We must pick, on theoretical or aesthetic grounds, a criterion for deciding which solution is the best.

7.7.1 Twomey's Solutions

Twomey[6] suggested the following approaches, among others. He did not assume that the variables had zero mean. So, for the rest of this section, let us assume that the mean value of $\mathbf{x}$ is $<\mathbf{x}> \neq 0$. One possible criterion of which is the "best" solution is to choose the *smoothest* vector that will solve equation (6).[7] We could measure the smoothness of the vector $\mathbf{x} = \mathrm{col}(x_1,\ldots,x_n)$ from the mean square difference between adjacent components of $\mathbf{x}$:

$$\eta^2 \equiv \sum_{i=1}^{n-1}\left(x_{i+1} - x_i\right)^2. \qquad (61)$$

This can be represented by a matrix operation.[8] Define the matrix

---

[5] In this context, round-off errors in the computations are rarely a significant problem.

[6] Twomey (1977), Chapter 6.

[7] Twomey (1963) based this solution on an article by Phillips (1962). Note that Foster's solution (1967), which I believe to be the fundamental work in this field, was then several years in the future.

[8] Twomey (1977, pp 124 *ff*).

$$\mathbf{K} \equiv \begin{pmatrix} 0 & 0 & 0 & 0 & \cdots \\ 1 & -1 & 0 & 0 & \cdots \\ 0 & 1 & -1 & 0 & \cdots \\ 0 & 0 & 1 & -1 & \cdots \\ \vdots & \vdots & \vdots & \vdots & \ddots \end{pmatrix}. \tag{62}$$

Then

$$\mathbf{Kx} = \mathrm{col}\left[0,(x_2 - x_1),\cdots,(x_n - x_{n-1})\right]. \tag{63}$$

Consequently,

$$\eta^2 = \left|\mathbf{Kx}\right|^2 = \mathbf{x}^t\mathbf{K}^t\mathbf{Kx} \tag{64}$$

To find the smoothest solution to equation (6), we find the value of $\mathbf{x}$ that minimizes $\lambda\eta + |\mathbf{A\hat{x}} - \mathbf{y}|^2$.[9] Here, λ is a constant that we introduce to make η and $\mathbf{A\hat{x}} - \mathbf{y}$ comparable quantities. Then

$$\chi^2 = (\mathbf{A\hat{x}} - \mathbf{y})^t(\mathbf{A\hat{x}} - \mathbf{y}) + \lambda\eta^2$$

$$= (\mathbf{A\hat{x}} - \mathbf{y})^t(\mathbf{A\hat{x}} - \mathbf{y}) + \lambda\,\mathbf{\hat{x}}^t\mathbf{K}^t\mathbf{K\hat{x}}. \tag{65}$$

(Note here that χ is a scalar.) Expand the first term on the r.h.s. of this equation:

$$\chi^2 = \mathbf{\hat{x}}^t\mathbf{A}^t\mathbf{A\hat{x}} - \mathbf{y}^t\mathbf{A\hat{x}} - \mathbf{\hat{x}}^t\mathbf{A}^t\mathbf{y} + \mathbf{y}^t\mathbf{y} + \lambda\,\mathbf{\hat{x}}^t\mathbf{K}^t\mathbf{K\hat{x}}. \tag{66}$$

Then take the derivative of χ^2 with respect to $\mathbf{\hat{x}}$ and set the result equal to 0 (using the method of §14.2):

$$\mathbf{\hat{x}}^t\mathbf{A}^t\mathbf{A} + \mathbf{A}^t\mathbf{A\hat{x}} - \mathbf{y}^t\mathbf{A} - \mathbf{A}^t\mathbf{y} + \lambda\mathbf{K}^t\mathbf{K\hat{x}} + \lambda\mathbf{\hat{x}}^t\mathbf{K}^t\mathbf{K} = 0. \tag{67}$$

We can rearrange this to get

$$\left(\mathbf{A}^t\mathbf{A\hat{x}} - \mathbf{A}^t\mathbf{y} + \lambda\mathbf{K}^t\mathbf{K\hat{x}}\right) + \left(\mathbf{A}^t\mathbf{A\hat{x}} - \mathbf{A}^t\mathbf{y} + \lambda\mathbf{K}^t\mathbf{K\hat{x}}\right)^t = 0. \tag{68}$$

This equation is satisfied if

---

[9] Twomey (1977, p 136) claims that he is using a Lagrange multiplier λ and minimizing η subject to the constraint $|\mathbf{Ax} - \mathbf{y}|^2 = \varepsilon^2$, where ε^2 is the noise variance. Notice the equality in this last equation. However, what he *says* he is doing does not make any sense and it does not seem to correspond to the equation he actually derives. Note that his solution does not, in fact, depend on ε^2.

$$\left(\mathbf{A^tA} + \lambda\mathbf{K^tK}\right)\hat{\mathbf{x}} = \mathbf{A^ty} \qquad (69)$$

or

$$\hat{\mathbf{x}} = \left(\mathbf{A^tA} + \lambda\,\mathbf{K^tK}\right)^{-1}\mathbf{A^ty}. \qquad (70)$$

Note that the last two terms in equation (68) are just the transpose of the first two.

Twomey[10] says that

> This is the equation for constrained linear inversion. The usual procedure for applying this equation is to choose several values for γ [= our λ], and then post-facto decide the most appropriate value for γ by computing the residual [$\mathbf{Ax} - \mathbf{y}$]; if this is appreciably larger than the overall error in [$\mathbf{x}$] due to all causes…then γ is too large….

I disagree with this philosophy. If we know the statistics of the problem at hand, we know how to pick λ to optimize the results. We should not have to pick the value of a parameter by trial and error. Also, this solution is not general enough: we should use equation (26) in most circumstances.

There are some subtle differences between equations (18) and (65). Firstly, the former equation involves an error *matrix*, while the latter involves a scalar (*i.e.*, one takes an outer product; the other, an inner product). The former method minimizes each component of error $|\mathbf{x} - \mathbf{R}\hat{\mathbf{x}}|^2$, which seems to me to be the logical thing to do. This explicitly includes the effects of measurement noise. The second method minimizes a scalar χ^2 given by equation (65). In this case, the noise does not enter explicitly. If we modified equation (65) to include a noise term, we would have to evaluate $\langle\mathbf{n}\cdot\mathbf{n}\rangle$, $\langle\hat{\mathbf{x}}\cdot\hat{\mathbf{x}}\rangle$, and $\langle\mathbf{n}\cdot\hat{\mathbf{x}}\rangle$. But since $\langle\mathbf{n}\cdot\hat{\mathbf{x}}\rangle \neq 0$, we cannot solve the problem as it was originally formulated: taking a derivative with respect to $\hat{\mathbf{x}}$ would now involve some terms of the form $\partial\hat{\mathbf{x}}/\partial\mathbf{n}$, which of course is unknown. Although minimizing χ^2 in equation (65) produces a familiar result, it is not clear why it actually makes sense to minimize this quantity.

There is nothing special about minimizing $\Sigma(x_i - x_{i-1})^2$. Twomey[11] suggests several other possibilities (even minimizing third differences). We could replace $\mathbf{K^tK}$ with $\mathbf{I}$; this is the equivalent of taking $\eta^2 = |\mathbf{x}|^2$. Therefore, the solution that minimizes $|\mathbf{x}|^2$ is

$$\hat{\mathbf{x}} = \left(\mathbf{A^tA} + \lambda\mathbf{I}\right)^{-1}\mathbf{A^ty}. \qquad (71)$$

By now, this equation should be an old friend.

---

[10] Twomey (1977), p 127.

[11] Twomey (1977).

Another solution that Twomey offers,[12] often called the Twomey-Phillips solution,[13] is

$$\hat{\mathbf{x}} = \langle\mathbf{x}\rangle + \mathbf{A}^t \left(\mathbf{A}\mathbf{A}^t + \beta\mathbf{I}\right)^{-1}\left(\mathbf{y} - \mathbf{A}\langle\mathbf{x}\rangle\right)$$

$$= \langle\mathbf{x}\rangle + \left(\mathbf{A}^t\mathbf{A} + \beta\mathbf{I}\right)^{-1}\mathbf{A}^t\left(\mathbf{y} - \mathbf{A}\langle\mathbf{x}\rangle\right). \tag{72}$$

Here[14] β is a scalar. Twomey arrives at this solution by minimizing $|\hat{\mathbf{x}} - \langle\mathbf{x}\rangle|$. Note the form that this will have if, as usual, we change variables so that $\langle\mathbf{x}\rangle = 0$. I introduce this form only because of its relevance to the discussion of iteration in Chapter 11.

The reader will note that Twomey offers a variety of solutions to equation (6), minimizing different things. He is correct in stating that the problem of finding solution to $\mathbf{y} = \mathbf{A}\mathbf{x}$ is not well defined until we pick a quantity to minimize. But he does not supply any *physical* insight to tell us what value of λ to use. I should emphasize that that choice should be determined by physical or statistical considerations.

7.7.2 What if we do not know the statistics of $\mathbf{x}$?

What happens if we do not know the statistics involved? In many situations, we do not know the covariance matrix of $\mathbf{x}$. This problem is always easily solved.[15]

First, let me point out that every remote-sensing instrument involves a sensor of some kind that measures the amount of radiation emitted or reflected by some object. If we cover the sensor, isolating it from the thing we want to measure, we have zero input and only the instrument noise enters our measurements. Therefore, we can easily find the noise covariance matrix $\mathbf{N}$ from such a set of measurements. Let us assume that there are several channels of data, and that the noise is uncorrelated between different channels. If necessary, we can rescale the measurements so that the noise is the same in every channel. Therefore, we can measure

$$\mathbf{N} = \sigma_N^2 \mathbf{I}. \tag{73}$$

Now since $\mathbf{y} = \mathbf{A}(\mathbf{x} + \mathbf{n})$, the covariance matrix of $\mathbf{y}$ is

---

[12] Twomey (1977), pp 137 *ff.*

[13] See Fleming (1976).

[14] As before, $\langle\mathbf{x}\rangle$ is the mean value of $\mathbf{x}$, not necessarily $= 0$.

[15] At least in theory.

$$\mathbf{S}_y \equiv \langle \mathbf{yy}^t \rangle = \langle \mathbf{A}(\mathbf{x}+\mathbf{n})[\mathbf{A}(\mathbf{x}+\mathbf{n})]^t \rangle$$

$$= \langle (\mathbf{Ax}+\mathbf{n})(\mathbf{Ax}+\mathbf{n})^t \rangle = \mathbf{AS}_x\mathbf{A}^t + \sigma_N^2\mathbf{I}. \qquad (74)$$

Here, $\mathbf{n}$ is a random noise vector that is independent of $\mathbf{x}$. So we measure the covariance matrix of the measurements $\mathbf{y}$. Furthermore, $\mathbf{AS}_x\mathbf{A}^t = \mathbf{S}_y - \sigma_N^2\mathbf{I}$. Since we know the matrix $\mathbf{A}$, we can determine the covariance matrix $\mathbf{S}_x$, at least in part.[16]

To find $\mathbf{S}_x$, we can write

$$\mathbf{A}^t\mathbf{AS}_x\mathbf{A}^t\mathbf{A} = \mathbf{A}^t\mathbf{S}_y\mathbf{A} - \sigma_N^2\mathbf{A}^t\mathbf{A}. \qquad (75)$$

If $(\mathbf{A}^t\mathbf{A})^{-1}$ exists, then

$$\mathbf{S}_x = (\mathbf{A}^t\mathbf{A})^{-1}\mathbf{A}^t\mathbf{S}_y\mathbf{A}(\mathbf{A}^t\mathbf{A})^{-1} - \sigma_N^2(\mathbf{A}^t\mathbf{A})^{-1}. \qquad (76)$$

Exercise 7-11: We have already discussed why $(\mathbf{A}^t\mathbf{A})^{-1}$ may be ill-conditioned. How does that affect our estimate of $\mathbf{S}_x$? How does it affect our ability to use equation (26) to meas $\mathbf{S}_y \equiv \langle \mathbf{yy}^t \rangle = \langle \mathbf{A}(\mathbf{x}+\mathbf{n})[\mathbf{A}(\mathbf{x}+\mathbf{n})]^t \rangle$ re $\mathbf{x}$? Remember that $\mathbf{S}_y$ is estimated from a large ensemble of measurements of $\mathbf{y}'$.

Suppose, for the sake of argument, that the situation is either black or white: each component of $\mathbf{x}$ either maps onto an observable component of $\mathbf{y}$ (*i.e.,* one where the variance of $\mathbf{x}$ changes $\mathbf{y}$ more than the noise does) or it maps to $\mathbf{0}$. As we have seen, we can always decompose $\mathbf{x}$ into orthogonal components, one that $\mathbf{A}$ maps to observable components of $\mathbf{y}$, and one that is contained in the null space (maps to $\mathbf{0}$). Call these components $\mathbf{x}_k$ and $\mathbf{x}_n$. We obviously cannot use this technique to measure the covariance matrix of $\mathbf{x}_n$, but we cannot measure $\mathbf{x}_n$ either—because it has no effect on the measurements—so we do not really care about its covariance matrix. However, we can measure the covariance matrix of $\mathbf{x}_k$, and that is exactly what we need to solve the inverse problem.

In the real world, there will be some shades of gray, and the individual researcher will have to decide where to place the boundaries. The definition of the null space has to be an operational one: in reality, *every* component of $\mathbf{x}$ will affect $\mathbf{y}$, at least to some extent, even if it is not readily measured. So the null space of $\mathbf{A}$ may not exist in the strict mathematical sense (although it will

---

[16] Note that we cannot do this if we do not know σ_N^2. Also, note that I have assumed that the noise covariance matrix is a scalar times the identity matrix.

have to if there are more unknowns than measurements). But, as we have seen, we can always rotate the bases of the independent-variable space and the measurement space so as to diagonalize $\mathbf{S}_x$ and $\mathbf{S}_y$, and therefore diagonalize $\mathbf{A}$ and $\mathbf{R}$, so that the components of $\mathbf{x}_k$ are each retrieved from a single component of $\mathbf{y}$, in these bases. As a rule, we should probably use all components of $\mathbf{x}$ that have signal-to-noise ratios $\lambda_i^{-1} \geq \sigma_x^2 / \sigma_N^2$.

7.8 Pseudoinverses

Here is another way to view the problem of inverting systems of linear equations. It is possible to find a *pseudoinverse*[17] of $\mathbf{A}$, denoted by $\mathbf{A}^\dagger$, that has the property that

$$\mathbf{A}^\dagger \mathbf{A} \approx \mathbf{I} \tag{77}$$

(I shall explain the symbol $\approx$ in a moment). There may be another pseudoinverse $\mathbf{B}$ such that $\mathbf{AB} \approx \mathbf{I}$, and where, in general, $\mathbf{A}^\dagger \neq \mathbf{B}$. That is, there may be distinct left and right pseudoinverses.

There are many ways to generate pseudoinverses, but one way that is of interest here is to use the singular value decomposition of $\mathbf{A}$. Remember that, for $n \geq m$, we can write an $n \times m$ matrix $\mathbf{A}$ in the form $\mathbf{A} = \mathbf{UDV}$, where $\mathbf{U}$ is an $n \times m$ matrix and the columns of $\mathbf{U}$ are orthogonal; $\mathbf{D}$ is an $m \times m$ diagonal matrix; and $\mathbf{V}$, an $m \times m$ orthogonal matrix. Let $\mathbf{D} = \mathrm{diag}(\lambda_1, \ldots, \lambda_m)$. Define the matrix $\mathbf{D}^\dagger$ to be

$$\mathbf{D}^\dagger = \mathrm{diag}\left(\lambda_1^{-1}, \ldots, \lambda_p^{-1}, 0, \ldots, 0\right), \tag{78}$$

where there are $m - p$ zeroes at the end. We choose p so that $\lambda_p^{-1} > c$, a constant chosen to limit the size of λ_i^{-1}. Note that

$$\mathbf{A}^\dagger \mathbf{A} = \mathbf{V}' \mathbf{D}^\dagger \mathbf{U}' \mathbf{UDV} = \mathbf{V}' \mathbf{D}^\dagger \mathbf{DV}, \tag{79}$$

and

$$\mathbf{D}^\dagger \mathbf{D} = \mathrm{diag}\left(\lambda_1^{-1}, \ldots, \lambda_p^{-1}, 0, \ldots, 0\right)$$

$$\times \mathrm{diag}\left(\lambda_1, \ldots, \lambda_p, \lambda_{p+1}, \ldots, \lambda_m\right) = \mathbf{I}_p, \tag{80}$$

where

---

[17] I have already discussed this briefly in §5.5.

$$\mathbf{I}_p = \mathrm{diag}(\underbrace{1, \ldots, 1}_{p \ 1's}, \underbrace{0, \ldots, 0}_{m-p \ 0's}) \tag{81}$$

Since $\mathbf{V}^t\mathbf{V} = \mathbf{I}$,

$$\mathbf{A}^{\dagger}\mathbf{A} = \mathbf{I}_p . \tag{82}$$

In other words, $\mathbf{A}^{\dagger}\mathbf{A}$ maps some components of $\mathbf{y}$ to $\hat{\mathbf{x}}$, but maps others to $\mathbf{0}$, depending on the size of each eigenvector. This eliminates the smallest eigenvectors of $\mathbf{A}$ and regularizes the solution.

Compare this with what we derived in §0.3. Here, the eigenvalues of $\mathbf{A}$ were shifted so that $\lambda_i \to \lambda_i + \sigma_N^2$. This transforms the problem so that no eigenvalue of $\mathbf{A}$ is less than σ_N^2, thus regularizing the problem in another way. In principle, these two approaches should lead to similar results.

7.9 Extension to quadratic terms

This section examines two methods for inverting systems of quadratic equations.[18] Up to now, it has been customary to linearize the equations that arise in remote sensing, although nonlinear methods have been applied in some areas like measuring the temperature of the atmosphere.[19] Here, I will extend the matrix methods used to solve systems of linear equations and show how to include quadratic terms explicitly. Including quadratic terms may allow us both to use more accurate inversion methods and to make better estimates of accuracy of proposed new remote-sensing instruments. Also, this method provides a way to estimate the *error* that we would make if we ignored quadratic terms. This last item by itself would make this analysis worthwhile.

Putting the problem in a general form, we are given a physical situation where some set of dependent variables, denoted $\mathbf{y} = \mathrm{col}(y_1, \ldots, y_m)$, is related in a known manner to a set of independent variables, $\mathbf{x} = \mathrm{col}(x_1, \ldots, x_n)$, by the system of equations

$$\mathbf{y} = f(\mathbf{x}) \quad \text{or} \quad y_i = f_i(\mathbf{x}), \tag{83}$$

where $f = \mathrm{col}(f_1, \ldots, f_m)$ is an m-valued function of $\mathbf{x}$. We want to infer the values $\mathbf{x}$ from measurements of $\mathbf{y}$. Assume that the problem is not underdetermined; *i.e.*, $m \geq n$; consideration of such systems would add nothing to the present discussion.

---

[18] This is a slight modification of Milman (1997).

[19] *E.g.*, see Twomey *et al.* (1977); Chahine (1970); and Rogers (1976); as well as the references in Houghton *et al.* (1984).

The measurements of $\mathbf{y}$ will contain noise that we represent by the random vector $\mathbf{n}$, which has zero mean. For the sake of simplicity, we shall assume that the components of $\mathbf{n}$ vary independently with variance σ_N^2. We seek the function $\boldsymbol{r}(\mathbf{y})$ that produces the best estimate of $\mathbf{x}$; *i.e.*,

$$\hat{\mathbf{x}} = \boldsymbol{r}(\mathbf{y} + \mathbf{n}), \tag{84}$$

where $\boldsymbol{r}$ is an n-valued function of $\mathbf{y}$. If $\boldsymbol{f}$ is a linear function, we can write equation (83) in the matrix form $\mathbf{y} = \mathbf{A}\mathbf{x}$. Then the inversion problem is one of finding a matrix[20] $\mathbf{R}$ to estimate $\hat{\mathbf{x}} = \mathbf{R}(\mathbf{y} + \mathbf{n})$.

Sometimes nonlinear problems are treated by linearizing equation (83). As an example, Kerekes[21] has analyzed the accuracy of temperature and water vapor soundings from infrared radiometers aboard satellites. He has treated only the linear part of the problem, although both the absolute humidity and the absorption coefficients of the atmospheric constituents depend strongly on temperature. It is not clear, until we calculate it, how large an error is involved in neglecting the quadratic terms.

There is a large literature on inversion of systems of linear equations. Some treat nonlinear equations; mostly, they invert them by iterating a linearized version of the problem.[22] It is hard to analyze the situation when there are no restrictions on the form of $\boldsymbol{f}$, and to determine whether or not an iterative process converges to a unique solution. If we treat $\boldsymbol{f}$ as a vector—a function measured at only a discrete number of points—and include measurement errors, then we know *a priori* that the solution is *not* unique.

I show here how to extend the linear analysis to include quadratic terms, while still minimizing the effects of random noise in the measurements. The method is general enough that it should apply to a wide class of problems: those where each f_i can be approximated by a quadratic equation. I shall offer two methods of dealing with quadratic equations: changing the independent variable and iteration of a nonlinear equation. Both of these methods have been applied to general nonlinear problems, and to some specific remote-sensing problems. But by concentrating on quadratic equations, we can do three things: estimate the *error* involved if we linearize the problem; devise an iterative scheme for improving on the linearized solution; and determine in a simple way whether the iteration converges to a unique solution. As far as I know, the methods presented below represent a new approach.

To study nonlinear functions, we can expand $\boldsymbol{f}$ in a Taylor series; replace $\mathbf{x}$ with $\mathbf{x} - <\mathbf{x}>$, and replace $\boldsymbol{f}(\mathbf{x})$ with $\boldsymbol{f}(\mathbf{x}) - \boldsymbol{f}(<\mathbf{x}>)$. Expanding to second order about $<\mathbf{x}> = \mathbf{0}$ and using component notation,

---

[20] I use lowercase, bold, italic letters for general functions and uppercase bold letters for matrices. Lowercase bold letters represent vectors.

[21] Kerekes, (1993).

[22] *E.g.*, see Parker (1977) or Houghton *et al.* (1984).

$$f_i(\mathbf{x}) = \sum_{j=1}^{n} \alpha_{ij}\, x_j \;+\; \sum_{j,k=1}^{n} \beta_{ijk}\, x_j\, x_k, \tag{85}$$

where

$$\alpha_{ij} = \frac{\partial f_i}{\partial x_j} \quad \text{and} \quad \beta_{ijk} = \frac{\partial^2 f_i}{\partial x_j\, \partial x_k}. \tag{86}$$

Note that, because f is not linear, $<f> \neq f(<\mathbf{x}>)$.

7.9.1 Quadratic scalar function

Before analyzing systems of nonlinear equations, I want to consider a scalar, nonlinear function of a vector $\mathbf{x}$. The point of this section is to form a heuristic bridge to the material that follows; it represents an underdetermined problem, which has no real application to physical systems.

Return to equation (85), but assume that $m = 1$. The coefficients $\alpha_{\beta ij}$ become a vector $\mathbf{a}$ and the coefficients β_{ijk}, a matrix $\mathbf{B}$, since there is but one value of i. Then equation (85) becomes

$$f(\mathbf{x}) = \mathbf{a} \cdot \mathbf{x} + (\mathbf{Bx}) \cdot \mathbf{x}. \tag{87}$$

Since the matrix $\mathbf{B}$ is symmetric—it represents the second-order partial derivatives of f—there exists a unitary matrix $\mathbf{U}$ such that

$$\mathbf{D} = \mathbf{UBU}^t, \tag{88}$$

where $\mathbf{D}$ is a diagonal matrix.[23] Then let $\mathbf{v} = \mathbf{Ux}$; this is a rotation of the basis vectors in n-space. Substituting this, along with equation (88), into equation (87),

$$f(\mathbf{v}) = (\mathbf{Ua}) \cdot \mathbf{v} + (\mathbf{Dv}) \cdot \mathbf{v}. \tag{89}$$

Write $\mathbf{D} = \text{diag}(\lambda_1, \dots, \lambda_n)$ and define $\mathbf{g} = \mathbf{Ua}$. We get a much simpler equation

$$f(\mathbf{v}) = \sum_{i=1}^{n} g_i v_i + \lambda_i v_i^2. \tag{90}$$

Transforming the independent variables further by

---

[23] $\mathbf{D}$ is a generic name for a diagonal matrix; I shall introduce a different matrix $\mathbf{D}$ in the next section.

$$s_i = v_i + \frac{\lambda_i}{g_i} v_i^2, \tag{91}$$

this reduces equation (87) to an equation that is linear in **s**:

$$f(\mathbf{s}) = \mathbf{g} \cdot \mathbf{s}. \tag{92}$$

Therefore, a nonlinear change of variables has transformed the quadratic equation into a linear one, and we need to solve only some quadratic equations to find **v**.

7.9.2 Quadratic vector function

When we now consider a nonlinear vector function

$$\mathbf{f}(\mathbf{x}) = \mathbf{A}\mathbf{x} + (\underline{\mathbf{B}}\mathbf{x}) \cdot \mathbf{x} \tag{93}$$

or, in component notation,

$$f_i(x) = \sum_j \alpha_{ij} x_j + \sum_{j,k} \beta_{ijk} x_j x_k. \tag{94}$$

Here, $\underline{\mathbf{B}}$ is a third-order tensor, denoted by an underscore, whose components are β_{ijk}. The last term in equation (94) serves as a definition for the notation $(\underline{\mathbf{B}}\mathbf{x}) \cdot \mathbf{x}$; note that this represents a vector quantity.

7.9.3 A transformation

We can transform the problem as follows; this will make it easier to see what is happening. Multiply each side of equation (93) by $(\mathbf{A}^t\mathbf{A})^{-1}\mathbf{A}^t$. Then

$$(\mathbf{A}^t\mathbf{A})^{-1}\mathbf{A}^t\mathbf{f} = \mathbf{x} + (\mathbf{A}^t\mathbf{A})^{-1}\mathbf{A}^t[(\underline{\mathbf{B}}\mathbf{x}) \cdot \mathbf{x}]. \tag{95}$$

There is a unitary matrix **U** that diagonalizes $\mathbf{S}_x = \langle \mathbf{x}\mathbf{x}^t \rangle$, just as we diagonalized **B** in equation (88). For any $n \times n$ matrix **U** and vectors **p** and **q** of length n, $(\mathbf{U}\mathbf{p})(\mathbf{U}\mathbf{p})^t = \mathbf{U}(\mathbf{p}\mathbf{q}^t)\mathbf{U}^t$, as we can easily prove by writing it in component notation. Define $\mathbf{v} = \mathbf{U}\mathbf{x}$. The covariance matrix of **v** is a diagonal matrix $\mathbf{U}\mathbf{S}\mathbf{U}^t = \mathbf{D}$. This transforms the basis vectors so that the components of **v** vary independently. We shall make use of this property shortly.

Multiply each side of equation (95) by **U** and substitute $\mathbf{U}^t\mathbf{v}$ for **x**. Also, define $\mathbf{Q} = \mathbf{U}(\mathbf{A}^t\mathbf{A})^{-1}\mathbf{A}^t$. In component notation, the ith element of the vector $\mathbf{Q}[(\underline{\mathbf{B}}\mathbf{U}^t\mathbf{v}) \cdot (\mathbf{U}^t\mathbf{v})]$ is

$$\sum_p q_{ip} \sum_{q,r} \beta_{pqr} \left(\sum_j u_{qj} v_j \right) \left(\sum_k u_{rk} v_k \right) = \sum q_{ip} \beta_{pqr} u_{qj} u_{rk} v_j v_k. \qquad (96)$$

Define $\underline{\mathbf{C}}$ by

$$c_{ijk} = \sum_{p,q,r} q_{ip} \beta_{pqr} u_{qj} u_{rk} \qquad (97)$$

where $\underline{\mathbf{C}}$, like $\underline{\mathbf{B}}$, is a third-order tensor; also, define $\mathbf{p} = \mathbf{Q}f$. Then we get the simple equation

$$\mathbf{p} = \mathbf{v} + \left[(\underline{\mathbf{C}}\mathbf{v}) \cdot \mathbf{v} \right]. \qquad (98)$$

Remember that $\mathbf{U}$ depends only on the statistics of $\mathbf{x}$, while $\mathbf{Q}$ depends both on $\mathbf{U}$ and the linear part of equation (85), the matrix $\mathbf{A}$. If this were a set of linear equations—if $\underline{\mathbf{C}}$ were $\mathbf{0}$—we would use the linear solution, with the covariance matrices modified appropriately. Since we transformed f to $\mathbf{p} = \mathbf{Q}f$, the noise covariance matrix becomes $<\mathbf{Q}\mathbf{n}\mathbf{n}^t\mathbf{Q}^t> = \sigma_N^2 \mathbf{Q}\mathbf{Q}^t$, since each measurement involves an error $\mathbf{Q}\mathbf{n}$. Furthermore, $\mathbf{A} = \mathbf{I}$. Then the solution to the linear equation $\mathbf{p} = \mathbf{v}$ is

$$\mathbf{v}' = \mathbf{D} \left[\mathbf{D} + \sigma_N^2 \mathbf{Q}\mathbf{Q}^t \right]^{-1} \mathbf{p}. \qquad (99)$$

This is the commonly used inverse for a linear problem. Equation (98) requires an extension of this idea to nonlinear equations.

In the above derivation, I assumed that the matrix $\mathbf{A}^t\mathbf{A}$ is not singular. If it is, then there is at least one component of $\mathbf{x}$ that cannot be retrieved from measurements of $\mathbf{y}$. If we delete the unmeasurable components of $\mathbf{x}$, $\mathbf{A}^t\mathbf{A}$ will no longer be singular.

7.10 Two approaches

So far, we have transformed equation (85) into a simpler form, but we still have no general solution to the quadratic problem. I can offer two different approaches. In each case, it is possible to estimate the errors due to noise in the measurements and to neglecting some quadratic terms.

7.10.1 Exact solution if all off-diagonal terms are zero

One possibility is to keep only the terms in equation (97) that are of the form c_{iii}, and setting all other terms to zero. If we are willing to make this approximation, then we can use equation (91) to transform the independent variable v; this will transform equation (98) into a linear equation that we can solve. To accomplish this, let

$$s_j = v_j + c_{jjj}v_j^2. \tag{100}$$

Then we can replace equation (98) with

$$\mathbf{p} = \mathbf{s} \tag{101}$$

plus terms that contain c_{jkl} where $k \neq j$ and $l \neq j$, which we ignore in this approximation.

If we do this, we should estimate the consequences of ignoring the other quadratic terms, which will contribute to the error in the inversion.[24] We can model this error as a kind of noise that arises from the nonlinear terms we have ignored. Define $c'_{ijk} = c_{ijk}$ if $i \neq j$ or $i \neq k$; and $c'_{iii} = 0$ (i.e., the terms we did not ignore are set to zero). These noise terms have the form

$$n'_i = \sum_{j,k} c'_{ijk} v_j v_k; \tag{102}$$

the covariance matrix has components $\sigma_{ij}^2 = <n_i'n_j'> - <n_i'><n_j'>$ given by

$$\sigma'^2_{ij} \equiv \sum_{k,m} c'_{ikm} \sum_{p,q} c'_{jpq} \langle v_k v_m v_p v_q \rangle - \left(\sum_{k,m} c'_{ikm} \langle v_k v_m \rangle \right) \left(\sum_{p,q} c'_{jpq} \langle v_p v_q \rangle \right). \tag{103}$$

Now the components of $\mathbf{v}$ vary independently; this means that $<v_i v_j> = 0$ if $i \neq j$. Let $<v_j^2> = \lambda_j$. If each component of $\mathbf{v}$ is a normally distributed random variable, then $<v_k v_m v_p v_q> = \lambda_k \lambda_p$ if $k = m$ and $p = q$; $<v_k v_m v_p v_q> = \lambda_k \lambda_m$ if $k = p$ and $m = q$ or $k = q$ and $m = p$; and $<v_k v_m v_p v_q> = 0$ otherwise. Using these relationships,

$$\sigma'^2_{ij} \equiv \sum_{k,p} c'_{ikk} c'_{jpp} \lambda_k \lambda_p + \sum_{k,m} c'_{ikm} c'_{jkm} \lambda_k \lambda_m$$

$$+ \sum_{k,m} c'_{imk} c'_{jkm} \lambda_k \lambda_m - \left(\sum_k c'_{ikk} \lambda_k \right) \left(\sum_p c'_{jpp} \lambda_p \right). \tag{104}$$

Finally, since $c_{jmk} = c_{mkm}$,

---

[24] I treated this problem (Milman, 1987) in the context of measuring wind speed over the ocean with microwave data.

$$\sigma_{ij}'^2 = 2\sum_{k,m} c_{ikm}' c_{jkm}' \lambda_k \lambda_m. \tag{105}$$

Note that the λ_i are just the eigenvalues of $\mathbf{S}_x$. To estimate the error in this inversion procedure, we can add the errors expressed in equation (105) to the noise covariance matrix $\mathbf{N}' = \sigma_N^2 \mathbf{QQ}^t$, find the inverse matrix using equation (99), and estimate the error from equation (32). This method has the advantage that we can evaluate the error in the nonlinear retrieval in a straightforward manner. It has the disadvantage that we have neglected some of the quadratic terms.

7.10.2 Iterated solution using all of the quadratic terms

As an alternative to the method in the last section, we can keep all of the quadratic terms and find an iterated solution to equation (98). To begin with, note that, if the quadratic terms were very small, we would have simply the solution given in equation (99). Using equation (32), the error in the estimated value of $\mathbf{v}$ will be

$$\chi^2 \approx \mathbf{D}\left[\mathbf{I} - \mathbf{D}\left(\mathbf{D} + \sigma_N^2 \mathbf{QQ}^t\right)^{-1}\right] = \sigma_N^2 \mathbf{DQQ}^t\left(\mathbf{D} + \sigma_N^2 \mathbf{QQ}^t\right)^{-1}. \tag{106}$$

The error that we would make by ignoring second-order terms is approximately equal to the variance of the term $(\underline{\mathbf{C}}\mathbf{v})\cdot \mathbf{v}$ that we neglected in the derivation of equation (99). It is approximately

$$\sigma_i'^2 \equiv \sum_{j,k} c_{ijk}\left(\sum_{l,m} c_{jlm} v_l v_m\right)\left(\sum_{p,q} c_{kpq} v_p v_q\right) \approx \sum_{j,k,l,p} c_{ijk} c_{jll} c_{kpp} \lambda_l \lambda_p. \tag{107}$$

That is, we neglect the interaction between different quadratic terms. Since we only need to calculate the error variances to within a factor of 2 or so, this approximation should be adequate.

Now if $\chi^2 < \sum_i \sigma_i'^2$, the accuracy of a linear inversion—*i.e.,* if we use only the linear part of equation (98)—is limited by having neglecting $\underline{\mathbf{C}}$. (When the reverse is true, the effects of quadratic terms will literally be lost in the noise, so there is no point in including them.)

To solve the quadratic problem, we first, define $\varepsilon = \mathbf{v}' - \mathbf{v}$. Then equation (98) becomes

$$\varepsilon = (\underline{\mathbf{C}}\mathbf{v}')\cdot\mathbf{v}' - 2(\underline{\mathbf{C}}\mathbf{v}')\cdot\varepsilon + (\underline{\mathbf{C}}\varepsilon)\cdot\varepsilon. \tag{108}$$

Define $\varepsilon^{(0)} = \mathbf{0}$ to be the initial estimate of the solution, and $\varepsilon^{(n+1)}$ to be the $(n+1)^{st}$ estimate, found from

$$\varepsilon^{(n+1)} = \left(\underline{\mathbf{C}}\mathbf{v}'\right)\cdot\left(\mathbf{v}' - 2\varepsilon^{(n)}\right) + \left(\underline{\mathbf{C}}\varepsilon^{(n)}\right)\cdot\varepsilon^{(n)}. \tag{109}$$

This process will converge to a unique solution if $\|\underline{\mathbf{C}}\varepsilon^{(n)}\| \ll 1$, where $\|\cdot\|$ represents the norm of the matrix.[25] Otherwise, the process may diverge. To see that it converges, consider the r.h.s. of equation (109) to be a function $\mathbf{h}(\varepsilon)$. Take the two vectors $\mathbf{x}_1$ and $\mathbf{x}_2$. The iteration will converge if $|\mathbf{h}(\mathbf{x}_1) - \mathbf{h}(\mathbf{x}_2)| < |\mathbf{x}_1 - \mathbf{x}_2|$, as long as $|\mathbf{x}_1|$ and $|\mathbf{x}_2|$ are not too large. But

$$\left|\mathbf{h}(\mathbf{x}_1) - \mathbf{h}(\mathbf{x}_2)\right| = \left|\left[\underline{\mathbf{C}}(\mathbf{x}_1 - \mathbf{x}_2 - 2\mathbf{v}')\right]\cdot(\mathbf{x}_1 - \mathbf{x}_2)\right| \tag{110}$$

which will be $< |\mathbf{x}_1 - \mathbf{x}_2|$ if the components of $\underline{\mathbf{C}}$ are small enough. When this is true, the series $\{\varepsilon^{(n)}\}$ converges, since at each step, $|\varepsilon^{(n+1)} - \varepsilon^{(n)}| = |\mathbf{h}(\varepsilon^{(n)}) - \mathbf{h}(\varepsilon^{(n-1)})| < |\varepsilon^{(n)} - \varepsilon^{(n-1)}|$. This is true, roughly speaking, when the second-order terms are small relative to the linear ones, although not so small that they can be neglected. Obviously, if they are not, the iteration will diverge, but then the linear approximation will not work either.

Note that we could combine these two approaches. From equation (100),

$$s_j s_k = \left(v_j + c_{jjj} v_j^2\right)\left(v_k + c_{kkk} v_k^2\right)$$

$$= v_j v_k + v_j v_k^2 c_{kkk} + v_k v_j^2 c_{jjj} + v_j^2 v_k^2 c_{jjj} c_{kkk}. \tag{111}$$

Since we assume that $c_{jjj} \ll 1$, to an adequate approximation, we can write

$$s_j s_k = v_j v_k \tag{112}$$

and

$$\mathbf{p} = \mathbf{s} + \left[(\underline{\mathbf{C}}'\mathbf{s})\cdot\mathbf{s}\right]. \tag{113}$$

Exercise 7-12: What is the error involved when we make this approximation?

References

Beteró, M., 'Regularization methods for linear inverse problems', in *Inverse Problems*, ed. by G. Talenti, Springer-Verlag, Berlin (1986).

Chahine, M. T., 'Inverse problems in radiative transfer: determination of atmospheric parameters,' *J. of the Atmospheric Sciences*, **27**, 960-967 (1970).

---

[25] The norm of a matrix $\mathbf{M}$, denoted by $\|\mathbf{M}\|$, is the supremum of $|\mathbf{M}\mathbf{a}| / |\mathbf{a}|$ over all vectors $\mathbf{a}$ with $|\mathbf{a}| > 0$.

Fleming, H. E., 'Comparison of linear inversion methods by examination of the duality between iterative and inverse matrix methods,' in *Inversion Methods in Atmospheric Remote Sensing, a workshop held at Langley Research Center, Hampton Virginia, December 15-17, 1976,* NASA Conference Publication CP-004, pp 325-360 (1976).

Foster, M., 'An application of the Wiener-Kolmogorov smoothing theory to matrix inversion,' *J. Soc. Indust. Appl. Math*, **9**, 387-392 (1961).

Graybill, F. A., *Introduction to Matrices with Applications to Statistics*, Wadsworth Publishing Company, Inc., Belmont, CA, (1969).

Groetsch, C. W., *The Theory of Tikhonov Regularization for Fredholm Equations of the First Kind*, Pitman Advanced Publishing Program, Boston (1984).

Houghton, J. T., F. W. Taylor, and C. D. Rodgers, *Remote Sounding of Atmospheres*, Cambridge University Press, Cambridge (1984).

Kerekes, J. P., 'A retrieval error analysis technique for passive infrared atmospheric sounders,' Lincoln Laboratory, MIT, Tech. Report 978 (1993).

Milman, A. S., 'How wind affects passive microwave measurements of sea surface temperature,' *IEEE Trans. Geosci. & Remote Sensing*, **GE-25**, 22-27 (1987).

Milman, A. S., 'Inversion of systems of quadratic equations,' *International J. of Remote Sensing*, **18**, 1365-1372 (1997).

Parker, R. L., 'Understanding Inverse Theory,' in F. A. Donath, F. G. Stehli, and G. W. Wetherill, Eds., *Ann. Rev. Earth and Planetary Science*, 35-64 (1977).

Phillips, D. L., 'A technique for the numerical solution of certain integral equations of the first kind,' *J. Association for Computing Machinery*, **9**, 84-97 (1962).

Rogers, C. D., 'Retrieval of atmospheric temperature and composition from remote measurements of thermal radiation,' *Reviews of Geophysics & Space Physics*, **14**, 609-624 (1976).

Sabatier, P. C. (Ed.), *Inverse Problems: An Interdisciplinary Study*, Academic Press, London (1987).

Twomey, S., B. Herman, and R. Rabinoff, 'An extension to the Chahine method of inverting the radiative transfer equation,' *J. of the Atmospheric Sciences*, **34**, 1085-1090 (1977).

Twomey, S., 'On the numerical solution of Fredholm integral equations of the first kind by the inversion of the linear system produced by quadrature,' *J. of the Association for Computing Machinery*, **10**, 97-101 (1963).

Twomey, S., 'Comparison of constrained linear inversion and an iterative nonlinear algorithm applied to the indirect estimation of particle size distributions,' *J. Computational Physics*, **18**, 188-200 (1975).

Twomey, S., *Introduction to the Mathematics of Inversion in Remote Sensing and Indirect Measurements*, Elsevier Scientific Pub. Co., Amsterdam (1977).

8. Fourier Transforms

A mathematician of the first rank, Laplace quickly revealed himself as only a mediocre administrator; from his first work we saw that we had been deceived. Laplace saw no question from its true point of view; he sought subtleties everywhere; had only doubtful ideas, and finally carried the spirit of the infinitely small into administration.

Napoleon (1769-1821)

Fourier[1] transforms are used to solve a wide variety of problems; this chapter will provide a brief introduction to the subject. *Fourier series*, which are similar in some ways to Fourier transforms, can be used to represent periodic functions. While no physical variable can be perfectly periodic—since nothing goes on forever—many systems are close enough to being periodic for the Fourier-series representation to be useful. Fourier series are also useful as a way of thinking about how certain devices work, or how certain processes behave. Optical instruments, including radiometers and radars, often have properties that depend on the Fourier transform of some physical quantity. We shall see, for instance, that the power pattern—the relative sensitivity to radiation coming from different directions—of an antenna is the Fourier transform of a quantity that is related to the size and shape of the antenna. In the study of integral equations, we shall see that there are special methods that are appropriate for solving integral equations that have the form of a Fourier transform.

Fourier transforms have been used since the nineteenth century. They were used to prove some important analytic results, but until relatively recently, performing even a simple numerical Fourier transform was an excruciatingly time-consuming task. Then, in 1965, Cooley and Tukey published a fast Fourier transform (or FFT) algorithm, making it possible to compute Fourier transforms very quickly.[2] Since that time, Fourier transforms have become much more widely used and understood. In the 1970s, the fast Fourier transform was a novelty; today, FFT software is as ubiquitous as personal computers.

The concept of a spectrum is that of the amount of power at each frequency that contributes to a random process. My long involvement in observ-

---

[1] Jean-Baptiste-Joseph Fourier (1768-1830) was a French mathematician, Egyptologist, and administrator. He published the *Théorie analytique de la chaleur* in 1822.

[2] Press *et al.* (1992, p 498) point out that other mathematicians had earlier discovered similar FFT methods, but their importance had not been recognized; the earliest was Gauss in 1805.

179

ing meteorological variables from spaceborne instruments has given me an appreciation that remote-sensing data are often used to study the statistics of weather phenomena. One example that is in the news today is that of global climate change: what evidence is there, if any, that the earth is becoming warmer? To determine this, we need measurements that cover both the whole globe and an appreciable time interval; unfortunately, reliable and systematic temperature measurements of any kind exist only for the last century or so. Global measurements made from satellites exist only for the last 18 years or so. So we are faced with a difficult task: if we see an upward trend in our measurements over the last 15 or 100 years, do we assume that we are seeing a long-term trend, or only a random fluctuation? In other words, are we seeing a permanent change in the *statistics* of the global temperature—a permanent change in the mean—or only a random fluctuation superimposed on unchanging statistics? This debate is going on right now, and it probably won't be resolved within the lifetimes of the present debaters.

The statistician's term for this is *stationarity*: a random process is stationary if its mean and variance do not change with time.[3] Global warming illustrates a general statistical problem: many statistical measures are not even defined if the underlying processes are not statistically stationary. It seems to me that, along with calculating spectra and autocorrelation functions of random processes, we need to consider some of these fundamental questions, as I do in this chapter and the next one.

We can look at the global-warming problem in the following light. Because of the short time record of temperature measurements, and because the way these measurements are made has changed over that time, we almost certainly will not be able to tell any time soon whether or not the earth is becoming warmer. In my view, the prudent scientist should say now that the data are inconclusive. However, there are compelling political interests in having an answer, and often people feel that there *must* be an answer to a question, even when there isn't. Not all problems can be solved. Nonetheless, there will be no shortage of scientists—on both sides of the question—who will claim that they have definite answers. They probably should know better. This is a rare example of a political question that hinges, ultimately, on some rather esoteric statistical arguments. Unfortunately, political institutions are rarely equipped to deal intelligently with scientific problems: we're bound to see more heat than light generated.

---

[3] Actually, there are several different kinds of stationarity; I will get to that later.

8.1 δ-Functions

Before going on to Fourier transforms proper, I should introduce the Dirac δ-function, which is an indispensable tool in Fourier analysis.[1] The function $\delta(x)$ is zero when its argument *is not* zero; it is infinite when its argument *is* zero. It is normalized so that

$$\int_{-\infty}^{\infty} \delta(x)\,dx = \int_{-\infty}^{\infty} \delta(x - x_0)\,dx = 1 \qquad (1)$$

for any value of x_0. It has the property that, for any function $f(x)$,

$$\int_{-\infty}^{\infty} f(x)\delta(x - x_0)\,dx = f(x_0). \qquad (2)$$

This is sometimes called the *sifting property*.

To define the δ-function rigorously, we can take it to be the limit of the following sequence of functions $h_\varepsilon(x)$:

$$h_\varepsilon(x) = 0 \qquad |x| > \varepsilon$$

$$h_\varepsilon(x) = \frac{1}{2\varepsilon} \qquad |x| \le \varepsilon. \qquad (3)$$

Clearly, for any $\varepsilon > 0$,

$$\int_{-\infty}^{\infty} h_\varepsilon(x)\,dx = 1, \qquad (4)$$

while $h_\varepsilon(x) = 0$ for all x outside a very small interval centered at 0. Taking the limit as $\varepsilon \to 0$,

$$\delta(x) = \underset{\varepsilon \to 0}{Lim}\, h_\varepsilon(x) \qquad (5)$$

The δ-function is one of a group of functions that are called *generalized* functions that have useful properties but cannot be defined in the usual way that functions are defined, either by specifying the value at each point or by

---

[1] δ-Functions are not, strictly speaking, functions: as we shall see, they are defined as the limit of a certain sequence of ordinary functions. They are usually called *generalized* functions, which are an indispensable part of a branch of mathematics called *operational calculus*. While the subject might interest some readers, I can only refer the reader to such books as Liverman (1964).

stating a rule for determining the value of the function from its argument.[2] The δ-function is defined instead as being the limit of a sequence of garden-variety functions. Clearly, it is not mathematically sufficient to define δ(x) by saying "it is zero when $x \neq 0$, and it is infinite when $x = 0$, and its integral = 1." One objection would be that such a thing might not even exist. So we define δ(x) and other generalized functions as limits of sequences of functions.

We can also define the δ-function as the limit of the family of Gaussian functions

$$g_\varepsilon(x) \equiv \frac{1}{\sqrt{2\pi}\varepsilon} e^{-x^2/2\varepsilon^2} . \tag{6}$$

Then

$$\delta(x) = \lim_{\varepsilon \to 0} g_\varepsilon(x). \tag{7}$$

Another very useful representation of δ(x) is

$$\delta(x) = \int_{-\infty}^{\infty} e^{-2\pi i \omega x} \, d\omega . \tag{8}$$

We shall use this equation frequently in the discussions of Fourier transforms.

Exercise 8-1: Show how δ(x) can be defined as a sequence of functions that leads to equation (8).

Another useful function is the *step function H(x)* defined by

$$H(x) = \begin{cases} 0 & x < 0 \\ 1 & x \geq 0 \end{cases} . \tag{9}$$

It is not hard to show that

$$\int_{-\infty}^{x} \delta(t) \, dt = H(x). \tag{10}$$

8.2 Fourier transforms and their properties

There are different ways that the Fourier transform can be defined, and I follow the notation of Bracewell (1965). I generally will use lowercase letters for

---

[2] *E.g.*, defining the function *f(x)* from the rule $f = \cos x$.

functions, and the corresponding uppercase letters for their Fourier transforms. Let $f(x)$ be a complex-valued function of the real variable x. The *Fourier transform* of $f(x)$ is[3]

$$F(\omega) = \int_{-\infty}^{\infty} f(x) e^{-2\pi i x \omega} \, dx. \tag{11}$$

It is not known what conditions are both necessary and sufficient for the Fourier transform of a function to exist.[4] However, a sufficient condition is that

$$\int_{-\infty}^{\infty} |f(x)|^2 \, dx < \infty; \tag{12}$$

i.e., that the integral exists and is finite. I shall assume from here on that all of the functions that we are discussing satisfy this condition. In practice, it would be very unusual to find a function that describes a real physical quantity that does *not* satisfy this condition.

I write Fourier transforms with integrals that extend from $-\infty$ to ∞, but these limits are only formal. Any physical process is limited: it has a beginning and an end. We can assume that any mathematical function that represents a physical process goes to zero outside some finite limits. If $f(x) = 0$ for $x < a$ and for $x > b$, then the integral from $-\infty$ to ∞ is the same as the integral from a to b.

For every Fourier transform, there is an inverse transform that restores the original function. The function $\hat{f}(x)$ defined by[5]

$$\hat{f}(x) = \int_{-\infty}^{\infty} F(\omega) e^{2\pi i x \omega} \, d\omega \tag{13}$$

is the inverse Fourier transform of F. Note that the sign is different in the exponential kernel for the Fourier transform and its inverse. This is called an inverse transform, reasonably enough, because $f = \hat{f}$. Strictly speaking, this is only correct when f is continuous; I will discuss this presently. To show that $f = \hat{f}$, substitute the definition of f' into the definition of F:

$$\hat{f}(x) = \int_{-\infty}^{\infty} \int_{-\infty}^{\infty} f(y) e^{-2\pi i \omega (y-x)} \, d\omega \, dy = \int_{-\infty}^{\infty} f(y) \delta(y-x) \, dy = f(x). \tag{14}$$

---

[3] Here, as elsewhere, I follow the notation of Bracewell (1965). Some authors define the Fourier transform with $+i$ in the exponential instead of $-i$.

[4] See Champeney (1987) pp 44 *ff.*

[5] If the forward transform has i in the exponent, the inverse transform must have $-i$.

Here, I have used equation (8) and the sifting property of the δ-function. This is a very common method of reducing multiple integrals (often four-fold or higher) to multiple integrals of lower order; this procedure reduces an *n*-fold integral to an (*n* - 2)-fold integral.

In the following sections, I shall denote the Fourier transforms of the complex functions $f(x)$, $g(x)$, and $h(x)$ by $F(\omega)$, $G(\omega)$, and $H(\omega)$, respectively; *a* is always a real number.

The Fourier transform is linear, since integration is linear. This means that, given the functions $f(x)$ and $g(x)$, and scalars α and β, let

$$h(x) = \alpha f(x) + \beta g(x). \tag{15}$$

Then,

$$H(\omega) = \alpha F(\omega) + \beta G(\omega). \tag{16}$$

The Fourier transform is defined even if $f(x)$ is not continuous. If f is not continuous at $x = \xi$, then

$$\int_{-\infty}^{\infty} F(\omega)e^{2\pi i\omega x}\, d\omega = \frac{1}{2}\left(\underset{x\to\xi^+}{Lim} f(x) + \underset{x\to\xi^-}{Lim} f(x) \right), \tag{17}$$

where the notation $x \to \xi^+$ means x approaches ξ from above, and $x \to \xi^-$ means x approaches ξ from below.[6]

Exercise 8-2: Show that the Fourier transform of a δ-function is a constant, and *vice versa*.

There are a few theorems that are relevant to Fourier transforms that I shall discuss here briefly.[7] These theorems are useful because they allow us to see the relationships between the Fourier transforms of different, but related, functions. This will help us to use what we know about the Fourier transforms of common functions to help us think about certain physical problems.

8.2.1 Definition

Many authors write Fourier transforms in the form

$$F(\omega) = \int_{-\infty}^{\infty} f(x)e^{-i\omega x}\, dx \tag{18}$$

---

[6] If $f(x)$ were continuous, then the two limits would be the same, so this equation would still be valid, if not very interesting.
[7] They are all discussed by Bracewell (1965).

and the inverse transform

$$f(x) = \frac{1}{2\pi} \int_{-\infty}^{\infty} F(\omega) e^{i\omega x} \, d\omega \, . \tag{19}$$

This form is slightly asymmetric. Furthermore, some theorems have to be changed slightly to accommodate this form. You need to know which form is being used.

8.2.2 Shift theorem

The Fourier transform of $f(x - a)$ is

$$\int_{-\infty}^{\infty} f(x-a) e^{-2\pi i\omega x} \, dx = \int_{-\infty}^{\infty} f(x) e^{-2\pi i\omega (x+a)} \, dx = e^{-2\pi i\omega a} \, F(\omega). \tag{20}$$

In other words, shifting the function f only changes the phase of its Fourier transform.

8.2.3 Similarity theorem

The Fourier transform of $f(ax)$ is

$$\int_{-\infty}^{\infty} f(ax) e^{-2\pi i\omega x} \, dx = \frac{1}{|a|} \int_{-\infty}^{\infty} f(x) e^{-2\pi ix(\omega/a)} \, dx = \frac{1}{|a|} F(\omega/a). \tag{21}$$

As an example, the Fourier transform of a Gaussian is another Gaussian:

$$\int_{-\infty}^{\infty} e^{-\pi(ax)^2} e^{-2\pi i\omega x} \, dx = \frac{1}{a} \int_{-\infty}^{\infty} e^{-\pi x^2} e^{-2\pi i\omega x/a} \, dx$$

$$= \frac{1}{a} \int_{-\infty}^{\infty} e^{-\pi x^2} e^{-2\pi i\omega x/a} \, dx = \frac{1}{a} e^{-\pi\left(\omega^2/a^2\right)} \int_{-\infty}^{\infty} e^{-\pi(x-i\omega/a)^2} \, dx$$

$$= \frac{1}{a} e^{-\pi\left(\omega^2/a^2\right)}. \tag{22}$$

So the Fourier transform of a fat Gaussian is a skinny one, and *vice versa*, as shown in Figure 8-1.

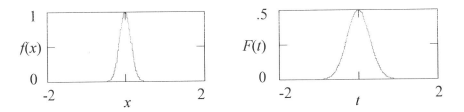

Figure 8-1. $f(x) = \exp\{-4\pi x^2\}$ and its Fourier transform $F(t) = \frac{1}{2}\exp\{-\frac{1}{4}\pi t^2\}$.

Exercise 8-3: What is the Fourier transform of $\exp\{-\frac{1}{2}(x - \mu)^2/\sigma^2\}$?

8.2.4 Rayleigh's theorem

Rayleigh's theorem[8] states that

$$\int_{-\infty}^{\infty} |f(x)|^2 \, dx = \int_{-\infty}^{\infty} |F(\omega)|^2 \, d\omega. \tag{23}$$

To prove this, use the definition of $F(\omega)$; then (the notation $f*$ means the complex conjugate of f)

$$|F(\omega)|^2 = F(\omega) F^*(\omega) = \int_{-\infty}^{\infty} f(x) e^{-2\pi i \omega x} \, dx \int_{-\infty}^{\infty} f^*(x') e^{2\pi i \omega x'} \, dx'; \tag{24}$$

therefore, after rewriting this as a double integral,

$$\int_{-\infty}^{\infty} |F(\omega)|^2 \, d\omega = \int_{-\infty}^{\infty} \int_{-\infty}^{\infty} \int_{-\infty}^{\infty} f(x) f^*(x') e^{-2\pi i \omega (x-x')} \, dx \, dx' \, d\omega. \tag{25}$$

Using equation (8),

$$\int_{-\infty}^{\infty} |F(\omega)|^2 \, d\omega = \int_{-\infty}^{\infty} \int_{-\infty}^{\infty} f(x) f^*(x') \, \delta(x - x') \, dx \, dx' \tag{26}$$

Rayleigh's theorem follows immediately from this. A similar theorem states that

$$\int_{-\infty}^{\infty} f(x) g^*(x) \, dx = \int_{-\infty}^{\infty} F(\omega) G^*(\omega) \, d\omega. \tag{27}$$

---

[8] Also known as *Plancherel's* or *Parseval's* theorem.

If $f(x) = g(x)$, this is the same as Rayleigh's theorem.

Often in physical situations, the integrands in these equations represent the amount of power in a certain process. These theorems show that the power measured by integrating the squared magnitude of a function is the same as the power measured by integrating the squared magnitude of its Fourier transform.

Exercise 8-4: Prove equation (27).

8.2.5 *The Convolution theorem*

Suppose that the following relationship exists between the functions f, g, and h:

$$h(x) = \int_{-\infty}^{\infty} f(u)\,g*(t-u)\,\mathrm{d}u. \tag{28}$$

This is called the *convolution* of f and g. We can use the definitions of the Fourier transform—equation (4)—and convolution to derive an important result, the convolution theorem:

$$H(\omega) = \int_{-\infty}^{\infty} h(x)\,e^{2\pi i \omega x}\,\mathrm{d}x = \int_{-\infty}^{\infty}\int_{-\infty}^{\infty} f(u)\,g*(x-u)\,e^{2\pi i \omega x}\,\mathrm{d}u\,\mathrm{d}x. \tag{29}$$

Use equation (11) to write this in terms of the Fourier transform functions F and G:

$$H(\omega) = \int_{-\infty}^{\infty}\int_{-\infty}^{\infty} \left[\int_{-\infty}^{\infty} F(\omega'')\,e^{2\pi i \omega'' x}\,\mathrm{d}\omega'' \right]$$

$$\left[\int_{-\infty}^{\infty} G(\omega')\,e^{-2\pi i \omega'(x-u)}\,\mathrm{d}\omega' \right]\mathrm{d}u\,\mathrm{d}x \tag{30}$$

Equation (30) simplifies to

$$H(\omega) = \int_{-\infty}^{\infty}\int_{-\infty}^{\infty}\int_{-\infty}^{\infty}\int_{-\infty}^{\infty} F(\omega'')\,e^{2\pi i \omega'' x}\,G(\omega')$$

$$\times\, e^{-2\pi i \omega'(x-u)}\,e^{-2\pi i \omega x}\,\mathrm{d}\omega'\,\mathrm{d}\omega''\,\mathrm{d}u\,\mathrm{d}x$$

$$= \int_{-\infty}^{\infty} \int_{-\infty}^{\infty} \int_{-\infty}^{\infty} F(\omega'') \, G(\omega') e^{2\pi i \omega' u} \, \delta(\omega + \omega' - \omega'') \mathrm{d}\omega' \mathrm{d}\omega'' \mathrm{d}u$$

$$= \int_{-\infty}^{\infty} \int_{-\infty}^{\infty} F(\omega'') \, G(\omega'' - \omega) e^{2\pi i (\omega'' - \omega) u} \, \mathrm{d}\omega'' \mathrm{d}u. \tag{31}$$

Here, I have used equation (8) several times to eliminate the variables of integration. Applying it once more, we get the simple result that

$$H(\omega) = F(\omega) G(\omega). \tag{32}$$

In simple words, the Fourier transform of the convolution of two functions is the product of their Fourier transforms, and *vice versa*.

The convolution of two functions is often written with the pentagram symbol $\star$; the convolution of the functions f and g would be written $f \star g$. A more compact form of the convolution theorem is

$$h = f \star g \iff H = FG. \tag{33}$$

8.2.6 Even and odd functions

A function $f(x)$ is *even* if $f(-x) = f(x)$; it is *odd* if $f(-x) = - f(x)$. For example, $\cos x$ is an even function; $\sin x$ is an odd function. It easy to show that any function can be written as the sum of an odd function and an even function. For any $f(x)$, define $f_e(x) = \frac{1}{2}[f(x) + f(-x)]$ and $f_o(x) = \frac{1}{2}[f(x) - f(-x)]$. Obviously, $f(x) = f_e(x) + f_o(x)$.

Exercise 8-5: Show that $f_e(x)$ and $f_o(x)$ are even and odd, respectively.

Exercise 8-6: Show that the product of two even or two odd functions is even; the product of an even and an odd function is odd.

Now for any even function and any constant a,

$$\int_{-a}^{a} f_e(x) \mathrm{d}x = 2 \int_{0}^{a} f_e(x) \mathrm{d}x, \tag{34}$$

while for any odd function,

$$\int_{-a}^{a} f_o(x) \mathrm{d}x = 0. \tag{35}$$

The Fourier transforms of even and odd functions have special properties. Using the relationship $e^{ix} = \cos x + i\sin x$,

$$\int_{-\infty}^{\infty} f_e(x)\left[\cos(2\pi\omega x) + i\sin(2\pi\omega x)\right]dx = 2\int_{0}^{\infty} f_e(x)\cos(2\pi\omega x)dx, \quad (36)$$

which is a real function, while

$$\int_{-\infty}^{\infty} f_o(x)\left[\cos(2\pi\omega x) + i\sin(2\pi\omega x)\right]dx = 2i\int_{0}^{\infty} f_o(x)\sin(2\pi\omega x)dx, \quad (37)$$

which is purely imaginary.

A complex function $g(x)$ is *Hermitian* if $g(-x) = g*(x)$. If $f(x)$ is real, its Fourier transform is Hermitian:

$$F(-\omega) = \int_{-\infty}^{\infty} f(x)\left[\cos(-2\pi\omega x) + i\sin(-2\pi\omega x)\right]dx$$

$$= \int_{-\infty}^{\infty} f(x)\left[\cos(2\pi\omega x) - i\sin(2\pi\omega x)\right]dx = F^*(\omega). \quad (38)$$

Similarly, let $f(x)$ be a real function of x, so $if(x)$ is purely imaginary. Then its Fourier transform has the property that

$$F(-\omega) = \int_{-\infty}^{\infty} if(x)\left[\cos(-2\pi\omega x) + i\sin(-2\pi\omega x)\right]dx$$

$$= \int_{-\infty}^{\infty} f(x)\left[i\cos(2\pi\omega x) + \sin(2\pi\omega x)\right]dx$$

$$= -\int_{-\infty}^{\infty} f(x)\left[i\cos(2\pi\omega x) - \sin(2\pi\omega x)\right]^* dx. \quad (39)$$

So $F(-\omega) = - F^*(\omega)$; such a function is called *skew Hermitian*.

Therefore, the symmetry of a function (even or odd) is intimately related to the symmetry of its Fourier transform (real or imaginary); real and imaginary functions have Fourier transforms that are Hermitian or skew Hermitian, respectively.

Exercise 8-7: Show that the Fourier transform of a Hermitian function is real; that of a skew-Hermitian function, imaginary.

8.2.7 Some useful functions

The function $\Pi(x)$ is defined by

$$\Pi(x) = \begin{cases} 1 & |x| < \frac{1}{2} \\ 0 & |x| \geq \frac{1}{2} \end{cases}. \tag{40}$$

Its Fourier transform is

$$\int_{-\infty}^{\infty} \Pi(x) e^{-2\pi i \omega x} \, dx = \int_{-\frac{1}{2}}^{\frac{1}{2}} \left[\cos(2\pi\omega x) + i\sin(2\pi\omega x) \right] dx. \tag{41}$$

Since $\sin x$ is an even function, it contributes zero to the integral. Consequently,

$$\int_{-\infty}^{\infty} \Pi(x) e^{-2\pi i \omega x} \, dx = 2\int_{0}^{\frac{1}{2}} \cos(2\pi\omega x) \, dx = \frac{\sin(\pi\omega)}{\pi\omega} \equiv \mathrm{sinc}\,\omega. \tag{42}$$

This useful function, $\mathrm{sinc}\omega$ (pronounced like "sink omega"), arises repeatedly when we use Fourier transforms. We will meet it again in Chapter 11.

If $\Pi(x)$ is convolved with itself, the resulting function is

$$\Lambda(x) = \int_{-\infty}^{\infty} \Pi(u)\Pi(x-u) \, du = \begin{cases} 1-|x|, & |x| < 1 \\ 0 & \text{otherwise} \end{cases}. \tag{43}$$

Using the convolution theorem, it is easy to show that the Fourier transform of $\Lambda(x)$ is

$$\int_{-\infty}^{\infty} \Lambda(x) e^{-2\pi i \omega x} \, dx = \mathrm{sinc}^2\omega. \tag{44}$$

Exercise 8-8: What are the Fourier transforms of $\cos x$ and $\sin x$? Can you find representations of $\delta(x)$ that involve only sines and cosines and that resemble equation (8)?

8.2.8 Fourier transforms in two dimensions

In two dimensions, the Fourier transform of $f(x,y)$ is

$$F(\xi,\eta) = \int_{-\infty}^{\infty}\int_{-\infty}^{\infty} f(x,y) e^{-2\pi i(\xi x + \eta y)} \, dx \, dy. \tag{45}$$

The formula for the inverse transform is similar. In the special case where $f(x,y)$ is radially symmetric, *i.e.*, $f(x,y) = f(r)$ where $r^2 = x^2 + y^2$, the Fourier transform is

$$\int_{-\infty}^{\infty} \int_{-\infty}^{\infty} f(x,y)e^{-2\pi i(\xi x + \eta y)}\, dx\, dy = i\pi \int_{-\infty}^{\infty} f(r)J_0(2\pi qr)r\, dr \qquad (46)$$

where

$$J_0(x) = \frac{1}{2\pi}\int_0^{2\pi} e^{ix\cos t}\, dt \qquad (47)$$

is the zero-th order Bessel function of the first kind.[9]

8.2.9 Derivatives and Integrals

The first derivative of $F(\omega)$ is

$$\frac{dF(\omega)}{d\omega} = \frac{d}{d\omega}\int_{-\infty}^{\infty} f(x)e^{-2\pi i x\omega}\, dx = \frac{2\pi}{i}\int_{-\infty}^{\infty} xf(x)e^{-2\pi i x\omega}\, dx. \qquad (48)$$

Similarly, the second derivative is

$$\frac{d^2 F(\omega)}{d\omega^2} = -4\pi^2 \int_{-\infty}^{\infty} x^2 f(x)e^{-2\pi i x\omega}\, dx. \qquad (49)$$

We can generalize this to include derivatives of non-integral order. For any $r > 0$, the rth derivative is

$$\frac{d^r F(\omega)}{d\omega^r} = \left(\frac{2\pi}{i}\right)^r \int_{-\infty}^{\infty} x^r f(x)e^{-2\pi i x\omega}\, dx; \qquad (50)$$

this is valid even if r is not an integer.

Integrals of $F(\omega)$ are also defined in a similar manner, and fractional integrals[10] can also be defined.

---

[9] See Bracewell (1965, p 247) for a proof.

[10] I will forebear from calling them *non-integral integrals*.

8.3 Fourier series

We can express any periodic function as the weighted sum of sine and cosine functions. This subject is closely related to Fourier transforms. Suppose $f(x)$ is a periodic function with period T. This means that

$$f(x + T) = f(x) \tag{51}$$

for all values of x. Then it is possible to express $f(x)$ as

$$f(x) = a_0 + \sum_{n=1}^{\infty} \left\{ a_n \cos\left(\frac{2\pi nx}{T}\right) + b_n \sin\left(\frac{2\pi nx}{T}\right) \right\}. \tag{52}$$

Throughout this discussion, I will assume that $f(x)$ is real. Note that $\cos(0) = 1$ and $\sin(0) = 0$, so the coefficient a_0 really represents $a_0\cos(2\pi 0x/T)$, while a multiplier of $\sin(2\pi 0x/T)$ would be pointless.

No physical quantity can be truly periodic in the mathematical sense. The universe had a beginning and presumably will have an end, so no time series can truly go from $-\infty$ to ∞. Similarly, nothing is infinite in spatial extent. Nonetheless, we can always imagine that any time series, for example, is multiplied by a very wide Gaussian and near $x = 0$, $f(x)$ is periodic to a very high degree of precision. The requirement that $f(x)$ must be periodic for it to possess a Fourier series expansion raises some difficult mathematical problems, but we need not pursue them here.

Given any periodic function $f(x)$, we can always define a new variable by $u = x/T$. Without any loss of generality, we can write

$$f(u) = a_0 + \sum_{n=1}^{\infty} \left\{ a_n \cos(2\pi nu) + b_n \sin(2\pi nu) \right\}. \tag{53}$$

While Fourier transforms are complex, Fourier series are usually real (but they do not have to be—see below). Whether or not this is an advantage depends on the user's viewpoint: complex notation is often more compact than real notation, but it may be less familiar. Also, while Fourier transforms generally involve negative as well as positive frequencies, Fourier series generally involve only positive frequencies. In equation (53), $f(x)$ is periodic with period 1; it is customary to regard x as being in the interval [0,1], although any other interval of unit length would also work.

Given $f(x)$, the coefficients a_n and b_n are given by

$$a_n = 2\int_0^1 f(x)\cos(2\pi nx)\,dx \tag{54}$$

and

$$b_n = 2\int_0^1 f(x)\sin(2\pi nx)\,dx. \tag{55}$$

Since $f(x)$ and $\sin(2\pi nx)$ or $\cos(2\pi nx)$ are periodic with period $T = 1$, the limits of these last two integrals could be $-\frac{1}{2}$ to $\frac{1}{2}$, or any numbers a and $a + 1$. Equations (54) and (55) come about as follows.

The functions $\sin(2\pi nx)$ and $\cos(2\pi nx)$ are orthonormal:

$$\int_0^1 \cos(2\pi nx)\cos(2\pi mx)\,dx = \tfrac{1}{2}\delta_{mn}; \tag{56}$$

$$\int_0^1 \sin(2\pi nx)\sin(2\pi mx)\,dx = \tfrac{1}{2}\delta_{mn}; \tag{57}$$

and

$$\int_0^1 \sin(2\pi nx)\cos(2\pi mx)\,dx = 0. \tag{58}$$

Remember that $\delta_{mn} = 1$ if $m = n$; $\delta_{mn} = 0$ otherwise.

Exercise 8-9: Prove these last three relationships, using the trigonometric identities

$$\cos\alpha\cos\beta = \tfrac{1}{2}\big[\cos(\alpha - \beta) + \cos(\alpha + \beta)\big] \tag{59}$$

$$\sin\alpha\sin\beta = \tfrac{1}{2}\big[\cos(\alpha - \beta) - \cos(\alpha + \beta)\big]. \tag{60}$$

To prove equations (54) and (55), take equation (53) and multiply each side by $\cos(2\pi nx)$ and integrate from 0 to 1. Since $\cos(2\pi nx)$ is orthogonal to $\sin(2\pi nx)$, all of the sine terms disappear. So do all of the terms proportional to $\cos(2\pi mx)$ if $m \neq n$. This establishes equation (54). Equation (55) is proved in a similar manner. Finally, since $<\cos 2\pi x> = <\sin 2\pi x> = 0$,

$$a_0 = \int_0^1 f(x)\,dx. \tag{61}$$

Sin$(2\pi nx)$ and cos$(2\pi nx)$ are not the only functions that we can use to represent f as a series expansion; there are many other commonly encountered families of orthogonal functions. The most common are the various families of orthogonal polynomials.[11]

A Fourier series may or may not converge. Consider a series defined with $a_n = 1$ for all n. At $x = 0$, the sum would be

$$f(0) = \sum_{n=0}^{\infty} 1, \tag{62}$$

while

$$f(\pi) = 1 - 1 + 1 - \cdots, \tag{63}$$

neither of which series converges. However, if $f(x)$ is integrable, *i.e.*,

$$\int_{-\frac{1}{2}}^{\frac{1}{2}} |f(x)| dx < \infty, \tag{64}$$

then the Fourier series for $f(x)$ will converge.

Exercise 8-10: So why doesn't this series for $f(\pi)$ converge? Can we modify it slightly so that it will converge?

If $f(x)$ and $g(x)$ both have Fourier series, the Fourier series of $f(x)g(x)$ is the product of the two series, provided that $f(x)g(x)$ is integrable. Also, the derivative of $f(x)$ given by equation (53) is given by

$$\frac{df(x)}{dx} = 2\pi \sum_{n=0}^{\infty} n \left\{ -a_n \sin(2\pi nx) + b_n \cos(2\pi nx) \right\}. \tag{65}$$

That is, a Fourier series can be differentiated term by term. It can also be integrated term by term.

8.3.1 Average values

As an aside, we can interpret the expression

$$\int_{0}^{1} \cos(2\pi nx) \cos(2\pi mx) dx = \langle \cos(2\pi nx) \cos(2\pi mx) \rangle \tag{66}$$

as being the average value of the product $\cos(2\pi nx) \cos(2\pi mx)$, with a similar equations applying to products of sines. It is straightforward to show that $<\cos x \cos(x+\phi)> = 0$ unless $\phi = 0$, 2π, 4π, ..., and similarly for sines, while

---

[11] See Abramowitz and Stegun (1965), Chapter 22.

<sinx cos(x+φ)> = 0 unless φ = ½kπ, where k is an odd integer [use equations (59) and (60)].

Exercise 8-11: Suppose we have a family of random functions $f_i(x)$ given by

$$f_k(x) = \sum_{n=1}^{\infty} \alpha_n \cos(2\pi nx + \phi_{kn}),$$ (67)

where ϕ_{kn} are random phases. What is the value of

$$\gamma_n = \langle f_k(x)\cos 2\pi nx \rangle,$$ (68)

where the average is taken over a random set $\{f_k\}$?

8.3.2 Derivation from Fourier transform

There is obviously a close connection between Fourier series and Fourier transforms. Lanczos[12] suggests that the Fourier transform is more fundamental than the Fourier series because the latter applies only to strictly periodic functions. I shall follow Lanczos' development here to show how Fourier series can be obtained from the Fourier transform.

Consider a periodic function $f(x)$ with period 1. If it had some other period T, we could make the obvious substitution for x so that the period became 1. Let N be some large integer (it will approach ∞ presently), and let

$$F_N(\omega) = \int_{-N}^{N+1} f(x)e^{-2\pi ix\omega}dx = \sum_{n=-N}^{N} e^{-2\pi in\omega} \int_0^1 f(x)e^{-2\pi ix\omega}dx.$$ (69)

We can see this by writing the integral as

$$\int_{-N}^{N+1} \cdot dx = \int_{-N}^{-N+1} \cdot dx + \int_{-N+1}^{-N+2} \cdot dx + \cdots + \int_{N}^{N+1} \cdot dx.$$ (70)

Now define

$$\Phi_N(\omega) = \sum_{n=-N}^{N+1} e^{-2\pi in\omega} = e^{-2\pi iN\omega} \sum_{n=0}^{2N+1} \left(e^{2\pi i\omega}\right)^n$$ (71)

which is a geometric series, and use the identity

---

[12] Lanzcos (1966), p 182.

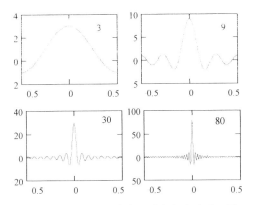

Figure 8-2. Plots of $\sin[N\pi x] / \sin(\pi x)$ for $N = 3, 8, 30,$ and 80.

$$\sum_{n=0}^{2N} r^n - r \sum_{n=0}^{2N} r^n = (1-r)\sum_{n=0}^{2N} r^n = 1 - r^{2N+1}. \qquad (72)$$

Then

$$\Phi_N(\omega) = e^{-2\pi N\omega} \sum_{n=0}^{2N+1} e^{2\pi i n\omega}$$

$$= \frac{e^{-2\pi N} - e^{2\pi i(N+1)\omega}}{1 - e^{2\pi i\omega}} = \frac{e^{-2\pi(N+\frac{1}{2})} - e^{2\pi i(N+\frac{1}{2})\omega}}{e^{-\pi i\omega} - e^{\pi i\omega}} = \frac{\sin\left[(2N+1)\pi\omega\right]}{\sin\pi\omega}. \qquad (73)$$

Consequently,

$$F_N(\omega) = \Phi_N(\omega)\int_0^1 f(x)e^{-2\pi i x\omega}\,dx. \qquad (74)$$

Now define an inverse transform by

$$f_N(x) = \int_{-\infty}^{\infty} F_N(\omega)e^{2\pi i x\omega}\,d\omega = \int_{-\infty}^{\infty} \Phi_N(\omega)e^{2\pi i\omega x}\int_0^1 f(x')e^{-2\pi i x'\omega}\,dx'\,d\omega. \qquad (75)$$

$f_N(\omega)$ is very large for $\omega = k$, where k is an integer, and very close to zero elsewhere, although

$$\underset{N\to\infty}{Lim}\int_{k-\varepsilon}^{k+\varepsilon} \Phi_N(\omega)\,d\omega = 1. \qquad (76)$$

I will give a heuristic argument to show that this is so; Lanczos (1966) gives a formal proof. Note that, as $\omega \to 0$, $\sin \pi \omega / \pi \omega = \operatorname{sinc} \omega \to 1$. Therefore, for large N and $\omega =$ an integer, $\Phi_N(k)$ has a height $(2N+1)$ and a width $1/(2N+1)$; see Figure 8-2. There is only a neg * mergeformatligible contribution to the integral outside the region where $|(2N+1)\omega| < \frac{1}{2}$. This (heuristically) establishes equation (8).

Using the definition of the Fourier coefficients a_n and b_n, and expressing the complex exponentials in equation (75) in terms of sines and cosines, we find in the limit as $N \to \infty$ that

$$f_N(x) \to a_0 + \sum_{n=0}^{\infty} \left(a_n \cos 2\pi nx + b_n \sin 2\pi nx \right). \tag{77}$$

This shows that taking the Fourier transform of a periodic function that extends from $-\infty$ to ∞ leads to the Fourier series.

8.3.3 Complex Fourier series

There is a complex form of the Fourier series that uses exponentials instead of sines and cosines. Let

$$c_n = a_n + ib_n, \qquad n < 0,$$

$$c_n = a_n - ib_n, \qquad n \geq 0. \tag{78}$$

Then equation (52) becomes

$$f(x) = \sum_{n=-\infty}^{\infty} c_n e^{2\pi inx}. \tag{79}$$

Also, we can find the coefficients c_m from

$$\int_{-\infty}^{\infty} f(x) e^{-2\pi imx} \, dx = \sum_n \int_{-\infty}^{\infty} c_n e^{2\pi inx} e^{-2\pi imx} \, dx = c_m. \tag{80}$$

Here, I have used the representation of a δ-function

$$\delta_{mn} = \int_{-\infty}^{\infty} e^{-2\pi i(m-n)x} dx. \tag{81}$$

8.4 Discrete Fourier transform

The Fourier transform is an integral transform. Except for a very few functions whose Fourier transforms can be evaluated analytically, we can only find the Fourier transform $F(s)$ of $f(x)$ by performing a numerical integration [*i.e.*, a Fast Fourier transform (FFT)]. Because FFT's work best with evenly sampled data, let us assume that we have measured some $f(x)$ between $x = 0$ and $x = x_{max}$. Suppose we take N points x_n, $n = 0, \cdots, N - 1$, and sample $f(x)$ at intervals $\delta x = x_{max} / N$. How large should N be? How many samples do we need of $f(x)$? We can also switch the question around: how large should δx be to ensure that we retain in $F(s)$ all of the information that was in $f(x)$? If we use an interval δx that is too large (N too small), we will distort $F(s)$. I shall show what the value of δx should be by comparing the Fourier transform to the Fourier series representation of $f(x)$.

In order to calculate the Fourier transform of $f(x)$, we replace the integral with a sum over x; this is called a discrete Fourier transform (DFT). That is,

$$F(s) = \int_{-\infty}^{\infty} f(x)e^{-2\pi i x s}\, dx \;\rightarrow\; \delta x \sum_{m=0}^{N-1} f(m\delta x)e^{-2\pi i s m \delta x}. \tag{82}$$

To see how the discrete Fourier transform is related to the Fourier series of $f(x)$, substitute equation (79) into this sum:

$$F(s) = \delta x \sum_{n=-\infty}^{\infty} \sum_{m=0}^{N-1} c_n e^{2\pi i m n \delta x} e^{-2\pi i s m \delta x} = \delta x \sum_{n=-\infty}^{\infty} \sum_{m=0}^{N-1} c_n e^{2\pi i (n-s) m \delta x}. \tag{83}$$

This sum is zero unless s is an integer. Now let $x_{max} = N \delta x$; *i.e.*, the infinite limits in the integral in equation (82) are replaced with 0 and x_{max}; then replace δx with x_{max} / N, so

$$F(s) = \delta x \sum_{n=-\infty}^{\infty} \sum_{m=0}^{N-1} c_n e^{2\pi i (n-s) m x_{max} / N}. \tag{84}$$

Now let $\delta s = 1 / x_{max}$ and $s = k\delta s$; the discrete Fourier transform has the simple form

$$F(k\,\delta s) \equiv F_k = \delta x \sum_{n=-\infty}^{\infty} \sum_{m=0}^{N-1} c_n e^{2\pi i (n-k) m / N}. \tag{85}$$

If $k' = k + N$, $F(k'\delta s) = F(k\delta s)$: F is periodic with period $N\delta s$. So there are only N different values of s that we can calculate, $s = 0,..., N\delta s$.

In terms of the Fourier series, the Fourier transform is

$$F(k\,\delta s) \ = \ \delta x \sum_{\ell=-\infty}^{\infty} c_{k+\ell N} \tag{86}$$

because of the periodicity in k. Therefore, the discrete Fourier transform will contain power not only at frequency $k\,\delta s$, but at $k +$ (integral multiples of N) unless $c_n = 0$ for all $n > N$. This effect of having power from higher frequencies contribute to terms in the discrete Fourier transform is called *aliasing*.

8.5 The sampling theorem

Another way to put this is that, if we are sampling at N points, and the highest frequency that contributes to $f(x)$ is s_{max}, it is necessary that $s_{max} = N / x_{max}$, so the sampling interval must not be larger than $\delta x = x_{max} / N = 1 / s_{max}$. On the other hand, Nyquist proved that there is nothing to be gained by making δx smaller than $1 / s_{max}$. Therefore, data should be sampled at an interval no larger than

$$\delta x \ = \ \frac{1}{s_{max}}, \tag{87}$$

which is called the *Nyquist frequency*.

Here, I have taken the spectrum to extend from 0 to s_{max}. If instead the spectrum includes negative frequencies,[13] and has nonzero values only between $-s_c$ and $+s_c$, the Nyquist frequency is

$$\delta x \ = \ \frac{1}{2s_c}. \tag{88}$$

Since $s_{max} = 2s_c$, the two definitions of Nyquist frequency are the same. In other words, what matters is the bandwidth of $f(x)$, not the starting point. The Nyquist frequency is the same, whether the band where $F(s)$ is nonzero extends from $-s_c$ to $+s_c$, or 0 to $s_{max} = 2s_c$, or any other interval of width $2s_c$; see Figure 8-3.

We can look at this business another way. Remember that the Fourier transform of a real function is Hermitian: $F(-s) = F^*(s)$. Therefore, the spectrum of $f(x)$ has the property that $P_f(-s) \equiv |F(-s)|^2 = P_f(s)$. For a real process, there must be equal amounts of power at positive and negative frequencies. Since most physical quantities are represented by real numbers, most spectra

---

[13] If x represents time, then s is truly a frequency. If x represents a distance, s is called a *spatial frequency* or *wavenumber*. It is customary, when discussing a variable x without any physical units attached to call the conjugate variable s a frequency.

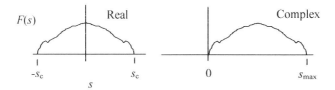

Figure 8-3. The spectrum runs from $-s_c$ to $+s_c$ (real process) or from 0 to $s_{max} = 2s_c$ (complex process).

will be symmetric about $s = 0$. In this case, the Nyquist sampling interval is $\delta x = 1/(2s_c)$.

Another way of stating Nyquist's theorem is that, if s_c is the largest frequency that contributes to $f(x)$ (including both positive and negative frequencies), the all of the information contained in $f(x)$ can be retained if we sample at a frequency $\delta x = 1/(2s_c)$. But on the other hand, if we use a larger sampling interval, we *will* lose some of the information in $f(x)$. When we calculate the spectrum of $f(x)$ from these data, we will encounter aliasing: power at frequencies $s_c + s$ will appear in the (undersampled) spectrum at the frequency $s_c - s$.

It is not hard to prove this theorem. First, we should note that

$$\int_{-\infty}^{\infty} \Pi(s) e^{-2\pi i x s} ds = \int_{-\frac{1}{2}}^{\frac{1}{2}} e^{-2\pi i x s} ds = \frac{\sin \pi x}{\pi x} \equiv \mathrm{sinc} x . \tag{89}$$

Later we will need to use the similarity theorem to find the Fourier transform of $\Pi(s/W)$:

$$\int_{-\infty}^{\infty} \Pi(s/W) e^{-2\pi i x s} ds = \frac{1}{W} \mathrm{sinc}(Wx) . \tag{90}$$

Now let $f(x)$ be a function of x defined on the real line, and $F(s)$ its Fourier transform. In many applications, $F(s)$ is limited by some physical constraint: we can only measure $F(s)$ between the limits, say $\pm\frac{1}{2}W$. We say that $f(x)$ is *bandwidth limited*, meaning that its Fourier transform is zero outside a given finite limit. The function f is the Fourier transform of F:

$$f(x) \equiv \int_{-\infty}^{\infty} \Pi(s/W) F(s) e^{-2\pi i s x} ds . \tag{91}$$

We can write $F(s)$ as a complex Fourier series:

$$F(s) = \Pi(s/W) \sum_{k=-\infty}^{\infty} c_k e^{2\pi i k s/W} , \tag{92}$$

where, again, $F(s) = 0$ for $s > \frac{1}{2}W$. The coefficients c_k are found from

$$c_k = \int_{-\infty}^{\infty} \Pi(s/W) F(s) e^{-2\pi iks/W} \, ds = f(k/W). \tag{93}$$

Therefore,

$$F(s) = \sum_{k=-\infty}^{\infty} e^{2\pi iks/W} f(k/W), \tag{94}$$

and consequently

$$f(x) = \sum_{k=-\infty}^{\infty} f(k/W) \int_{-\infty}^{\infty} \Pi(s/W) e^{2\pi iks/W} e^{-2\pi isx} \, ds. \tag{95}$$

Note that f, by definition, is the Fourier transform of a function $F(s)$ that is identically zero for $|s| > \frac{1}{2}W$. Now using the definition of sinc(x),

$$f(x) = \frac{1}{W} \sum_{k=-\infty}^{\infty} f(k/W) \, \text{sinc}[W(k/W - x)]; \tag{96}$$

a slightly more useful form might be

$$f(x/W) = \frac{1}{W} \sum_{k=-\infty}^{\infty} f(k/W) \text{sinc}(k - x). \tag{97}$$

We see, therefore, that we can reconstruct $f(x)$ using only samples spaced at an interval of $1/W$. Therefore, all of the information contained in $f(x)$ [or $F(s)$] is contained in these samples; sampling at a finer interval cannot produce more information.

Some instruments measure the phase of a signal, as well as its magnitude. Many radars do this; call it a complex process. Other instruments measure only the amplitude; call this a real process. An ordinary camera falls into this class. If we cannot measure the phase of a signal, we cannot distinguish between positive and negative frequencies, so if we measure the spectrum of a real process, we measure frequencies between 0 and s_c, taking them to be positive; since the spectrum must be an even function of frequency, we can extend it to negative frequencies in the obvious way. So for real processes, we take the bandwidth to be $2s_c$.

For complex processes, we can distinguish between positive and negative frequencies. Furthermore, we can multiply the processes by $\cos(f_0)$, where f_0 is a carrier frequency, and shift the spectrum from the range $(-s_c, s_c)$ to the range

$(f_0 - s_c, f_0 + s_c)$.[14] From this viewpoint, it is more natural to think of the bandwidth $s_{max} = 2s_c$.

8.5.1 The uncertainty principle

I mentioned the uncertainty principle in quantum mechanics in Chapter 2. Here, I will discuss an uncertainty principle that is more general in its application. It says, roughly, that the width of a function times the width of its Fourier transform cannot be smaller than a certain value. This means that, if f_W is a bandwidth-limited function, where W is the bandwidth, then the width (or variance) of f_W must become larger as W becomes smaller.

Let $f(x)$ be the Fourier transform $F(s)$ and have the property that

$$\lim_{x \to \infty} \sqrt{x} f(x) = 0, \tag{98}$$

which we need to ensure that the integrals will converge. Also, we shall normalize f by

$$\int_{-\infty}^{\infty} \left| f(x) \right|^2 dx = 1; \tag{99}$$

we shall assume that this integral exists also.

We can measure how concentrated each function is by evaluating the integrals

$$W_x^2 \equiv \int_{-\infty}^{\infty} x^2 f^2(x) dx \tag{100}$$

and

$$W_s^2 \equiv \int_{-\infty}^{\infty} s^2 F^2(s) ds. \tag{101}$$

Both f and F cannot be narrow functions. The uncertainty principle states that

$$W_x W_s \geq \frac{1}{4\pi}, \tag{102}$$

where equality obtains only if $f(x) = \kappa F(s)$ for some constant κ, and consequently that f and F are both Gaussians.[15]

---

[14] This is the process called *heterodyning*.

[15] The proof here follows that of Papoulis(1962) pp 62 *ff*, and Papoulis (1977) pp 273-275.

We shall use Schwarz's inequality (see §14.5), whence we get

$$\left| \int_{-\infty}^{\infty} xf(x)\frac{df}{dx}dx \right|^2 \leq \left[\int_{-\infty}^{\infty} \left| xf(x) \right|^2 dx \right]\left[\int_{-\infty}^{\infty} \left| \frac{df}{dx} \right|^2 dx \right]. \tag{103}$$

From Rayleigh's theorem [see equation (23)], we know that

$$\int_{-\infty}^{\infty} \left| f(x) \right|^2 dx = \int_{-\infty}^{\infty} \left| F(s) \right|^2 ds. \tag{104}$$

Integrate the l.h.s. of equation (103) by parts: we get

$$\int_{-\infty}^{\infty} xf(x)\frac{df}{dx}dx = \tfrac{1}{2}xf^2 \Big|_{-\infty}^{\infty} - \tfrac{1}{2}\int_{-\infty}^{\infty} \left| f(x) \right|^2 dx = -\tfrac{1}{2}. \tag{105}$$

In this last equation I have used equation (98) to eliminate the first term on the r.h.s. Now use the relationship that df/dt is the Fourier transform of $2\pi isF(s)$ [see equation (48)]. Then

$$\int_{-\infty}^{\infty} \left| \frac{df}{dx} \right|^2 dx = 4\pi^2 \int_{-\infty}^{\infty} \left| sF(s) \right|^2 ds. \tag{106}$$

Combining equations (106), (105), and (103), we find that

$$\left[\int_{-\infty}^{\infty} \left| xf(x) \right|^2 dx \right]\left[\int_{-\infty}^{\infty} \left| sF(s) \right|^2 ds \right] = W_x^2 W_s^2 = \geq \frac{1}{16\pi^2}, \tag{107}$$

thus proving equation (102). Furthermore, equality requires that $f(x) = \kappa F(s)$, which can happen only if f is a Gaussian [*cf.* equation (103)].

8.6 Fast Fourier transform

Even with a computer, it is very time-consuming to calculate the Fourier transform of a function directly from the definition given in equation (11) or equation (82): there would be too many complex exponentials to evaluate. The number of operations needed to perform a discrete Fourier transform of length n from this definition is proportional to n^2. Fortunately there are methods, called *Fast Fourier Transforms* (known universally as FFTs)[16] that we can use to evaluate Fourier transforms efficiently. The first published version was by Cooley and Tukey in 1965. Using an FFT, the number of operations is propor-

---

[16] Unfortunately, the verb *to FFT* has also entered the language; it is usually used where *to Fourier transform* would be more appropriate.

tional only to $n\log_2 n$. One place to find both a good explanation of the methods and canned software to do FFTs is in *Numerical Recipes in Fortran* or *Numerical Recipes in C*, both by Press *et al.*

Looking at the DFT that is defined in equation (82), we would expect that a DFT of length N would require $O(N^2)$ multiplications. By rearranging the order of the operations and some clever factoring, FFT algorithms reduce this to $O(N \log_2 N)$ multiplications, which is a very substantial savings when N is large.

Although the Fourier transform is defined by an integral, the actual calculation is performed at only a finite number of points, so the integral is being approximated by a sum. Usually, $f(x)$ is measured at discrete values of x; assume that they are evenly spaced at $x_j = j\delta x$. Let $f_j = f(x_j)$ and N be the number of points where x is measured. Then, as we saw before,

$$F(s_j) = F_j = \sum_{k=0}^{N-1} e^{-2\pi ijk/N} f_k .$$ (108)

The spacing δs of F (*i.e.,* the spectral resolution) is

$$\delta s = \frac{1}{N\delta x} = \frac{1}{\Delta x},$$ (109)

where $\Delta x \equiv N\delta x$. The inverse transform is

$$f_k = \frac{1}{N} \sum_{j=0}^{N-1} e^{+2\pi ijk/N} F_j .$$ (110)

Note the factor $1/N$ in this equation.

Exercise 8-12: Why is this factor $1/N$ needed?

8.6.1 Convolution

Suppose we want to evaluate the convolution

$$h(x) = \int_{-\infty}^{\infty} f(y-x)g(y)\,dy$$ (111)

Finding the convolution $f \star g$ using the definition is very time consuming; a much faster method uses the convolution theorem. To find $f \star g$, perform the following steps:

1. Use an FFT to find F and G.

2. Take their product.

3. Perform an inverse FFT.

Exercise 8-13: Work out the details of how to evaluate $f \star g$ using Fourier transforms.

8.6.2 Zero-padding

FFT routines work most efficiently when the number of data points is an integral power of 2. Many of them also work reasonably well if the number N of data points is the product of powers of small primes: $N = 2^p 3^q 5^r$, perhaps, where p, q, and r are integers. But what if the number of data points is not one of these numbers?

Suppose there are M points that are input to the FFT. The easiest way to deal with the problem is to *zero-pad* the data; *i.e.*, add $N - M$ zeros to the data so that there are $N = 2^n$ data points. This does result in the Fourier transform being sampled more frequently than necessary, but this is not a significant problem.

8.7 Other transforms

There are other integral transforms that are similar to Fourier transforms and are worth mentioning in passing. Most of them are formed by replacing the complex exponential kernel $e^{-2\pi i \omega t}$ with a different kernel. Many of the theorems about Fourier transforms have analogues with these other transforms. You can find many examples of transform pairs in Råde and Westergren (1990).

8.7.1 Laplace transform

The Laplace transform is similar to the Fourier transform, with the complex exponential kernel replaced with a real exponential kernel. Let $f(x)$ be a function such that $\lim\limits_{\varepsilon \to 0} \int\limits_{-\varepsilon}^{\infty} f(t) e^{-st} dt$ exists for all $s \geq s_0$, and f is continuous, except that it may be discontinuous at finitely many points. Then the Laplace transform of $f(x)$ is

$$F(s) = \lim\limits_{\varepsilon \to 0} \int\limits_{-\varepsilon}^{\infty} e^{-sx} f(x) dx ; \tag{112}$$

the inverse transform is given by

$$f(x) = \lim_{b \to \infty} \frac{1}{2\pi i} \int_{a - ib}^{a + ib} e^{sx} F(s) \, ds \qquad a \geq s_0 . \tag{113}$$

Laplace transforms have wide applications to differential equations and circuit analysis.

Exercise 8-14: Which of the theorems about Fourier transforms apply to Laplace transforms?

8.7.2 Hankel transform

The Hankel transform uses a Bessel function as the kernel: for $n > -\frac{1}{2}$,

$$F_n(s) = \int_0^\infty x f(x) J_n(sx) \, dx . \tag{114}$$

The inverse is

$$f(x) = \int_0^\infty s F_n(s) J_n(xs) \, ds . \tag{115}$$

In these equation, J_n is the Bessel function of the first kind of order n. These transform relations also hold when J_n is replaced by Y_n, which is the Bessel function of the second kind. If we define the Hankel functions as $H_n^{(1)} \equiv J_n + iY_n$ and $H_n^{(2)} \equiv J_n - iY_n$, we get a transform pair

$$F_n(s) = \int_0^\infty x f(x) H_n^{(1)}(sx) \, dx \tag{116}$$

and

$$f(x) = \int_0^\infty s F_n(s) H_n^{(2)}(xs) \, ds . \tag{117}$$

We have already encountered the zeroth-order Hankel transform in §8.2.8.

Exercise 8-15: Which of the theorems about Fourier transforms apply to Hankel transforms?

8.7.3 Z-transform

The z-transform[17] is similar to a DFT. Define a function $x(n)$, where n is a positive integer. Then define the transform

$$X(z) \equiv \sum_{n=0}^{\infty} x(n) z^{-n} \qquad (118)$$

for any complex number z such that this series converges. The inversion formula is

$$x(n) = \frac{1}{2\pi i} \oint_C X(z) z^{n-1} \, dz , \qquad (119)$$

where C is the unit circle. We can apply the theory of residues to solve this integral.

We can make this look more like a DFT by letting $z = e^{2\pi i/N}$. Then $z^{-n} = e^{-2\pi i n/N}$, and

$$X(z^k) \equiv X_k = \sum_{n=-\infty}^{\infty} x_n e^{-2\pi i nk/N} . \qquad (120)$$

Here, I have used the more suggestive notation $x_n = x(n)$. In this case, the z-transform associates each sequence of numbers $x_0, x_1, x_2, \ldots$, to a function defined on the unit circle at equal intervals of $2\pi/N$. Compare this to the definition of the complex Fourier series (8).

Exercise 8-16: What are the interesting theorems about z-transforms?

8.8 A Fourier transform coloring book

It is often helpful to be able to visualize some common Fourier transforms pairs. The following list is taken from Bracewell (1965), where the reader can find a more complete set of illustrations. Campbell and Foster (1961)[18] have a much more extensive table of Fourier transform pairs, although the notation seems somewhat dated. Oberhettinger (1973 and 1990) also have extensive tables of Fourier transforms and Fourier series. In addition, *BETA Mathematics Handbook* (Råde and Westergren, 1990), is a useful compendium of mathematical formulas, including sections on Fourier and other transforms and examples both of more Fourier transforms and of certain Fourier series.

---

[17] See Papoulis (1977), pp 31-37.
[18] Campbell (1961).

Some Fourier Transform Pairs[19]

$f(t) = \int\limits_{-\infty}^{\infty} F(\omega)e^{2\pi it\omega}\,d\omega$	$F(\omega) = \int\limits_{-\infty}^{\infty} f(t)e^{-2\pi i\omega t}\,dt$
$f(t) = 1$	$\delta(w)$
$e^{-\pi t^2}$	$e^{-\pi\omega^2}$
$te^{-\pi t^2}$	$i\omega\,e^{-\pi\omega^2}$
$\Pi(t)$	$\mathrm{sinc}(\omega)$
$\Lambda(t)$	$\mathrm{sinc}^2(\omega)$
$H(t)$	$\tfrac{1}{2}\delta(\omega) - i/(2\pi\omega)$
$J_0(2\pi t)$	$\Pi(\tfrac{1}{2}\omega)\,(1-\omega^2)^{-\tfrac{1}{2}}$
$2\cos\pi t$	$\delta(\omega+\tfrac{1}{2}) + \delta(\omega-\tfrac{1}{2})$
$2\sin\pi t$	$i\delta(\omega+\tfrac{1}{2}) - i\delta(\omega-\tfrac{1}{2})$
$2\Pi(t)\cos\pi t$	$\mathrm{sinc}(\omega+\tfrac{1}{2}) + \mathrm{sinc}(\omega-\tfrac{1}{2})$
$2\Pi(t)\sin 2\pi t$	$i\,\mathrm{sinc}(\omega+1) - i\,\mathrm{sinc}(\omega-1)$
$e^{-|t|}$	$2\,[1+(2\pi\omega)^2]^{-1}$
$te^{-a|t|}\ (a>0)$	$-4ia\omega/(a^2+\omega^2)^2$
$|t|e^{-a|t|}\ (a>0)$	$2(a^2-\omega^2)/(a^2+\omega^2)^2$

---

[19] Some of these formulas come from Råde and Westergren (1990).

$$f(t) = \int_{-\infty}^{\infty} F(\omega)e^{2\pi it\omega} \, d\omega \qquad \bigg| \qquad F(\omega) = \int_{-\infty}^{\infty} f(t)e^{-2\pi i\omega t} \, dt$$

$$e^{-a|t|\text{sgn } t} \ (a > 0) \qquad \bigg| \qquad -2i\omega \, / \, (a^2 + \omega^2)$$

$$\sin at^2 \ (a > 0) \qquad \bigg| \qquad \sqrt{\frac{\pi}{a}} \cos\left(\frac{\omega^2}{4a} + \frac{\pi}{4}\right)$$

$$\cos at^2 \ (a > 0) \qquad \bigg| \qquad \sqrt{\frac{\pi}{a}} \cos\left(\frac{\omega^2}{4a} - \frac{\pi}{4}\right)$$

$$\frac{i}{2(n-1)!}(it)^{n-1}e^{ict}\,\text{sgn}\,t \qquad \bigg| \qquad \frac{1}{(\omega - c)^n} \quad (c \text{ real}, n = 1,2,\ldots.)$$

$$\frac{1}{2a}e^{-a|t|} \quad (a > 0) \qquad \bigg| \qquad \frac{1}{2a}e^{-a|t|} \quad (a > 0)$$

$$\frac{i}{2}e^{-a|t|}\,\text{sgn}\,t \quad (a > 0) \qquad \bigg| \qquad \frac{\omega}{\omega^2 + a^2}$$

Functions Used in Conjunction with Fourier Transforms

$$\Pi(x) = \begin{cases} 1 & |x| \le \frac{1}{2} \\ 0 & \textit{otherwise} \end{cases} \qquad\qquad \Lambda(x) = \begin{cases} 1 - |x| & |x| \le 1 \\ 0 & \textit{otherwise} \end{cases}$$

$$\text{sinc}(x) = \frac{\sin(\pi x)}{\pi x} \qquad\qquad H(x) = \begin{cases} 1 & x \ge 0 \\ 0 & x < 0 \end{cases}$$

$$\text{sgn}(x) = \begin{cases} 1 & x \ge 0 \\ -1 & x < 0 \end{cases}$$

A Pictorial Guide to Fourier Transforms

$$f(t) = \int_{-\infty}^{\infty} F(\omega)e^{2\pi i t\omega}\, d\omega \qquad\qquad F(\omega) = \int_{-\infty}^{\infty} f(t)e^{-2\pi i\omega t}\, dt$$

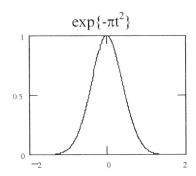

$\exp\{-\pi t^2\}$

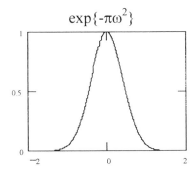

$\exp\{-\pi\omega^2\}$

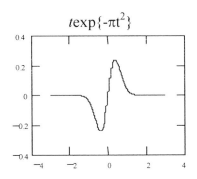

$t\exp\{-\pi t^2\}$

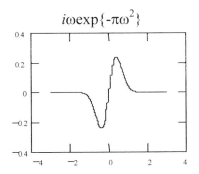

$i\omega\exp\{-\pi\omega^2\}$

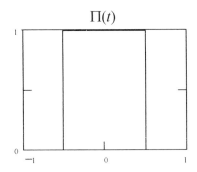

$\Pi(t)$

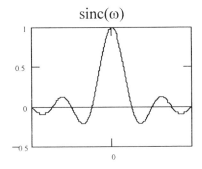

$\mathrm{sinc}(\omega)$

$$f(t) = \int_{-\infty}^{\infty} F(\omega)e^{2\pi it\omega} \, d\omega \qquad\qquad F(\omega) = \int_{-\infty}^{\infty} f(t)e^{-2\pi i\omega t} \, dt$$

$\Lambda(t)$

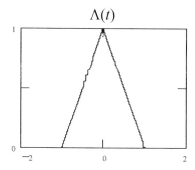

$\text{sinc}^2(\omega)$

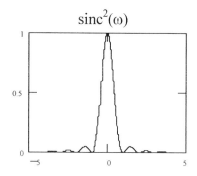

$J_0(2\pi t)$

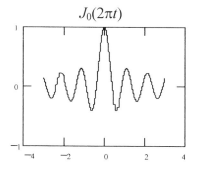

$\Pi(\omega)(1 - \omega)^{-\frac{1}{2}}$

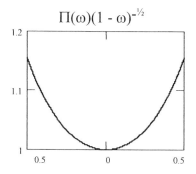

$\cos \pi t^2$

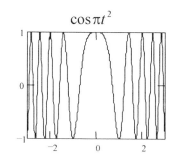

$\cos\left(\dfrac{\omega^2}{4\pi} - \dfrac{\pi}{4}\right)$

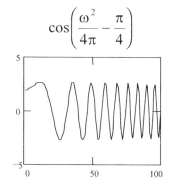

$$f(t) = \int_{-\infty}^{\infty} F(\omega) e^{2\pi i t \omega} \, d\omega \qquad\qquad F(\omega) = \int_{-\infty}^{\infty} f(t) e^{-2\pi i \omega t} \, dt$$

$\sin \pi t^2$

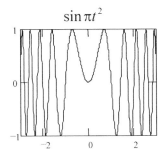

$\cos\left(\dfrac{\omega^2}{4\pi} + \dfrac{\pi}{4}\right)$

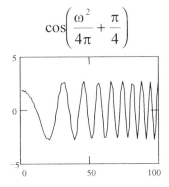

$e^{-|t|}$

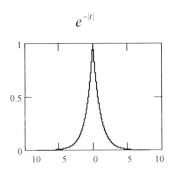

$\dfrac{2}{1+\omega^2}$

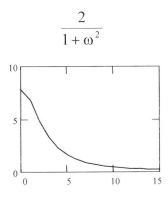

$t e^{-|t|}$

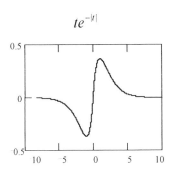

$\dfrac{-4i\omega}{\left(1+\omega^2\right)^2}$

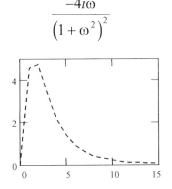

References

Abramowitz, M., and I. Stegun, *Handbook of Mathematical Functions*, Dover, New York, (1968).

Bracewell, R., *The Fourier Transform and Its Applications*, McGraw-Hill Book Co., New York, 381 pp (1965).

Campbell, G. A., and R. M. Foster, *Fourier Integrals for Practical Applications*, D. Van Nostrand Company, Inc., Princeton, NJ (1961).

Champeney, D. C., *A Handbook of Fourier Theorems*, Cambridge University Press, Cambridge, 185 pp (1987).

Cooley, J. W., and J. W. Tukey, 'An algorithm for the machine calculation of complex Fourier Series,' *Mathematics of Computation*, **19**, 297-301 (1965).

Lanczos, C., *Discourse on Fourier Series*, Hafner Publishing Co., New York (1966).

Liverman, T. P. G., *Generalized Functions and Direct Operational Methods*, Prentice-Hall, Inc., Elglewood Cliffs (1964).

Oberhettinger, F., *Fourier Expansions, a Collection of Formulas*, Academic Press, New York (1973).

Oberhettinger, F., *Tables of Fourier Transforms and Fourier Transforms of Distributions*, Springer-Verlag, Berlin (1990).

Papoulis, A., *The Fourier Integral and Its Applications*, McGraw-Hill Book Company, New York (1962).

Papoulis, A., *Signal Analysis*, McGraw-Hill Book Company, New York (1977).

Press, W. H., S. A. Teukolsky, W. T. Vettering, and B. R. Flannery, *Numerical Recipes in Fortran, the Art of Scientific Computing*, Second Edition, Cambridge University Press, Cambridge (1992).

Råde, L. and Westergren, B., *BETA Mathematics Handbook*, Second Edition, CRC Press, Boca Raton, 494 pp (1990).

9. Autocorrelation Functions and Spectra

In the index to the six hundred odd pages of Arnold Toynbee's A Study of History, *abridged version, the names of Copernicus, Galileo, Descartes and Newton do not occur yet their cosmic quest destroyed the medieval vision of an immutable social order in a walled-in universe and transformed the European landscape, society, culture, habits and general outlook, as thoroughly as if a new species had arisen on this planet.*

Arthur Koestler (1905-1983)

We are all familiar with the spectrum of sunlight. As Newton first showed, a prism can separate white light into many different colors. We know today that these colors correspond to different wavelengths of the electromagnetic radiation. We can say equivalently that the spectrum separates light into waves with different frequencies; or separates it into photons with different energies. The energy of each photon is $E = h\nu$, where h is Planck's constant and ν is the frequency. Another way to view this is that we are taking a complicated electromagnetic field and representing it as a sum of pure colors, which correspond to independent oscillations with different frequencies.

In this chapter, I shall discuss spectra of random processes. This is a generalization of the process of breaking sunlight down into its component colors with a prism. Spectral analysis is a method of representing a random process in terms of a set of basis functions—usually sines and cosines, but other sets of functions are also used—that have some desirable properties. Many physical phenomena obey second-order differential equations whose general solutions are sines and cosines; $\sin\omega t$ and $\cos\omega t$ represent possible realizations of such processes. For example, when a piano string is struck by one of the felt hammers in the piano, it vibrates in a sinusoidal manner, producing both a fundamental tone and harmonics at higher frequencies. Here, as in many other physical processes, something is physically vibrating in a manner that is naturally described by sines and cosines. Any harmonic oscillator—where the restoring force is proportional to the displacement—vibrates sinusoidally.

There is an important difference between physics and pure mathematics. We can take *any* random physical process and decompose it—mathematically—into pure sinusoidal components. But although this always is possible mathematically, it may or may not make any sense physically. Even though there is a spectrum, a given physical system may or may not have actual components that vibrate. We should keep this in mind when we are considering the spectrum of a physical process. I explore an example of the difference between the mathematical and physical aspects of spectra in §§9.5 and 9.6.

9.1 Random variables

The term *random variable* means the following. Suppose that the function $f_0(x)$ represents the value of some physical quantity f at each location (or time) x. This function is the result of some set of measurements; if we perform a second set of measurements using exactly the same instruments and procedures, we will find a new function $f_1(x)$ that will be different, both because we have a new realization of the noise in the measurements, and also because the thing we are measuring may have changed. If the quantity being measured changes randomly, we can say that each time, we are measuring a different realization of a random process. We can say that there is a parent population of possible measurements, or possible realizations of the process: each time we measure something, we are randomly selecting a different realization.

As we shall see, there are two different definitions of the spectrum; they are based on the work of Wiener and Khintchine, respectively. Khintchine developed a statistical theory of spectra, based on defining the autocorrelation function (see below) as an average over an ensemble of realizations of a random process. Wiener based his theory on properties of a single function $f(x)$ which is taken to extend from $-\infty$ to ∞. His theory is in no way probabilistic, although for many classes of functions, the two definitions are equivalent.

9.1.1 Stationary processes

A random process is said to be *stationary* if, roughly speaking, its statistics are the same everywhere. Let x be a spatial variable, and $f(x)$ a random function of x. Define the nth moment of f to be[1]

$$\mu_n = E\left[x^n f(x)\right]. \tag{1}$$

In general, each moment might depend on the position x. However, for many random processes, we would expect that, if the physical properties of a system are independent of the position x, then the moments of a property f of the system should not depend on position either. For example, consider again the sur-

---

[1] $E[f]$ denotes the *expected value* of f. The expected value is, roughly, the average over the entire parent population of possible realizations of the random function f.

face of the ocean. Because of wind-generated waves, the surface moves randomly up and down at each point on the surface. Suppose that $f(x)$ represents the height of the surface at the location x. For a large ocean and values of x covering only a small part of that ocean, we would expect the moments of the surface height distribution to be the same, independent of x.

A random process $f(x)$ is *strictly stationary* if all the moments of f are independent of x. In particular, the mean and the autocorrelation function $E[f^*(x)f(x+\tau)]$—which I shall define presently—will be independent of x. A random process is called *wide sense stationary* (often abbreviated WSS) if the mean and second moment are independent of position, even though the higher moments do depend on position. Any strictly stationary process is necessarily WSS.

Exercise 9-1: Show that in the special case where the process is Gaussian, a random process that is wide sense stationary is also strictly stationary.

If a random process is not stationary, at least in the wide sense, then the autocorrelation function as such does not exist. We shall assume that all random processes that we are considering in this book are wide sense stationary, except for §9.3, where I shall briefly discuss random processes with stationary increments.

If f is a random process and x is a spatial variable, rather than a time variable, it is customary to say that $f(x)$ is *homogeneous* (rather than stationary) if $E[f(x)]$ and $E[f^*(x)f(x+\tau)]$ are independent of x. However, I may also use the term *stationary* for homogeneous processes.

We should note that it is generally not possible to tell whether or not a particular random process $f(x)$ is stationary. Suppose we make measurements for $x_0 < x < x_1$. We cannot tell, from that set of measurements, whether or not the statistical properties of $f(x)$ will be different for $x < x_0$ or $x > x_1$. We might divide the measurements we have into two equal intervals, calculate the statistical properties for each interval, and see how closely they agree. If the difference of the means of the two parts is larger than the variance of either part, for example, we might conclude that the process is nonstationary. But usually such tests give only equivocal results, and we must simply take it as an article of faith that the process is stationary.

9.1.2 Ergodic processes

We can define $E[f(x)f^*(x+\tau)]$ in one of two ways. On the one hand, we can hold x fixed and perform an average over an ensemble of realizations of f. Or we could take a single realization of f and perform an average over x. Let us define the so-called *ensemble average* by

$$E_e[f(x)] \equiv \lim_{N \to \infty} \frac{1}{N} \sum_{n=1}^{N} f_n(x), \tag{2}$$

where f_n is the nth realization of the random process f. The *spatial average* is

$$E_s[f(x)] = \lim_{T \to \infty} \frac{1}{2T} \int_{-T}^{T} f(x)\, dx, \tag{3}$$

where f is a single realization of that process. In general, these two averages will not be the same, but for many random systems, they are. A random process f is said to be *ergodic* if $E_e[f(x)] = E_s[f(x)]$. That is, the spatial and ensemble averages are the same.

In general, there is no way to demonstrate a physical process is ergodic, since that would require evaluating an infinite amount of data. We may assume that a particular process is ergodic, for lack of any other information about it, or to make a particular problem tractable, but we cannot prove it.

9.1.3 Autocorrelation function and spectrum

There are two ways to define the spectrum of a random process, depending on whether there is a single realization of the process available or an ensemble of realizations. We can imagine a random process as being one where there is a parent population that contains every possible realization of the process; an experiment that we do selects one realization at random. In some cases, we can perform the experiment repeatedly, drawing each time from the same parent population. In other cases, we can perform the experiment only once. *E.g.*, we could measure the temperature at a certain airport every day for 100 years—say, every day of the twentieth century. We cannot repeat this experiment, since the measurements for some other 100-year interval might be systematically different from the ones for the twentieth century. Similarly, measurements made somewhere else would not necessarily have the same statistical properties: would measurements made in London, England, come from the same parent population as those made in Mexico City?

To define the spectrum of $f(x)$ rigorously, we start with the autocorrelation function: as we shall see, the spectrum is defined to be the Fourier transform of an autocorrelation function. Consider the situation where we can repeat an experiment many times. Let $f(x)$ be a realization of the random variable; we define its autocorrelation function to be the expected value over a set of realizations of[2]

---

[2] The maximum value of $C_f(\tau)$ occurs at $\tau = 0$. Some authors define the autocorrelation function to be $C_f(\tau) / C_f(0)$.

$$C_f(\tau) = E_e[f^*(x)f(x+\tau)]. \tag{4}$$

Equation (4) is the basis of a statistical theory of spectra.[3] This definition only makes sense for processes that are at least wide-sense stationary. The autocorrelation function does not exist for processes that are not stationary, since its value would depend on x as well as τ. The variable τ is often called the *lag*, and $f^*(x)f(x+\tau)$ called a *lagged product*.

By definition, the *spectrum* of this random process is[4]

$$S_f(\omega) = \int_{-\infty}^{\infty} C_f(\tau)e^{-2\pi i\omega\tau}\,d\tau. \tag{5}$$

Equation (4) represents an estimate of average over the parent population; the autocorrelation function is a property of the parent population, as is $S_f(\omega)$. It is not exactly represented by any single realization of $f(x)$, or even by any finite set of realizations. Since the autocorrelation function is Hermitian (it is even if it involves a real process), the spectrum is real. For a real-valued process f, $S_f \geq 0$.

Now suppose we are given only a single realization of $f(x)$, but we still want to know its spectrum. We can estimate its autocorrelation function from the following limit:

$$\mathcal{C}_f(\tau) = \lim_{T\to\infty} \frac{1}{2T}\int_{-T}^{T} f^*(x)f(x+\tau)\,dx. \tag{6}$$

In order for this limit to exist,

$$\int_{-\infty}^{\infty} |f(x)|^2\,dx < \infty. \tag{7}$$

This must be true for any physical process, since the energy involved is always finite. This form of the autocorrelation function is different from equation (4), for while $C_f(\tau)$ is defined as being a property of a random process, $\mathcal{C}_f(\tau)$ is defined in terms of a specific function $f(x)$ or a specific realization of

---

[3] This is the form used by Khintchine (1934), as referenced in Champeney (1987) pp 103-104. See also Gardner (1990).

[4] I should admit to the reader that I have one bad habit: when writing Fourier transforms, I reflexively denote the limits of integration as being $\pm\infty$. For abstract discussions, this may be adequate, but we should realize that, when we are discussing actual data (or potential data), we may have to use finite limits.

a random process. Wiener[5] developed his spectral theory from this form of the autocorrelation function. Unlike the Khintchine version [equation (4)], this autocorrelation function always exists, whether or not f is stationary.[6]

In the Wiener theory, the spectrum of a single realization is the Fourier transform of $\mathscr{C}_f(\tau)$:

$$\mathscr{S}_f(\omega) = \int_{-\infty}^{\infty} \mathscr{C}_f(\tau) e^{-2\pi i \tau \omega} \, d\tau \ . \tag{8}$$

9.1.4 The Wiener-Khintchine theorem

We have two definitions of the autocorrelation function. Are they the same thing? Remember that a random process is ergodic if the autocorrelation functions defined by equations (4) and (6) are the same. That is, if the average over an ensemble of realizations is equal to the average over x. Most physical systems that we are interested in are ergodic. Some are not, however, and in these cases, the two kinds of autocorrelation function will be different. However, in many situations, it is not possible to evaluate both forms of the autocorrelation function for the same physical system, so we simply *assume* that the system is ergodic. We make that assumption, but it may or may not be true—there is no way to tell for sure without having an infinite amount of data.

In one sense, the spectrum of a random variable is a property of the parent population. We can estimate its value from one or several sets of measurements. Whether we consider $f(x)$ to be simply a single realization of a random process or a function in its own right, we can estimate its spectrum from its *periodogram*, which is the squared magnitude of its Fourier transform:

$$P_{f,T}(\omega) \equiv \left| \int_{-T}^{T} f(x) e^{-2\pi i \omega x} \, dx \right|^2 = \left| F_T(\omega) \right|^2 \tag{9}$$

where

---

[5] Wiener (1964), as referenced by Champeney (1987) pp 102 *ff.* See also Gardner (1990) §3.2. Wiener's theory is actually not a statistical theory at all, since it does not matter whether or not f is a random variable.

[6] Note that, when we assume that $f(x)$ is stationary, and that $f(x)$ is not identically zero over some interval $[a,b]$ where $a > b$, then $f(x)$ must continue to fluctuate even as $x \to \pm\infty$. Needless to say, nothing goes on forever, so no physical process can be exactly stationary. The present discussion of spectra has to be heuristic, rather than mathematically rigorous: see Gardner (1990, Chapter 10) for a rigorous treatment.

$$F_T(\omega) \equiv \int_{-T}^{T} f(x)e^{-2\pi i x \omega}\,dx. \tag{10}$$

To see how the periodogram is related to the autocorrelation function defined in equation (6) and the spectrum in equation (8), substitute the definition of $F(\omega)$ into equation (9):

$$\mathcal{C}_f(\tau) = \frac{1}{2T}\int_{-T}^{T}\int_{-\infty}^{\infty}\int_{-\infty}^{\infty} F^*(\omega)\,F(\omega')e^{2\pi i \omega' \tau}\,e^{2\pi i(\omega-\omega')x}\,d\omega\,d\omega'\,dx \tag{11}$$

(note that the limits of the x-integral are $\pm\,T$, not $\pm\infty$). Note also that I have used the shift theorem here to represent the Fourier transform of $f(x+\tau)$. Using the representation of the δ-function

$$\delta(\omega) = \int_{-\infty}^{\infty} e^{-2\pi i \omega x}\,dx \approx \int_{-T}^{T} e^{-2\pi i \omega x}\,dx \tag{12}$$

from equation (8) of Chapter 8, this reduces to

$$\mathcal{C}_f(\tau) \approx \int_{-\infty}^{\infty}\int_{-\infty}^{\infty} F^*(\omega)\,F(\omega')e^{2\pi i \omega' \tau}\,\delta(\omega-\omega')\,d\omega\,d\omega'$$

$$\approx \int_{-\infty}^{\infty} |F(\omega)|^2\,e^{2\pi i \omega \tau}\,d\omega$$

$$= \int_{-\infty}^{\infty} P_{f,T}(\omega)e^{2\pi i \omega \tau}\,d\omega \tag{13}$$

in the limit as $T\to\infty$. So the periodogram is

$$P_f(\omega) = \int_{-T}^{T} \mathcal{C}_f(\tau)e^{2\pi i \omega \tau}\,d\tau. \tag{14}$$

Finally, the spectrum is[7]

$$\mathcal{S}_f(\omega) = \lim_{T\to\infty} E\left[\frac{1}{2T}\int_{-T}^{T}\mathcal{C}_f(\tau)e^{-2\pi i \omega \tau}d\tau\right] = \lim_{T\to\infty} E[P_{f,T}(\omega)]. \tag{15}$$

---

[7] I use the notation $P_{f,T}$ here because the spectrum $\mathcal{S}$ is a property of the parent population of f, while the periodogram P is a calculated quantity.

Using either equation (5) or equation (8) as its definition, the spectrum is the Fourier transform of the autocorrelation function. This is known as the *Wiener-Khintchine theorem*.

Why don't we use equation (9) as a definition of the spectrum? We shall see here and in a subsequent section that we must be careful of how we define the limiting processes involved. The autocorrelation function defined in equation (6) always exists, which is why we started there. The equivalent definition of the spectrum would involve the following limit. Suppose that we define the partial periodogram to be

$$P_{0,T}(\omega) = \left| \int_0^T f(t) e^{-2\pi i t \omega} dt \right|^2. \tag{16}$$

We run into a problem if we try to define the spectrum to be

$$\mathcal{S}_f(\omega) \stackrel{?}{=} E\left[\lim_{T \to \infty} P_{0,T}(\omega) \right]. \tag{17}$$

Papoulis[8] points out that, for this expression to be valid, it would be necessary both that $\mathcal{S}_f(\omega) \to P_{0,T}(\omega)$ and that the variance of $\mathcal{S}_f(\omega) \to 0$. He shows that the first requirement is met as long as

$$\int_{-\infty}^{\infty} \tau C_f(\tau) d\tau < \infty. \tag{18}$$

However, the variance of $\mathcal{S}_f(\omega)$ does not approach zero as $T \to \infty$. In §9.2.3 I show that the variance of $P_{0,T}(\omega)$ remains constant as T increases. To see why this should be so, let $\delta\omega$ be the spacing between points where we evaluate $P_{0,T}(\omega)$. Since $\delta\omega \propto 1/T$, the number of points where we evaluate $P_{0,T}$ increases in proportion to T. So as T increases, the number of independent estimates between 0 and ω_{max}, say, increases, but the accuracy does not. Therefore, the variance of $S_f(\omega)$ does not approach 0 as $T \to \infty$.

Therefore, we define the spectrum by

$$\mathcal{S}_f(\omega) = \lim_{T \to \infty} E\left[P_{0,T}(\omega) \right]. \tag{19}$$

This limit always exists.

Exercise 9-2: Is a stationary random process always ergodic? Is an ergodic process always stationary?

---

[8] Papoulis (1965), pp 343-344; Middleton (1960, pp 140 *ff*); or Bendat and Piersol (1986, Chapter 5).

Exercise 9-3: Explain why equation (4) is only valid if the random process in question is at least wide-sense stationary.

9.1.5 Cross-correlation and cross-spectrum

The *cross-correlation* of two functions f and g

$$C_{fg}(\tau) = E\left[f^*(x)g(x+\tau)\right].$$ (20)

Given one realization each of $f(x)$ and $g(x)$, we can estimate the cross-correlation from

$$\mathcal{C}_{fg}(\tau) = \underset{T \to \infty}{Lim} \frac{1}{2T} \int_{-T}^{T} f^*(x)g(x+\tau)\,dx$$ (21)

if the limit exists. Again, note that equation (20) involves an *ensemble* of functions f and g, taken from different populations, while equation (21) involves a single realization of each random process.

Just as the spectrum of f is defined to be the Fourier transform of its auto-correlation function, the *cross spectrum* of f and g is the Fourier transform of the cross-correlation function [equation (21)]:

$$S_{fg}(\omega) = \int_{-\infty}^{\infty} C_{fg}(\tau)e^{-2\pi i\tau\omega}\,d\tau.$$ (22)

Let F and G be the Fourier transforms of f and g. Define the *cross-periodogram* to be

$$P_{fg,T}(\omega) = F_T^*(\omega)G_T(\omega),$$ (23)

where F_T was defined in equation (8). How are the cross-spectrum and the cross-periodogram related? Substitute the definitions of F and G into equation (23) and take the Fourier transform:

$$\mathcal{S}_{fg}(\omega) = \int_{-\infty}^{\infty}\int_{-\infty}^{\infty} \left[\int_{-\infty}^{\infty} F^*(\omega')e^{-2\pi i\omega'x}\,d\omega'\right]$$

$$\left[\int_{-\infty}^{\infty} G(\omega'')e^{2\pi i\omega''(x+\tau)}\,d\omega''\right]e^{-2\pi i\tau\omega}\,dx\,d\tau$$

$$= \int_{-\infty}^{\infty}\int_{-\infty}^{\infty}\int_{-\infty}^{\infty}\int_{-\infty}^{\infty} F^*(\omega')G(\omega'')e^{-2\pi i(\omega'-\omega'')x}e^{-2\pi i(\omega-\omega'')\tau}\, d\omega'\, d\omega''\, dx\, dt \, . \tag{24}$$

Rearranging this and using equation (3),

$$\mathscr{S}_{fg}(\omega) = \int_{-\infty}^{\infty}\int_{-\infty}^{\infty} F^*(\omega')G(\omega'')\delta(\omega'-\omega'')\delta(\omega-\omega'')d\omega'd\omega''$$

$$= F^*(\omega)G(\omega). \tag{25}$$

We can see that, analogously with equation (21),

$$P_{fg,T}(\omega) = \int_{-T}^{T}\mathscr{C}_{fg}(\tau)e^{-2\pi i\omega\tau}d\tau \, . \tag{26}$$

This result is analogous to the Wiener-Khintchine theorem. The *definition* of the cross-spectrum is [*cf.* equation (29)].

$$\mathscr{S}_{fg}(\omega) \equiv \lim_{T\to\infty} E\big[P_{fg,T}(\omega)\big]. \tag{27}$$

Exercise 9-4: Let $f(x)$ be a random process, and let $F(\omega)$ be its Fourier transform. Why can't we define an average Fourier transform $F_a = <F>$ and find the spectrum $\mathscr{S}_f$ from that? What fundamental difference is there between averaging complex numbers and real numbers?

9.1.6 Summary

To summarize: there are two ways to define the autocorrelation function, the spectrum, and the cross-spectral density:

$$C_f(\tau) = E_e\big[f^*(x)f(x+\tau)\big], \tag{28}$$

$$S_f(\omega) = \int_{-\infty}^{\infty} C_f(\tau)e^{-2\pi i\omega\tau}\, d\tau, \tag{29}$$

$$C_{fg}(\tau) = E_e\big[f^*(x)g(x+\tau)\big], \tag{30}$$

$$S_{fg}(\omega) = \int_{-\infty}^{\infty} C_{fg}(\tau)e^{-2\pi i\tau\omega}\,d\tau, \tag{31}$$

$$P(\omega) = \left|F(\omega)\right|^2, \tag{32}$$

$$\mathscr{C}_f(\tau) = \underset{T\to\infty}{Lim}\frac{1}{2T}\int_{-T}^{T} f^*(x)\,f(x+\tau)\,dx, \tag{33}$$

$$\mathscr{S}_f(\omega) = \int_{-\infty}^{\infty} \mathscr{C}_f(\tau)e^{-2\pi i\omega\tau}\,d\tau, \tag{34}$$

$$\mathscr{C}_{fg}(\tau) = \underset{T\to\infty}{Lim}\frac{1}{2T}\int_{-T}^{T} f^*(x)g(x+\tau)\,dx, \tag{35}$$

$$\mathscr{S}_{fg}(\omega) = \int_{-\infty}^{\infty} \mathscr{C}_{fg}(\tau)e^{-2\pi i\tau\omega}\,d\tau, \tag{36}$$

$$P_{fg}(\omega) = F^*(\omega)G(\omega). \tag{37}$$

I hope that placing these equations here will make these relationships easier to comprehend. Note again that equation (28) is defined in terms of a population of random functions f, while equation (33) involves a single realization of this random process.

　　Each pair of definitions involves an averaging process of some kind. In equation (28) there is an average over an ensemble of realizations; in equation (33), taking the limit as $T\to\infty$ is also a kind of averaging process. Only for ergodic processes are these two definitions exactly equivalent.

9.2 Estimating the spectrum

Given a random process $f(x)$, we can estimate its spectrum from the periodogram $P_f(\omega) = |F(\omega)|^2$, which is a noisy estimate of $\mathscr{S}_f$. It turns out that the variance of the periodogram at each frequency is equal to its expected value at that frequency, which leads to very spiky spectra.[9] If more than one realization is available, we can reduce the noise by averaging the spectra.

---

[9] See Press *et al.* (1992).

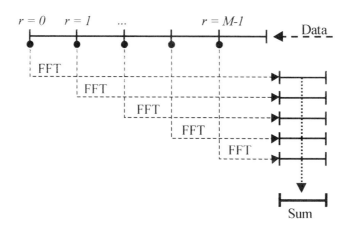

Figure 9-1. Divide the data into M parts; calculate the spectrum of each part; and sum them.

When only one realization is available, we can reduce the nosiness of our estimate of $\mathscr{S}_f$ by dividing the time series into M equal parts; finding the spectrum of each part; and then averaging them—see Figure 9-1. We can call this procedure *segmenting* the data. The price we pay doing this is reduced spectral resolution. Using equation (85) of §8.5, we can see that the spectral resolution is now $M\delta\omega$ instead of $\delta\omega$.

Instead of dividing $f(x)$ into M parts, we can calculate the periodogram using all of the data; the analog of taking the M individual spectra and averaging them is taking M consecutive values of P_f and averaging *them* (see Figure 9-2).[10] Call this method *smoothing* the periodogram. This produces a periodogram with N/M values.

The two operations are mathematically equivalent. As a practical matter, which way you do it probably doesn't matter. I shall explore the two methods in the next two sections, and show that they lead to exactly the same results.

9.2.1 Find the periodograms of M parts and sum them, or…

We can prove that the two processes, smoothing the periodogram or segmenting the data, are equivalent. Suppose that we are given a series $f(x_i)$, $i = 0, ..., N - 1$. Furthermore, let N be the product of two integers: $N = MP$. We want to show that dividing the data into M equal subsets of P points each is the equivalent of calculating the periodogram from all of the data and then summing over P adjacent points. The DFT of f is

---

[10] See Brockwell and Davis (1991, §10.4) or Press *et al.* (1992, pp 544 *ff*).

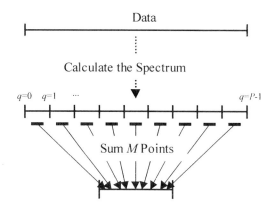

Figure 9-2. Calculate the spectrum; then sum each group of M consecutive points.

$$F_k = \sum_{j=0}^{N-1} f_j e^{-2\pi ijk/N}, \quad k = 0, \cdots, N-1, \tag{38}$$

where $f_j = f(x_j)$. The periodogram is

$$P_k = |F_k|^2 = \sum_{j=0}^{N-1}\sum_{j'=0}^{N-1} f_j f_{j'} e^{-2\pi i(j-j')k/N}. \tag{39}$$

If we let $j'' = j' - j$, then

$$P_k = \sum_{j''=0}^{N-1} e^{-2\pi ij''k/N} \sum_{j=0}^{N-1} f_j f_{j+j''}. \tag{40}$$

But the summation over j is just the autocorrelation function of f at lag j''. Using the definition

$$C_{j''} \equiv \sum_j f_j f_{j+j''}, \tag{41}$$

we see that

$$P_k = \sum_{j''=0}^{N-1} e^{-2\pi ij''k/N} C_{j''}. \tag{42}$$

Note that the lag j'' runs from 0 to N. This is just the Wiener-Khintchine theorem. Note that, since $C_{j''} = C_{-j''}$, $P_k = P_{-k}$, and P_k is real.

In many cases, the autocorrelation is zero for lags larger than some maximum (nonzero) lag. Suppose that $C_j = 0$ for $j \geq P$. Then we can eliminate the part of the sum from P to N without losing any information. If we do this, we can find the periodogram of each of $M \equiv N / P$ (we will assume that P divides N: usually N and P are powers of 2). So, let $j'' = rP + q$ and $j = r'P + q'$ and substitute this into equation (39). Then, instead of using equation (42), we can estimate the periodogram from

$$P_k = \sum_{q=0}^{P-1} e^{-2\pi i q k / N} \sum_{r'=0}^{M-1} \sum_{j'=0}^{P-1} f_{r'M+j'} f_{r'M+j'+q}$$

$$= \sum_{q=0}^{P-1} e^{-2\pi i q k / N} \sum_{r'=0}^{M-1} \mathscr{C}_{r',q} = \sum_{q=0}^{P-1} e^{-2\pi i q k / N} M \langle \mathscr{C}_q \rangle, \qquad (43)$$

where $\mathscr{C}_{r,q}$ denotes the autocorrelation function calculated from the rth part of the series—in other words,

$$\mathscr{C}_{r,q} \equiv \sum_{j=0}^{P-1} f_{rM+j} f_{rM+j+q} \qquad (44)$$

—and $M \langle \mathscr{C}_j \rangle$ is the sum of the M autocorrelation functions of length P. Note that $k = 0, 1, \ldots, P\text{-}1$ because $\langle \mathscr{C}_j \rangle$ has only P values. This is the result of dividing the series f_j into M parts and summing the periodograms.

Why has the noise been reduced? Not because we have averaged more combinations of f_i—in fact, we have used fewer. What we have accomplished is *not* adding the parts of the autocorrelation function $\mathscr{C}_k$ where $k \geq P$, where we have assumed that $f_j f_{j+k} = 0$ for $k \geq P$. If we had added them into the average, we would have introduced extra noise, but no extra information.

Exercise 9-5: What happens if $\langle f_j f_{j+k} \rangle \neq 0$ for some values of $k \geq P$?

9.2.2 …Calculate one periodogram and smooth it

Instead of dividing the data into M parts and summing the periodograms, we can sum M adjacent values of (the original) periodogram P. To this end, let $j = rP + q$, $r = 0, \ldots, M\text{-}1$ and $q = 0, \ldots, P\text{-}1$. Also, let $k = \ell P + s$. Then, substituting these expressions into equation (42),[11]

$$P_{\ell M + s} = \sum_{r=0}^{M-1} \sum_{q=0}^{P-1} e^{-2\pi i (rP + q)(\ell M + s) / MP} \mathscr{C}_{rP + q}. \qquad (45)$$

---

[11] We use a sum, not an average, to preserve the normalization.

Define

$$P_\ell \equiv \sum_{s=0}^{M-1} P_{\ell M + s} = \sum_{s=0}^{M-1} \sum_{r=0}^{M-1} \sum_{q=0}^{P-1} e^{-2\pi i (rP + q)(\ell M + s)/MP} \mathcal{C}_{rP+q} . \tag{46}$$

Rearranging the sums and remembering that factors of $\exp\{-2\pi i r\} = 1$ for any integer r, this becomes

$$P_\ell = \sum_{r=0}^{M-1} \sum_{q=0}^{P-1} e^{-2\pi i q\ell/P} \mathcal{C}_{rP+q} \sum_{s=0}^{M-1} e^{-2\pi i (rP+q)s/MP} . \tag{47}$$

As before, assume that $\mathcal{C}_q = 0$ for $q \geq P$. Then $\mathcal{C}_{rP+q}$ does not depend on r: $\mathcal{C}_{rP+q} = \mathcal{C}_{r'P+q}$ for any r and r'. So

$$P_\ell = \sum_{q=0}^{P-1} e^{-2\pi i q l/P} \langle \mathcal{C}_q \rangle \sum_{r=0}^{M-1} \sum_{s=0}^{M-1} e^{-2\pi i (rP+q)s/MP}$$

$$= \sum_{q=0}^{P-1} e^{-2\pi i q l/P} \langle \mathcal{C}_q \rangle \sum_{r=0}^{M-1} \sum_{s=0}^{M-1} e^{-2\pi i rs/M} e^{-2\pi i qs/MP} . \tag{48}$$

But

$$\left(1 - e^{-2\pi i s/M}\right) \sum_{s=0}^{M-1} e^{-2\pi i rs/M} = 1 - e^{-2\pi i s} = 0 \tag{49}$$

unless $s = 0$. In that case, we get the simple result that

$$P_\ell = M \sum_{q=0}^{P-1} e^{-2\pi i q\ell/P} \langle \mathcal{C}_q \rangle ; \tag{50}$$

the factor M comes because of the sum over r, which adds 1 to itself M times. Comparing this with equation (43), we see that we get the same result whichever way we calculate the smoothed periodogram.

9.2.3 *Variance of the autocorrelation function and the spectrum*

Consider the following limits that define the autocorrelation function and the spectrum.[12] Equation (6) defines $\mathcal{C}_f(\tau)$ in terms of a limiting process. Suppose that we start off with finite limits to the integration: say, from $-T_0$ to $+T_0$. If then we increase T to a larger value, say $T_1 > T_0$, we will have more samples

---

[12] See Jenkins and White (1969), pp 222-223.

of the random process at each lag τ, and the random errors will be reduced. Therefore, the autocorrelation function approaches a limit as $T \to \infty$.

But consider what happens if we try to do the same thing with the spectrum. We estimate the spectrum $\mathcal{S}_f(\omega)$ by calculating the periodogram, which is defined to be the magnitude squared of the Fourier transform of f—see equation (9). We already saw, in §9.1.4, that we needed to define the spectrum as the limiting value of the expectation of $\mathcal{S}_{0,T}$, rather than the expectation of the limit. Should we define it to be

$$\mathcal{S}_f(\omega) = \underset{T \to \infty}{Lim} \left| \int_{-T}^{T} f(x) e^{-2\pi i \omega x} \, dx \right|^2 = \underset{T \to \infty}{Lim} P_{f,T}(\omega)? \tag{51}$$

That is, from a computational standpoint, can we envision the spectrum as the limit of the periodogram as we take more and more data? If we do, we should notice something peculiar: no matter how large T is, the variance of $P_f(\omega)$ remains the same: $\text{var}[P_f(\omega)] = E[P_f(\omega)]$. Unlike the autocorrelation function, which approaches a limiting value as $T \to \infty$, $P_f(\omega)$ does not. The reason for this is that as T becomes larger, the step size $\delta\omega = 1 / (2T)$ in the Fourier transform decreases (or the number of points in the DFT increases) in proportion. In other words, the number of samples *per interval* $\delta\omega$ remains constant. Therefore, the variance of $P_f(\omega)$ remains constant. On the other hand, if we are satisfied with a certain spectral resolution $\delta\omega_0$, corresponding to some $T_0 = 1 / (2\delta\omega)$, then as T increases further, we can sum adjacent values of $P_f(\omega)$, thus reducing the variance (as we have seen in the previous section). Alternatively, if we had many realizations of f available, we could average the spectra over those realizations and reduce the variance that way. This is one reason why we define the spectrum to be the Fourier transform of the autocorrelation function, even though we use the periodogram to *estimate* the spectrum.

9.3 The structure function

The *structure function* is closely related to the autocorrelation function. It can be used to define the spectrum of certain kinds of nonstationary processes. Consider a random process $f(x)$. Define the structure function to be

$$D_f(\tau) \equiv E\left[\left| f(x + \tau) - f(x) \right|^2 \right]. \tag{52}$$

That is, the structure function is the variance of the *change in f* over a distance τ. Note that $D_f(0) = 0$ identically, and D_f is non-negative definite. If the variance of $f(x) = \sigma_f^2$ is finite, then $D_f(\infty) = \sigma_f^2$. If $f(x)$ is real,

$$D_f(\tau) \equiv E\left[f^2(x) - 2f(x)f(x+\tau) + f^2(x+\tau)\right]. \tag{53}$$

If f(x) *is stationary,* then $E[f^2(x)] = E[f^2(x+\tau)]$ and

$$D_f(\tau) \equiv 2E\left[f^2(x)\right] - 2E\left[f(x)f(x+\tau)\right]. \tag{54}$$

Using equation (28), the structure function of a stationary process is

$$D_f(\tau) \equiv 2\left[C_f(0) - C_f(\tau)\right]. \tag{55}$$

Exercise 9-6: Prove that the structure function of a stationary process is even. Is the structure function of a nonstationary process necessarily even?

We can extend the class of random processes that possess well-defined structure functions.[13] If $f(x)$ is stationary, then the structure function [equation (52)] will be independent of x. However, it might be that even though $f(x)$ is *not* stationary [*i.e.*, the autocorrelation function defined in equation (28) *does* depend on x], it still may happen that the statistics of the differences $f(x+\tau) - f(x)$ are independent of x. Such a process is called a process with *stationary increments*.[14] As with the definition of stationarity in §9.1.1, a process has stationary increments if and only if all of the moments of $f(x+\tau) - f(x)$ are independent of x. Even though the autocorrelation function does not exist for this process, the structure function does. Just as, for a stationary process, we define the spectrum to be the Fourier transform of the autocorrelation function [see equation (5)], we can define the spectrum of an SI process in terms of the structure function.

To start, we need to consider the following property of SI processes. Suppose that $g(x)$ is a stationary process. Then

$$f(x) = \int_{x_0}^{x} g(t)\,dt + c \tag{56}$$

is an SI process, where c is any constant.[15] To see this, note that

$$E\left[\left|f(x+\tau) - f(x)\right|^2\right] = E\left[\left|\int_x^{x+\tau} g(t)\,dt\right|^2\right] = E\left[\int_x^{x+\tau} g(t)\,dt \int_x^{x+\tau} g(t')\,dt'\right]$$

---

[13] The following discussion comes from Monin and Yaglom (1975, pp 86 *ff*); see also Yaglom (1987, §23) and Gardner (1990, §3.2).

[14] In order to simplify the text, I shall refer to a random process with stationary increments as an *SI process.*

[15] See Monin and Yaglom (1975, p 80).

$$= E\left[\int_0^\tau \int_0^\tau g(t+x)g(t'+x)\,dt'\,dt\right] = \int_0^\tau \int_0^\tau E\left[g(t+x)g(t'+x)\right]dt'\,dt, \quad (57)$$

where I have substituted $t + x$ for t, and similarly for t'. But by the definition of a stationary process,

$$C_g(t - t') = E\left[g(t+x)g(t'+x)\right] \tag{58}$$

(the autocorrelation function of g) depends only on $t - t'$; in particular, it is independent of x. Therefore,

$$D_f(\tau) = E\left[\left|f(x+\tau) - f(x)\right|^2\right] = \int_0^\tau \int_0^\tau C_g(t - t')\,dt\,dt' \tag{59}$$

Since $g(x)$ is stationary, it has a spectrum. Define $G(\omega)$ to be the Fourier transform of $g(x)$, so

$$g(x) = \int_{-\infty}^\infty G(\omega)e^{2\pi i \omega x}\,d\omega. \tag{60}$$

Substituting this into equation (56), and this into (59), we get

$$D_f(\tau) = \int_0^\tau \int_0^\tau \int_{-\infty}^\infty e^{2\pi i(t-t')\omega}\left|G(\omega)\right|^2 d\omega\,dt'\,dt$$

$$= \frac{i}{2\pi} \int_0^\tau \int_{-\infty}^\infty \frac{\left(e^{-2\pi i\omega\tau} - 1\right)}{\omega} e^{2\pi i\omega t}\left|G(\omega)\right|^2 d\omega\,dt$$

$$= \frac{1}{4\pi^2} \int_{-\infty}^\infty \frac{\left(e^{-2\pi i\omega\tau} - 1\right)\left(e^{2\pi i\omega\tau} - 1\right)}{\omega^2}\left|G(\omega)\right|^2 d\omega. \tag{61}$$

Finally,[16]

$$D_f(\tau) = \frac{1}{\pi^2} \int_0^\infty (1 - \cos 2\pi\omega\tau)\frac{\left|G(\omega)\right|^2}{\omega^2}\,d\omega. \tag{62}$$

---

[16] Using $e^{ix} = \cos x + i\sin x$.

This is a generalization of the Wiener-Khintchine theorem. For SI processes, the structure function is related to the "spectrum" $|G(\omega)|^2 / \omega^2$ by a transform that is very similar to a Fourier transform.

Exercise 9-7: Does $D_f(\tau)$ have reasonable values at $\tau = 0$? at $\tau = \infty$?

9.4 Filters

A filter is a device—either physical or mathematical—that modifies a given function $f(x)$ by modifying its spectrum. In electrical engineering, a filter is a device that may block the low-frequency part of a signal,[17] allowing only high frequencies to pass. Or it may work the other way around, blocking high frequencies and passing low frequencies. Or it may block both high and low frequencies, passing only a band of intermediate frequencies.

A filter that blocks low frequencies is called a *high-pass* filter; one that blocks high frequencies, a *low-pass* filter. A filter that blocks both high and low frequencies, passing only frequencies between two limits ω_1 and ω_2, is called a *bandpass* filter. A filter is a passive device, so the amount of energy coming out of the filter at each frequency is proportional to the amount of power that went in at that frequency; the constant of proportionality will depend on frequency.

One important use of filters is to reduce the noise in a given set of data. If we know *a priori* that the process we are measuring only contains frequencies between $\omega = 0$ and $\omega = \omega_c$, say, but there is noise at all frequencies up to some $\omega_k > \omega_c$, we can improve the signal-to-noise ratio by lowpass-filtering the data with a cutoff frequency ω_f; *i.e.*, it passes all frequencies $\omega < \omega_f$ (see Figure 9-3).

Mathematically, we can represent a filter by a function $h(x)$. The action of filtering is achieved by a convolving $f(x)$ with $h(x)$. Let $F(\omega)$ and $H(\omega)$ be the Fourier transforms of $f(x)$ and $h(x)$, respectively. Then the output of the filter is

$$g(u) = \int_{-\infty}^{\infty} f(x)h(u - x)\,dx. \tag{63}$$

Using the convolution theorem, the Fourier transform of $g(u)$ is

$$G(\omega) = F(\omega)H(\omega). \tag{64}$$

Lastly, the spectrum $S(\omega)$ is

---

[17] By *signal* here, I mean any function $f(t)$.

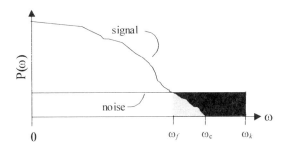

Figure 9-3. Signal and noise power.

$$S(\omega) \;=\; \left|F(\omega)\right|^2 \left|H(\omega)\right|^2. \tag{65}$$

Figure 9-4a shows a random function with a spectrum proportional to the frequency ω. Figures 9-4b and -c show the same function filtered with low-pass and high-pass filters, respectively. Figure 9-4d shows the original spectrum (light line), and the low- and high-pass filtered spectra.

In light of the comment at the end of the last section, we should note that any real filter—i.e., one where $h(x)$ is real—must behave the same way at positive and negative frequencies. As an example, suppose we are given a real filter $h(x)$ whose Fourier transform, we are told, is $H(\omega)$, a Gaussian centered on the frequency ω_0:

$$H(\omega) \;=\; e^{-\pi(\omega-\omega_0)^2}. \tag{66}$$

Then the result of using $h(x)$ to filter $f(x)$ is

$$g(x) \;=\; \int\limits_{-\infty}^{\infty} F(\omega)\,H(\omega)\,e^{2\pi i x \omega}\,d\omega \tag{67}$$

which in this case gives

$$g(x) \;=\; \int\limits_{-\infty}^{\infty} F(\omega)\,e^{-\pi(\omega-\omega_0)^2}\,e^{2\pi i x \omega}\,d\omega$$

$$=\; e^{2\pi i x \omega_0} \int\limits_{-\infty}^{\infty} F(\omega-\omega_0)\,e^{-\pi\omega^2}\,e^{2\pi i x \omega}\,d\omega\;; \tag{68}$$

note that I have used the shift theorem here.

The problem here, of course, is that while $g(x)$ is obviously real—both $f(x)$ and $h(x)$ are real—it looks like equation (68) will give a complex result. What went wrong?

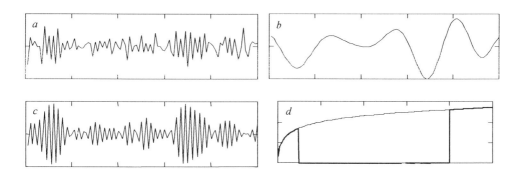

Figure 9-4. *a*) Noise time series; *b*) low-pass filtered series; *c*) high-pass filtered series; *d*) original spectrum (dashed line) and filtered spectra (solid lines on left and right).

We should have remembered that both F and H are Hermitian. Consequently, equation (66) should have been

$$H(\omega) = e^{-\pi(\omega - \omega_0)^2} + e^{-\pi(\omega + \omega_0)^2}. \tag{69}$$

And equation (68) should have read

$$g(x) = \int_{-\infty}^{\infty} F(\omega) e^{2\pi i x \omega} \left[e^{-\pi(\omega - \omega_0)^2} + e^{-\pi(\omega + \omega_0)^2} \right] d\omega$$

$$= \int_{-\infty}^{\infty} \left\{ e^{-\pi(\omega - \omega_0)^2} \left[F(\omega) e^{2\pi i x \omega} + F(-\omega) e^{-2\pi i x \omega} \right] \right\} d\omega$$

$$= 2 \int_{-\infty}^{\infty} e^{-\pi(\omega - \omega_0)^2} \left[F_r(\omega) \cos(2\pi x \omega) - F_i(\omega) \sin(2\pi x \omega) \right] d\omega \tag{70}$$

which obviously *is* real. Here, $F_r(\omega)$ and $F_i(\omega)$ are the real and imaginary parts of $F(\omega)$; *i.e.*, $F(\omega) = F_r(\omega) + iF_i(\omega)$. Because $F(\omega)$ is Hermitian, $F_r(-\omega) = F_r(\omega)$ and $F_i(-\omega) = -F_i(\omega)$. Therefore both terms under the integral in equation (70) are even functions of ω.

Exercise 9-8: If $H(\omega)$ is given by equation (69), what is $h(x)$?

9.5 Propagating waves

In this section, I shall give an example of how we can use some of the ideas about spectra that I have been discussing. It concerns the identification of propagating waves in the earth's atmosphere, as they might be revealed in satellite data; here I introduce a new mathematical method for identifying such waves. The reason for doing this is that meteorologists want to study the dynamics of the earth's atmosphere, which are very complicated. If we can identify waves of some sort propagating through the atmosphere, *and they have a physical significance,* in the sense that something is actually oscillating, we will be better able to understand the dynamics of the atmosphere. In addition, we will be able to make better forecasts. However, we must be very careful not to *create* waves that have only mathematical, not physical, significance. While the impetus for this study was my interest in meteorological data, these ideas apply to the general problem of analyzing two-dimensional random data, and of identifying the underlying *physical* processes that gave rise to those data.

To this end, we must be careful to include two types of motions in our model. One is propagating waves: disturbances that move from location x_1 to x_2, x_3, ..., and so forth. The other component is due to changes at each location that are statistically independent at different locations. If we ignore this second component, the data analysis will create false propagating waves that will, in general, show up either at wavelengths where there are no propagating waves, or as waves moving in the wrong direction. Without this added level of analysis, we will tend to identify all of the energy with propagating waves, even when the physical processes involved are spatially statistically independent.

As an aside, we should note that if a space-time random process is spatially statistically independent, the values at any two locations will be uncorrelated. However, if the process is not spatially statistically independent, values at some pairs of locations will be highly correlated, while values at other pairs of locations will still be uncorrelated. As a simple example, suppose that we are measuring a random process that is a superposition of waves of the form $f_i(x,t) = \cos[2\pi(x/\lambda - \omega t) + \phi_i]$, where ϕ_i is a random phase. If we make measurements at locations x_1 and x_2, they will be perfectly correlated if $(x_2 - x_1) / \lambda = 1$; they will be uncorrelated if $(x_2 - x_1) / \lambda = 0.5$.

Exercise 9-9: Suppose there is a random process that I can represent as

$$F(x,t) = \sum_{i=1}^{N} a_i \cos\left[2\pi(x/\lambda - \omega t) + \phi_i\right], \tag{71}$$

where the ϕ_i are random phases. What is the two-dimensional autocorrelation function of F? Evaluate $\langle F(x_1,t)F(x_2,t)\rangle$, where the average is taken over time.

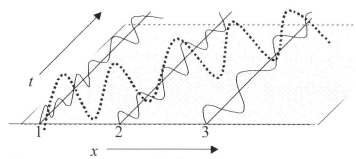

Figure 9-5. A wave propagating from point 1 to point 3 (heavy dotted line), and the noiselike components that are localized at each point (light solid lines).

My interest in this subject arose when I was studying a method for finding propagating waves in the atmosphere.[18] Many authors claim to be able to identify waves in the atmosphere by doing the following:

1. Take a time series of data: *e.g.*, the surface pressure at Canton Island in the Pacific for each day;

2. Filter the data with a bandpass filter—one that passes waves with periods, say, between 30 and 90 days;

3. Note the resultant sinusoidal appearance of the filtered time series.

4. Try to interpret the *physical* significance of these results.

The filtered data appear to be sinusoidal because of the way we processed them: even if we have a white-noise signal, we will still create a sinusoidal behavior simply by filtering the time series. A sinusoidal component in the filtered data might have some mathematical significance, but it has no physical significance: those components were created by our data analysis. Note in particular that random, uncorrelated physical events will appear to have a coherence that, in fact, has no physical basis. It was to improve on this situation that I developed the following ideas.

To identify propagating waves, I introduce the concept of a random process that contains a mixture of propagating waves, presumed to be sinusoidal in nature, and random fluctuations that I call *noiselike* because they are uncorrelated both in space and time, even though they are a real part of the variable under discussion, and are *not* due to measurement error. I show this schematically in Figure 9-5. Many authors[19] have tried to explain *all* of the variability in a geophysical variable in terms of propagating waves—sines and cosines— only. However, in order to use this kind of analysis—where all of the fluctuations are assumed to be due solely to physical waves in the atmosphere—we

---

[18] See, for example, Madden and Julian (1971, 1972), and Julian (1973).

[19] *E.g.*, see Hayashi 1971, 1977*a*, and 1977*b*.

would need to assume a degree of statistical stationarity that is, in fact, not present. That is why I take a different path and assume that every geophysical variable is determined by two components, only one of which contains only propagating waves,[20] while the other is random and uncorrelated from place to place—or time to time. Without this random component, we would be forced to interpret all of the variation as due to propagating waves. Although this might be mathematically correct, it is not a physically reasonable description of the earth's atmosphere.

Even with the addition of these noiselike terms to the discussion, the separation of the fluctuations of a variable into propagating waves and noise-like components is still ill-determined. Since the surface of the earth is not uniform, it is unreasonable to expect that a wave would propagate around the earth with its wavelength, frequency, and amplitude unaffected by things like the presence or absence of continents. Therefore, waves will grow or diminish in amplitude; few, if any, of them will propagate entirely around the earth unchanged. Therefore, even with this two-component model that assumes that the waves are in a steady state, the decomposition of the data into waves may not necessarily correspond to true waves in the atmosphere.

In the discussion that follows, assume that $v(x,t)$ is a geophysical variable that is measured at locations x and times t, where the locations span a significant part of the earth's surface and t spans many years. We would treat the 3-dimensional case similarly (that is, a variable that depends on latitude, longitude, and time).

9.5.1 Cross-correlation and propagating waves

Any function $v(x,t)$ can be represented as a Fourier series, and consequently as a sum of sine and cosine components. While this is mathematically true, it may be of no *physical* significance. Imagine, if you will, a particle undergoing Brownian motion: its position is affected by random, uncorrelated impulses. Plot its position along the x-axis as a function of time. Now we can, if we want, represent this plot as a Fourier series, but we also have to realize that, while there are sines and cosines involved in the expansion, there is nothing that is actually waving. There are no waves involved. None. It is in this context, then, that I ask the question: do propagating waves contribute to $v(x,t)$?

Note that I must assume in this analysis that the phenomenon we are studying is statistically stationary both in time and space. The discussion that follows refers to stationary processes only.

---

[20] I shall use the term *propagating waves* to include standing waves also.

9.6 Separating waves from noiselike features

Suppose that the random process in question is composed of two types of phenomena. One is random and very localized in time and space, while the other occurs in the form of propagating waves. We want to separate the propagating waves from the localized phenomena. Assume that we have N time series that represent measurements made along a line at positions x_i, $i = 1, \ldots, N$. The measurements are represented by

$$v_i(t) = w_i(t) + n_i(t), \tag{72}$$

where

$$w_i(t) = \int_0^\infty \mu(\omega) \cos\left[2\pi\omega\, t + \phi_i(\omega)\right] d\omega \; ; \tag{73}$$

$\mu^2(\omega)$ is the spectral density of w, and $\phi_i(\omega)$ is a phase that depends on ω. We assume that w_1 and w_2 have the same spectral density. The terms $n_i(t)$ are random processes that are uncorrelated with each other or with w_i. Let

$$\left\langle n_i^2(t) \right\rangle = \sigma_i^2 \,, \tag{74}$$

for $i = 1, 2, \ldots N$; while, by hypothesis,

$$\left\langle n_i(t) n_j(t) \right\rangle = 0 \qquad \text{if } i \neq j . \tag{75}$$

Remember that I call $n_i(t)$ *noiselike* components because each one is uncorrelated with anything else; they are not like noise in the sense of being only a nuisance that interferes with measurement of something else that is really important. In particular, do not confuse them with measurement errors; they are assumed to be real phenomena that are important in their own right. The *propagating part* of the signal is w_i. While n_i might have periodic components, we assume that these periodic components are not correlated between different n_i. Also, it is not necessary to assume that n_i is δ-correlated in τ.

The cross-correlation function is[21]

$$C_{ij}(\tau) \equiv \left\langle v_i(t) v_j(t + \tau) \right\rangle = \lim_{T \to \infty} \frac{1}{2T} \int_{-T}^{T} v_i(t) v_j(t + \tau) dt . \tag{76}$$

First, we can calculate the cross-correlation of the w_i:

---

[21] Note the slight change in notation.

$$R_{ij}(\tau) = \left\langle \int_0^\infty \int_0^\infty \mu(\omega)\mu(\omega')\cos\left[2\pi\omega't + \phi_i(\omega')\right] \right.$$

$$\left. \cos\left[2\pi\omega(t+\tau) + \phi_j(\omega)\right]d\omega\, d\omega' \right\rangle. \qquad (77)$$

To evaluate this, we can use the statistical relationship that

$$\left\langle \cos\left[2\pi\omega't + \phi_i(\omega)\right]\cos\left[2\pi\omega(t+\tau) + \phi_j(\omega)\right]\right\rangle = 0 \qquad (78)$$

if $\omega \neq \omega'$, and the trigonometric identity

$$\left\langle \cos\left[2\pi\omega t + \phi_i(\omega)\right]\cos\left[2\pi\omega(t+\tau) + \phi_j(\omega)\right]\right\rangle$$

$$= \left\langle \cos(2\pi\omega t)\cos\left[2\pi\omega(t+\tau) + \Delta_{ij}(\omega)\right]\right\rangle$$

$$= \left\langle \cos(2\pi\omega t)\left\{\cos\left[2\pi\omega(t+\tau)\right]\cos\Delta_{ij}(\omega) - \sin\left[2\pi\omega(t+\tau)\right]\sin\Delta_{ij}(\omega)\right\}\right\rangle$$

$$= \left\langle \cos(2\pi\omega t)\left\{\cos\Delta_{ij}(\omega)\left[\cos(2\pi\omega t)\cos(2\pi\omega\tau) - \sin(2\pi\omega t)\sin(2\pi\omega\tau)\right]\right.\right.$$

$$\left.\left. - \sin\Delta_{ij}(\omega)\left[\cos(2\pi\omega t)\sin(2\pi\omega\tau) + \sin(2\pi\omega t)\cos(2\pi\omega\tau)\right]\right\}\right\rangle$$

$$= \tfrac{1}{2}\left[\cos\Delta_{ij}(\omega)\cos(2\pi\omega\tau) - \sin\Delta_{ij}(\omega)\sin(2\pi\omega\tau)\right]. \qquad (79)$$

where $\Delta_{ij}(\omega) \equiv \phi_j(\omega) - \phi_i(\omega)$ is the relative phase; $\Delta_{ii} \equiv 0$. Note that I have used the orthogonality of $\cos 2\pi x$ and $\sin 2\pi x$, as well as $<\cos^2 t> = \tfrac{1}{2}$. This finally reduces to

$$R_{ij}(\tau) = \tfrac{1}{2}\int_0^\infty \mu^2(\omega)\left[\cos(2\pi\omega\tau)\cos\Delta_{ij} - \sin(2\pi\omega\tau)\sin\Delta_{ij}\right]d\omega. \qquad (80)$$

Because n_i and w_i are uncorrelated,

$$C_{ij}(\tau) = R_{ij}(\tau) + c_{ij}(\tau), \qquad (81)$$

where

$$c_{ij}(\tau) \equiv \langle n_i(t) n_j(t + \tau) \rangle. \tag{82}$$

Since for $i \neq j$, $c_{ij}(\tau) = 0$ by hypothesis, $\mathcal{C}_{ij}(\tau)$ will contain a contribution only from the wave part of the signal. Define

$$C_{ij}^o(\tau) \equiv \tfrac{1}{2}\left[C_{ij}(\tau) - C_{ij}(-\tau) \right] = -\tfrac{1}{2}\sin(2\pi\omega\tau)\int_0^\infty \mu^2(\omega)\sin[\Delta_{ij}(\omega)]\,d\omega. \tag{83}$$

Note that a similar expression for $\mathcal{C}_{ii}{}^o(\tau)$ would be identically zero. However, by taking the Fourier transform of $\mathcal{C}_{ij}{}^o(\tau)$, we can find the product $\mu^2(\omega)$ $\sin[\Delta_{ij}(\omega)]$.

The phase $\Delta_{ij} = 2\pi(\lambda/L_{ij})$, where λ is the wavelength at frequency ω and $L_{ij} \equiv x_i - x_j$. Although we know that the speed of propagation[22] is $c = \lambda / \omega$, we do not know what c is, and we do not even know that c is independent of λ.

The even part of R_{ij} is also independent of the noise. Define

$$C_{ij}^e(\tau) \equiv \tfrac{1}{2}\left[C_{ij}(\tau) + C_{ij}(-\tau) \right] = \tfrac{1}{2}\cos(2\pi\omega\tau)\int_0^\infty \mu^2(\omega)\cos[\Delta_{ij}(\omega)]\,d\omega. \tag{84}$$

If we restrict the range of ω to some interval $[\omega_i, \omega_j]$, say, we can determine $\cos[\Delta_{ij}(\omega)]$ and $\sin[\Delta_{ij}(\omega)]$. The phase difference will be

$$\tan[\Delta_{ij}] = \frac{C_{ij}^o(\tau)}{C_{ij}^e(\tau)}. \tag{85}$$

Presumably, this ratio is independent of τ.

The spacings L_{ij} determine what wavelengths we can measure. If $\lambda \gg L_{ij}$, Δ_{ij} will be small and we will have very little sensitivity to those waves. On the other hand, if $\lambda \ll L_{ij}$, $\Delta_{ij} \gg 2\pi$ and the waves will not be coherent, even over a small bandwidth $\delta\omega$, so we will be averaging over different phases. Therefore the choice of L_{ij} acts as a kind of filter with a maximum response at $\lambda = L_{ij}$.

I should reiterate the two central ideas here:

1. In order to avoid overinterpreting the data, we must include, along with any propagating waves, noiselike terms due to the variance of each w_i at each point i that are uncorrelated with anything else: such processes always happen in nature.
2. The spectral density of w_i is best determined from the cross-correlation function.

---

[22] Do not confuse c, the speed of propagation, with $c_{ij}(\tau)$.

Exercise 9-10: In reality, we only can have measurements over some finite interval, say from $t = 0$ to $t = T_0$. Then we should replace equation (76) with

$$C_{ij}(\tau) \;=\; \frac{1}{2T_0} \int_0^{T_0} v_i(t)\, v_j(t + \tau)\, dt .$$ (86)

How does this affect the formalism developed here?

Exercise 9-11: How will measurement noise (as opposed to the noiselike parts of the actual variation of w) affect our estimates of $\tan[\Delta_{ij}]$?

9.6.1 Relationship to the 2-D Fourier transform

It is instructive to compare the results of the last section with that of taking the 2-D Fourier transform of the data.[23] Instead of having only two points, consider measurements made at $2N$ different locations x_j. Assume that $x_j = j\delta x$, where δx is the interval between spatial measurements. We can consider the $2N$ functions $v_j(t)$ to be equivalent to a two-dimensional function $v(x_j,t)$, which I will write simply as $v(x,t)$ so that I can use the formalism of integral transforms. Let $L \equiv 2N\delta x$. Then we can write this in the form of a Fourier series:

$$v(x,t) \;=\; \sum_{j=0}^{N-1} \left[a_j(t)\cos(2\pi j x / L) + b_j(t)\sin(2\pi j x / L) \right],$$ (87)

where $b_0(t) \equiv 0$. The analogs of $\mathcal{C}_{ij}(\tau)$ are

$$p_j(\tau) \;\equiv\; \operatorname*{Lim}_{T \to \infty} \frac{1}{2T} \int_{-T}^{T} a_j(t)\, b_j(t + \tau)\, dt ,$$

$$q_j(t) \;\equiv\; \operatorname*{Lim}_{T \to \infty} \frac{1}{2T} \int_{-T}^{T} a_j(t)\, a_j(t + \tau)\, dt ,$$

and

$$r_j(t) \;\equiv\; \operatorname*{Lim}_{T \to \infty} \frac{1}{2T} \int_{-T}^{T} b_j(t)\, b_j(t + \tau)\, dt .$$ (88)

---

[23] This development parallels that of Hayashi (1971, 1977-a, 1977-b, 1982).

The Fourier coefficients are given by

$$a_j(t) = \int_{-\frac{1}{2}L}^{\frac{1}{2}L} v(x,t)\cos(2\pi jx / L)\,dx$$

and

$$b_j(t) = \int_{-\frac{1}{2}L}^{\frac{1}{2}L} v(x,t)\sin(2\pi jx / L)\,dx \,. \tag{89}$$

Let us look at the cross-correlation function $p_j(\tau)$ for an ensemble of waves as given in equations (87). In this case, the phases ϕ_j correspond to $2\pi jx$, where the wavenumber j is an integer. Here too, we can represent $v(x,t)$ as the sum of propagating waves and a noiselike component:

$$v(x,t) = \int_0^\infty \sum_k \mu(k,\omega)\cos\left[2\pi(k\,x / L - \omega\,t) + \Delta_k\right]d\omega + n(x,t), \tag{90}$$

where $n(x,t)$ is a noiselike random variable—with zero mean—that is uncorrelated between different points. Then[24]

$$a_j(t) = \int_{-\frac{1}{2}L}^{\frac{1}{2}L} \cos(2\pi jx / L)$$

$$\left\{\int_0^\infty \sum_k \mu(k,\omega)\cos\left[2\pi(k\,x / L - \omega\,t) + \Delta_k\right]d\omega + n(x,t)\right\}dx \tag{91}$$

Expand the $\cos x$ term in the usual way; using the orthogonality of the cosines and sines,

$$a_j(t) = \int_{-\infty}^\infty \sum_m \mu(j,\omega)\cos 2\pi\left[\omega\,t - \Delta_m\right]d\omega + n_j(t)\,. \tag{92}$$

Here,

$$n_j(t) \equiv \int_{-\frac{1}{2}L}^{\frac{1}{2}L} n(x,t)\cos(2\pi jx / L)\,dx \tag{93}$$

which is another random variable whose variance may depend on j (the noise may not be spatially spectrally white). Similarly,

---

[24] We allow the possibility that $c_j(\tau) \equiv \langle n_j(t)\, n_j(t + \tau)\rangle \neq 0$ for some $\tau > 0$.

$$b_j(t) = \int\limits_{-\infty}^{\infty} \sum_m \mu(j,\omega)\sin 2\pi[\omega t - \Delta_m]\,d\omega + n_j'(t), \tag{94}$$

where n_j' is defined in a manner similar to equation (93). Therefore,

$$p_j(\tau) = \int\limits_{-T}^{T}\int\limits_{0}^{\infty}\int\limits_{0}^{\infty} \sum_{m,n} \mu(j,\omega)\mu(j,\omega')$$

$$\times \cos[2\pi\omega t - \Delta_m]\sin[2\pi\omega'(t+\tau) - \Delta_n]\,d\omega\,d\omega'\,dt + c_{jj}(\tau), \tag{95}$$

where $c_{jj}(\tau)$ is the autocorrelation function of $n_j(t)$. Since the only contribution to this average comes where $\omega = \omega'$ and $\Delta_m = \Delta_n$, this reduces to[25]

$$p_j(\tau) = -\tfrac{1}{2}\int\limits_{0}^{\infty} \mu^2(j,\omega)\sin(2\pi\omega\tau)\,d\omega + c_{jj}(\tau). \tag{96}$$

The function $q_j(\tau)$ is evaluated in the same way;[26] the result is

$$q_j(\tau) = \tfrac{1}{2}\int\limits_{0}^{\infty} \mu^2(j,\omega)\cos(2\pi\omega\tau)\,d\omega + c_{jj}(\tau). \tag{97}$$

Obviously, $r_j(\tau)$ measures the same thing.

The spectral power at each frequency is given by the Fourier transforms of equations (96) and (97). In the simple case where the noise $n(x_j,t)$ has the same spectrum at each point x_j,

$$\int\limits_{-\infty}^{\infty} c_{jj}(\tau)e^{2\pi i\omega\tau}\,d\tau \equiv S_N(\omega), \tag{98}$$

and the spectrum of p_j is

$$S_j(\omega) = \left| \int\limits_{-\infty}^{\infty} [p_j(\tau) + iq_j(\tau)]e^{2\pi i\omega\tau}\,d\tau \right|^2$$

$$= \mu_j^2(\omega) + S_N(\omega). \tag{99}$$

This shows that random components will contribute to the presumed spectrum of waves moving in the + and − directions. This is the disadvantage of using a

---

[25] $\cos(x)\,\sin(y) = \tfrac{1}{2}\{\sin(x+y) - \sin(x-y)\}.\grave{}$

[26] $\cos(x)\,\cos(y) = \tfrac{1}{2}\{\cos(x+y) + \cos(x-y)\}.$

2-D Fourier transform: since everything looks like a propagating wave, there is no capacity for distinguishing between noiselike random phenomena and true propagating waves. Yet these noiselike features are a common—if not omnipresent—feature of the state of the real world.

These ideas show, if nothing else, just how difficult it may be to identify propagating waves in 2- or 3-dimensional data. If all problems look like nails, the only tool you ever use will be a hammer. In this case, when trying to identify propagating waves, we cannot blindly use Fourier analysis: we must account explicitly for other (*i.e.*, noiselike) phenomena that are definitely not nails (call them screws) and be careful to use a screwdriver where it is more appropriate than an hammer.

Exercise 9-12: Suppose there is a single wave propagating from x_1 to x_2, of the form $\phi(x,t) = \cos(2\pi x/\lambda - \omega t)$. There are also (uncorrelated) noiselike components $n_1(t)$ and $n_2(t)$. We measure $v_1(t) = \cos(2\pi x_1/\lambda - \omega t) + n_1(t)$ and $v_2(t) = \cos(2\pi x_2/\lambda - \omega t) + n_2(t)$. What is the autocorrelation function of v_1? Of v_2? What is the cross-correlation function of v_1 and v_2? Given a set of measurements, what will we estimate the form of ϕ to be? How will this estimate depend on the quantity $(x_2 - x_1)/\lambda$?

Exercise 9-13: Suppose we use the analysis of this section. Use a computer program to model this situation, using a component with wavelength λ and including a noiselike component at each measurement point. Then compute the power spectrum of these waves, assuming that all of the energy is really in the form of propagating waves (*i.e.*, forget that you introduced the noiselike components). Do you see waves at wavelengths other than λ? Where did these components come from? Now perform an analysis using the method of the previous section. How do the results differ?

References

Bendat, J. S., and A. G. Piersol, *Random Data Analysis and Measurement Procedures*, John Wiley & Sons, New York (1986).

Brockwell, P. J., and R. A. Davis, *Time Series: Theory and Methods*, Second Edition, Springer-Verlag, New York (1991).

Champeney, D. C., *A Handbook of Fourier Theorems*, Cambridge University Press, Cambridge, 185 pp (1987).

Gardner, W. A., *Introduction to Random Processes with Applications to Signals & Systems*, 2nd Edition, McGraw-Hill Publishing Co., New York, 456 pp (1990).

Hayashi, Y., 'A generalized method of resolving disturbances into progressive and retrogressive waves by space Fourier and time cross-spectral analyses,' *J. Meteorological Society of Japan*, **49**, 125-128 (1971).

Hayashi, Y., 'On the coherence between progressive and retrogressive waves and a partition of space-time power spectra into standing and traveling parts,' *J. Meteorological Society of Japan*, **16**, 338-373 (1977b).

Hayashi, Y., 'Space-time power spectral analysis using the maximum entropy method,' *J. Meteorological Society of Japan*, **55**, 415-420 (1977a).

Hayashi, Y., 'Space-time spectral analysis and its applications to atmospheric waves,' *J. Meteorological Society of Japan*, **60**, 156-170 (1982).

Jenkins, F. A., and H. E. White, *Fundamentals of Optics*, third edition, McGraw-Hill Book Co., New York (1957).

Julian, P. R., 'Comments on the determination of significance levels of the coherence statistic,' *J. of the Atmospheric Sciences*, **32**, 836-837 (1973).

Khintchine, A., *Math. Annalen.* **109**, 604-615 (1934).

Madden, R. A., and P. R. Julian, 'Description of global-scale circulation cells in the tropics with a 40-50 day period,' *J. of the Atmospheric Sciences*, **29**, 1109-1123 (1972).

Madden, R. A., and P. R. Julian, 'Detection of a 40-day oscillation in the zonal wind in the Tropical Pacific,' *J. of the Atmospheric Sciences*, **28**, 702-708 (1971).

Middleton, D., *An Introduction to Statistical Communication Theory*, McGraw-Hill Publishing Co., New York (1960).

Milman, A. S., 'A comment on the use of bandpass filtering to discover atmospheric oscillations,' *International J. of Remote Sensing*, **19**, 2275-2282 (1998).

Monin, A. S., and A. M. Yaglom, *Statistical Fluid Mechanics: Mechanics of Turbulence*, Volume 2, The MIT Press, Cambridge, MA (1975).

Papoulis, A., *Probability, Random Variables, and Stochastic Processes*, McGraw Hill, New York (1965).

Press, W. H., S. A. Teukolsky, W. T. Vettering, and B. R. Flannery, *Numerical Recipes in Fortran, the Art of Scientific Computing*, Second Edition, Cambridge University Press, Cambridge, England (1992).

Wiener, N., 'Generalized Harmonic Analysis', in *Selected Papers,* MIT Press, Cambridge (1964).

Yaglom, A. M., *Correlation Theory of Stationary and Related Random Functions*, Volume I: Basic Results, Springer-Verlag, New York (1987).

10. Integral Equations

In previous chapters, I treated discrete problems that are represented by a matrix equation $\mathbf{y} = \mathbf{Ax}$. Implicit in this formulation is the notion that there is really a *function* $g(x)$ that is sampled at a set of points $\{x_i\}$, and a corresponding function $f(y)$ sampled at a set of points $\{y_j\}$. We identify the vector $\mathrm{col}[g(x_i)]$ with $\mathbf{x}$, and similarly with $f(y)$. We could, alternatively, treat f and g as continuous functions and write the integral equation (1) below instead of the corresponding matrix equation. We shall see that there is a duality between matrix equations and integral equations, in that one is a discretized form of the other; in practice, we generally need to approximate an integral equation with a matrix equation before we can actually perform computations. On the other hand, there are ways of thinking about some problems that are more fruitful in the integral-equation setting.

10.1 Hilbert spaces

I discuss Hilbert spaces in §14.8. We saw that a matrix equation is a mapping from a vector space $\mathcal{V}$ to another vector space $\mathcal{W}$: the matrix $\mathbf{A}$ maps each vector $\mathbf{x} \in \mathcal{V}$ to a new vector $\mathbf{y} \in \mathcal{W}$. Similarly, an integral equation has a kernel—a function that I denote $K(x,y)$—and it maps each function f in a Hilbert space $\mathcal{F}$ to another function g in a Hilbert space $\mathcal{G}$. I can write the relationship

$$g(x) = \int_a^b K(x,y)f(y)\mathrm{d}y \tag{1}$$

symbolically as

$$g = Kf . \tag{2}$$

247

The comparison with matrix equations is straightforward: the functions f and g correspond to the vectors $\mathbf{x}$ and $\mathbf{y}$; the kernel $K(x,y)$ to the matrix $\mathbf{A}$; and integration replaces matrix multiplication.

In order to discuss solutions to integral equations, we need to define the norm of a function and the distance between two functions. All of this is done within the framework of Hilbert spaces.

When we discuss vectors, we take for granted that any vector can be written as a linear sum of a finite number of basis vectors. We would write

$$\mathbf{x} = \sum_{i=1}^{n} \alpha_i \mathbf{e}_i \tag{3}$$

for an element $\mathbf{x}$ of the n-dimensional vector space E^n, where $\{\mathbf{e}_i\}$ is a basis of E^n, and $\{\alpha_i\}$ a set of scalars.[1] Similarly, we shall see that we can write a function $f \in \mathcal{F}$ as a sum over basis functions ϕ:

$$f(y) = \sum_{i=1}^{M} \alpha_i \phi_i(y). \tag{4}$$

Here, M is often infinite, which brings up questions of convergence. Remember that a Cauchy sequence[2] is a sequence of elements of $\mathcal{F}$, $f_0, f_1, \ldots$, where, for any positive ε, there is an integer N such that $d(f_j, f_k) < \varepsilon$ for all $j, k > N$. Here, $d(\cdot,\cdot)$ is the distance between two functions. A Hilbert space $\mathcal{F}$ is *complete*, in that every Cauchy sequence $\{f_i : f_i \in \mathcal{F}\}$ converges to a limit in $\mathcal{F}$. In this way, we can make use of infinite series to represent the functions that arise in the study of integral equations.

There are at least four ways to solve integral equations. I will treat several in this chapter.

1. Since integration of $f(y)$ can be approximated by a finite sum of $f(y_i)$, we can use an approximate quadrature[3] formula to replace an integral equation with a matrix equation that we already know how to solve.

2. We can solve integral equations by iteration. If we use a quadrature formula that varies the points where the function is evaluated, depending on the value of the function, we are in effect using a different matrix to approximate the integration for each new function.

---

[1] Remember that the notation $\{\}$ denotes a set.

[2] See §14.8.4.

[3] The numerical evaluation of an integral is called *quadrature* because, before the time of Newton, the problem of finding the area under a curve was stated as being that of finding a square whose area was equal to the area under a given curve. The name dates from a time when mathematicians were much more geometrical in their thinking than we are today.

3. For some kernels, the integral equation represents a convolution; we can use Fourier transforms to solve such equations.

4. If the kernel is periodic, we can find an orthogonal set of basis functions $\{\phi_i\}$ that are also eigenfunctions of the kernel; we can solve the integral equation by expressing f and g as sums of these functions.

Perhaps I should have entitled this chapter "Linear Integral Equations," since, in general, integral equations can be nonlinear in a variety of ways. Although our main interest is in linear equations, because those are the ones we can solve most readily, most of the integral equations that arise in remote sensing are really nonlinear. As far as I know, these equations can only be solved by iteration; I shall discuss that in the next chapter.

10.2 Fredholm equations

Consider the following problem. Let $f(y)$ and $g(x)$ be continuous functions that are related by the equation (1).[4] $K(x,y)$, called the *kernel* of the transformation, is a known function of x and y. The limits a and b are any real numbers—with the obvious restriction that $a \neq b$. The limits a or $b = \pm\infty$ are also allowed. The problem is, given $g(x)$, to find the corresponding $f(y)$. In practical problems, we are given the values of $g(x_i)$, $i = 1,\ldots,N$; we need to find a set of values of $f(y_i)$ that solve equation (1).

Equation (1) is linear in f:

$$\int_a^b \left[rf_1(y) + sf_2(y) \right] K(x,y)\,dy = rg_1(x) + sg_2(x), \qquad (5)$$

where g_1 and g_2 are related to f_1 and f_2 by equation (1) and r and s are any complex numbers.

Equation (1) is called a *Fredholm[5] equation of the first kind.* Such equations are notoriously ill-conditioned and unstable when they arise in physical situations. The problem is that it is possible that there are two or more— actually, an infinite number of—functions $f(y)$ that produce almost indistinguishable functions $g(x)$. When there is some noise in the measurement of $g(x)$, it may be impossible to determine which of these functions $f(y)$ actually gave rise to it. This means that a small error in the measurement of $g(x)$ can produce a very large error in the estimated value of $f(y)$. For instance, if we were measuring vertical temperature profiles in the earth's atmosphere, we might, because of this, infer some negative temperatures—which we know

---

[4] See Delves and Mohamed (1985) and Groetsch (1984) for discussions of these topics.

[5] Ivar Fredholm (1866-1927) was a Swedish mathematician; he invented the theory of integral equations.

quite well can never happen. The problem might, of course, be our choice of which measurements to make, but it is probably the way we estimated $f(y)$ from the measured data. In this chapter, we shall investigate the properties of these equations and develop ways to find stable solutions.[6]

I discuss functions that belong to $L^2[0,1]$ in §14.8.5. We can generalize this notation to $L^2[a,b]$ to indicate the class of functions that have the property that

$$\int_a^b |f(x)|^2 dx < \infty. \tag{6}$$

The functions $f \in L^2[a,b]$ form a Hilbert space.

10.2.1 Boundedness condition

Throughout this discussion, we shall assume that the kernel has the property that

$$\int_a^b |K(x,y)|^2 dy < M_K, \tag{7}$$

where M_K is a positive constant independent of x. We shall place a similar restriction on f:

$$\int_a^b |f(x)|^2 dx < M_f, \tag{8}$$

where M_f is another positive constant. These conditions will always be met in practice because there is only a finite amount of power available for any real physical process; we shall assume from now on, without further comment, that these conditions are always met.

10.2.2 Fredholm equations of the second kind

A *Fredholm equation of the second kind* has the form

$$\phi(x) - \lambda \int_a^b K(x,y)\phi(y)dy = f(x), \tag{9}$$

---

[6] See Groetsch (1984) and Delves and Mohamed (1985) for further discussions of Fredholm equations of the first kind.

where $f(x)$ is a given function, λ is a parameter, and we seek functions $\phi(x)$ that satisfy equation (9). I shall not pursue this kind of equation further here; there is a vast literature on these equations that the interested reader can consult.

10.2.3 Relationship to matrix equations

I shall start with definitions that relate some properties of complex functions to similar properties of vectors. Part of the connection arises because, when we actually solve equations like equation (1) based on measured data, we may replace the integral with a summation and we approximate the function $f(y)$ with a vector $\mathbf{f}$ specified at certain points. We can identify each component of $\mathbf{f}$ with the value of f at some specified y_i:

$$\mathbf{f} = \mathrm{col}\big[f(y_1), f(y_2), \cdots, f(y_n)\big]. \tag{10}$$

I discussed the basis of a vector space in §5.2.2: it is a set of mutually orthogonal vectors that span the vector space. We saw that the eigenvectors of a matrix form an orthonormal basis; we shall see in §10.7 that the eigenfunctions of equation (1) form a basis of the Hilbert space.

A comment on the relationship between integral equations and matrix equations: as I noted earlier, we may evaluate an integral numerically with a quadrature formula. By necessity, this means that we are replacing the integral equation with a matrix equation where the matrix has a finite number of dimensions. With integral equations, however, a kernel will possess, in general, an infinite number of eigenfunctions; we can write any function in a Hilbert space as a sum of these eigenfunctions. When we approximate the integral equation with a matrix equation, we necessarily limit ourselves to a finite number of eigenvectors. Instead of considering the elements f of a Hilbert space $\mathscr{F}$, we are limited to the elements $\mathbf{x}$ of some vector space $\mathscr{X}$. If we express f with equation (4) and $\mathbf{x}$ with equation (3), it is clear that many of the functions that exist in the original Hilbert space $\mathscr{F}$ have no counterpart in the vector space $\mathscr{X}$. Therefore an integral equation may have many solutions in $\mathscr{F}$ that do not exist in the corresponding vector space $\mathscr{X}$; this may cause problems when we approximate an integral equation with a matrix equation.

10.3 Convolution integrals

The Fourier-transform method applies to the special case where the kernel depends only on the difference between x and y, so we can write the kernel in equation (1) as $K(x,y) = q(x - y)$. Such kernels smooth $f(x)$, which, as we shall see, means that they remove high-frequency components from its spectrum. As always, we shall consider what effect that noise has on the estimates $\hat{f}$ of f.

If there were *no* noise in the measurement of g, we would measure

$$g(x) = \int_{-\infty}^{\infty} f(y)q(y-x)\,dy. \tag{11}$$

Here, I assume that $q(x)$ does not change when $f(y)$ changes. (Later, we shall encounter a radiation transfer problem where $f(y)$ represents the temperature of the atmosphere at altitude y and the kernel $q(x,y)$, which includes the effects of absorption and emission of radiation, also depends on the temperature $f(y)$. This is a much more difficult problem to solve.) We shall aslo assume that the spectrum of f exists and does not change with time; we shall seek a fixed function $D(v)$ that will solve equation (11).

Since equation (11) is a convolution, we can make use of the convolution theorem. Let $G(v)$ and $F(v)$ be the Fourier transforms of g and f. Let $Q(v)$ be the Fourier transform of $q(x)$, which we take to be a fixed, known function. Then the convolution theorem states that

$$G(v) = Q(v)F(v). \tag{12}$$

So, to solve equation (11), we might want to write $F(v) = G(v)/Q(v)$ and take the Fourier transform of both sides; the solution would be

$$\hat{f}(y) = \int_{-\infty}^{\infty} \frac{G(v)}{Q(v)} e^{2\pi i y v}\,dv. \tag{13}$$

So far, so good. We seem to have a recipe for estimating $\hat{f}$. This is illustrated schematically in Figure 10-1. The vertical arrows represent Fourier transforms.

There is one obvious problem, however. As a consequence of equation (7), the integral $\int_{-\infty}^{\infty} Q(v)\,dv$ is finite, so $Q(v) \to 0$ as $v \to \infty$. Therefore, there is a limit v_c such that, for all $v > v_c$, $Q(v)$ is indistinguishable from zero. For these frequencies, equation (13) involves division by zero, in which case the noise becomes infinitely large.

Exercise 10-1: Suppose that $q(x)$ is a Gaussian; then $Q(v)$ is a Gaussian also. Also, suppose that $F(v)$ has spectral components outside the region where $Q(v)$ is very close to zero. What effect does noise have on a retrieval that uses equation (13)?

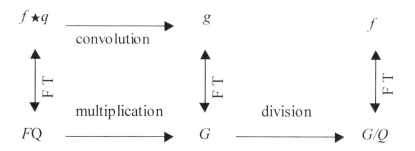

Figure 10-1. Relationships between the different functions. Vertical arrows represent Fourier Transforms (F.T.).

10.4 Noise

When we include the effects of noise explicitly in our analysis, it becomes apparent what is wrong. I will only consider *white noise*, so called because its spectrum is constant. It is called white noise by analogy with white light, which contains all of the colors of the spectrum (although not in equal quantities). Let $n(x)$ be the (random) noise signal with zero mean and variance σ_N^2. Its power per unit bandwidth is constant:[7]

$$\left|N(v)\right|^2 = \sigma_N^2. \tag{14}$$

The total noise power within a band Δv is

$$N_{\text{tot}}^2 = \sigma_N^2 \Delta v. \tag{15}$$

Usually, a device that makes a physical measurement has incorporated into it, either by design or necessity, a component that acts as a low-pass filter: there is a cutoff frequency such that only frequencies v with $|v| < v_c$ contribute to the output. Therefore the total noise power is

$$\int_{-\infty}^{\infty} \left|N(v)\right|^2 \, dv = \int_{-v_c}^{v_c} \sigma_N^2 \, dv = 2v_c \sigma_N^2 \tag{16}$$

(The factor of 2 arises because frequency can be positive or negative.) Instead of a cutoff frequency, we can use an equivalent integration time

$$T_i = \tfrac{1}{2}v_c \tag{17}$$

---

[7] $N(v)$ is the Fourier transform of $n(x)$.

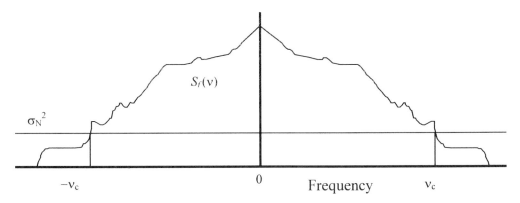

Figure 10-2. Noise spectrum N and power spectrum of F.

that is, roughly speaking, the amount of time devoted to a particular measurement.[8] The longer the integration time, the smaller v_c is, and the less noise is present.

Suppose $F(v)$ is monotonically decreasing in v. We can choose as the cutoff frequency that frequency where the $S_f(v)$ has the same power as the noise: $S_f(v) = \sigma_N^2$, as I show schematically in Figure 10-2. If we choose v_c too large (too short an integration time), we will include more noise than necessary; if we choose it too small, we will lose information about $F(v)$. Of course other choices for v_c are possible, but this one should be as good as any.

In order to use this method, we solve for the Fourier transform of $f(x)$. I discussed the notion of the spectrum of a random variable in §9.1. If we assume that that $f(y)$ comes from a parent population that is ergodic, then we can assume that the spectrum of $f(y)$ is the average value of its periodogram:

$$S_f(v) \equiv \left\langle \left| F(v) \right|^2 \right\rangle, \qquad (18)$$

where $F(v)$ is the Fourier transform of $f(y)$. Here, the average is taken over realizations of f. Also, the spectrum of $n(x)$ is $|N(v)|^2 = \sigma_N^2$.

10.5 Solution when noise is included

To solve equation (1) we are going once again to solve for $F(v)$ and then transform it to get $f(y)$. We measure the quantity

$$g'(x) = g(x) + n(x). \qquad (19)$$

Since

---

[8] If the measurements are discrete, like measuring the length of a table with a ruler, it represents the number of independent measurements that were made.

$$g'(x) = \int\limits_{-\infty}^{\infty} f(y)\, q(y-x) \mathrm{d}y + n(x), \tag{20}$$

the Fourier transform of g' is

$$G'(\nu) = F(\nu)Q(\nu) + N(\nu). \tag{21}$$

We invert equation (21) by estimating $F(\nu)$ from

$$\hat{F}(\nu) = D(\nu)G'(\nu) \tag{22}$$

where[9]—as I shall show presently—

$$D(\nu) = \frac{S_f(\nu)Q(\nu)}{S_f(\nu)|Q(\nu)|^2 + \sigma_N^2} \qquad |\nu| \le \nu_c. \tag{23}$$

Remember that here, S_f is the average periodogram of f. Outside the interval $[-\nu_c,\nu_c]$, $D(\nu) = 0$. The function $D(\nu)$ is called a Wiener filter, after the mathematician Norbert Wiener.[10] I will prove equation (23) shortly; I want to discuss it first.

Due to the presence of σ_N^2, the denominator in equation (23) never goes to zero, and

$$\alpha = \int\limits_{-\nu_c}^{\nu_c} |D(\nu)|^2\, \mathrm{d}\nu \tag{24}$$

is finite; I call α the *noise amplification factor*. I shall explain the significance of α presently. When we Fourier transform $\hat{F}$, using suitable limits, we get

$$\hat{f}(y) = \int\limits_{-\nu_c}^{\nu_c} D(\nu)G'(\nu)e^{2\pi i \nu y}\, \mathrm{d}\nu. \tag{25}$$

The choice of $\pm\nu_c$ as the limits of integration is usually a reasonably good one, but other limits could also be used. However, it is clear that it is pointless to make the interval much wider, since F contains only noise at frequencies higher than ν_c. This shows how important it is to know the spectrum of f before we try to devise an algorithm for retrieving it from measurements of g. If we increase the limit past $\pm\nu_c$, only noise, not information, is added to $\hat{f}$.

---

[9] Remember that q is a known function.
[10] See Wiener (1949) and Gardner (1990, §11.7).

Using equation (22),

$$\hat{f}(y) = \int_{-v_c}^{v_c} D(v) [F(v)Q(v) + N(v)] e^{2\pi i v y} \, dv. \tag{26}$$

This is the sum of two terms, one containing information, the other, noise. Of course, we cannot, for any individual realization of f, separate these two components (if we could, we could eliminate noise entirely), but we can separate their spectra. Since $|N(v)|^2$ is independent of v, and using the definition of α [equation (24)],

$$\int_{-v_c}^{v_c} \left| D(v)\, N(v)\, e^{2\pi i v x} \right|^2 \, dv = \sigma_N^2 \int_{-v_c}^{v_c} |D(v)|^2 \, dv = \alpha\, \sigma_N^2. \tag{27}$$

Then the spectrum of $\hat{f}$ is:

$$\left| \hat{F}(v) \right|^2 = \left| D(v)\, F(v)\, Q(v) \right|^2 + \alpha\, \sigma_N^2. \tag{28}$$

In light of equation (23),

$$\left| \hat{F}(v) \right|^2 = \frac{S_f^2(v) |Q(v)|^4 |F(v)|^2}{\left[S_f(v) |Q(v)|^2 + \sigma_N^2 \right]^2} + \alpha\, \sigma_N^2$$

$$= \frac{|F(v)|^2}{\left[1 + \dfrac{\sigma_N^2}{S_f(v) |Q(v)|^2} \right]^2} + \alpha\, \sigma_N^2. \tag{29}$$

Note that, as $v \to \pm\infty$, $S_f(v) \to 0$, so $\hat{F}(v) \to 0$ also. If it were not for the term σ_N^2 in the denominator of equation (23), $\hat{F}(v)$ would become very large wherever $S_f(v)$ became small. In Chapter 12, I will show that there is a tradeoff between the magnitude of α and the *resolution* of an inverse.

One notable feature of equation (29) is that, at frequencies where $S_f(v)|Q(v)|^2 \gg \sigma_N^2$, $\hat{F}(v) \approx F(v)$, while when $S_f(v)|Q(v)|^2 \ll \sigma_N^2$, $\hat{F}(v) \to 0$.

10.5.1 Derivation of the Wiener filter

We can prove that equation (23) is the optimal solution (in the least-squares sense) as follows. Since

$$\hat{F}(v) = D(v)G'(v) = D(v)[F(v)Q(v) + N(v)], \tag{30}$$

we can find the function $D(v)$ that minimizes the squared error[11]

$$\chi^2 \equiv \left\langle \int_{-v_c}^{v_c} \left|\hat{F}(v) - F(v)\right|^2 dv \right\rangle = \left\langle \int_{-v_c}^{v_c} \left|D(v)[F(v)Q(v) + N(v)] - F(v)\right|^2 dv \right\rangle$$

$$= \left\langle \int_{-v_c}^{v_c} \left|F(v)[D(v)Q(v) - 1] + D(v)N(v)\right|^2 dv \right\rangle. \tag{31}$$

Here, I have used Rayleigh's theorem (§8.2.4), to conclude that

$$\int_{-\infty}^{\infty} \left|\hat{f}(y) - f(y)\right|^2 dy = \int_{-\infty}^{\infty} \left|\hat{F}(v) - F(v)\right|^2 dv. \tag{32}$$

Expand equation (31) to get [since $D(v) = 0$ if $v > v_c$, we can make the formal limits of integration $\pm\infty$]

$$\chi^2 = \left\langle \int_{-\infty}^{\infty} F(v)F^*(v)[D(v)Q(v) - 1][D^*(v)Q^*(v) - 1]dv \right\rangle$$

$$+ \left\langle \int_{-\infty}^{\infty} F(v)[D(v)Q(v) - 1]D^*(v)N^*(v)dv \right\rangle$$

$$+ \left\langle \int_{-\infty}^{\infty} F^*(v)[D^*(v)^*Q(v) - 1]D(v)N(v)dv \right\rangle$$

$$+ \left\langle \int_{-\infty}^{\infty} D(v)D^*(v)N(v)N^*(v)dv \right\rangle. \tag{33}$$

The second and third terms are both zero, since $N(v)$ is a random variable that is uncorrelated with $F(v)$. Also, D and Q are fixed functions, so they do not change from one realization to the next. Therefore, using equations (18) and (14),

---

[11] Here, <·> represents an average over an ensemble of observations.

$$\chi^2 = \int\limits_{-\infty}^{\infty} S_f(v)\big[D(v)Q(v)-1\big]\big[D^*(v)Q^*(v)-1\big]dv$$

$$+ \sigma_N^2 \int\limits_{-\infty}^{\infty}\big|D(v)\big|^2 dv. \tag{34}$$

After some further manipulation, this becomes

$$\chi^2 = \int\limits_{-\infty}^{\infty} S_f(v)\Big[\big|D(v)\big|^2\big|Q(v)\big|^2 - D(v)Q(v) - D^*(v)Q^*(v) + 1\Big]dv$$

$$+ \sigma_N^2 \int\limits_{-\infty}^{\infty}\big|D(v)\big|^2 dv. \tag{35}$$

In order to minimize χ^2, replace $D(v)$ with $D(v) + \varepsilon h(v)$, where $h(v)$ is an arbitrary function not identically equal to zero and ε is small positive constant. Just as we find an extremum of a real function $h(t)$ by taking the derivative with respect to t and setting it equal to zero, we can replace D with $D + \varepsilon h$ and differentiate with respect to ε:

$$\chi^2 = \int\limits_{-\infty}^{\infty} S_f(v)\Big\{\big|D(v)+\varepsilon h(v)\big|^2\big|Q(v)\big|^2 + 1\Big\}dv$$

$$- \int\limits_{-\infty}^{\infty} S_f(v)\Big\{\big[D(v)+\varepsilon h(v)\big]Q(v) + \big[D^*(v)+\varepsilon h^*(v)\big]Q^*(v)\Big\}dv$$

$$+ \sigma_N^2 \int\limits_{-\infty}^{\infty}\big|D(v)+\varepsilon h(v)\big|^2 dv. \tag{36}$$

Now

$$\big|D(v)+\varepsilon h(v)\big|^2 = \big|D(v)\big|^2 + \varepsilon D^*(v)h^*(v)$$

$$+ \varepsilon D(v)h(v) + \varepsilon^2\big|h(v)\big|^2. \tag{37}$$

Since ε is very small, we ignore terms proportional to ε^2. Some more manipulation yields

$$\chi^2 = \int_{-\infty}^{\infty} S_f(v) \left\{ |Q(v)|^2 \left[|D(v)|^2 + \varepsilon h(v) D(v) + \varepsilon h^*(v) D^*(v) \right] + 1 \right\} dv$$

$$- \int_{-\infty}^{\infty} S_f(v) \left\{ \left[D(v) + \varepsilon h(v) \right] Q(v) + \left[D^*(v) + \varepsilon h^*(v) \right] Q^*(v) \right\} dv$$

$$+ \sigma_N^2 \int_{-\infty}^{\infty} \left\{ |D(v)|^2 + \varepsilon h(v) D(v) + \varepsilon h^*(v) D^*(v) \right\} dv. \tag{38}$$

At an extremum, $d\chi^2/d\varepsilon = 0$:

$$0 = \int_{-\infty}^{\infty} S_f(v) \left\{ |Q(v)|^2 \left[h(v) D(v) + h^*(v) D^*(v) \right] \right\} dv$$

$$- \int_{-\infty}^{\infty} S_f(v) \left\{ h(v) Q(v) + h^*(v) Q^*(v) \right\} dv$$

$$+ \sigma_N^2 \int_{-\infty}^{\infty} \left\{ h(v) D(v) + h^*(v) D^*(v) \right\} dv. \tag{39}$$

Since this must be zero for any function $h(v)$ whatsoever, it is necessary that the integrand vanish everywhere, so

$$\left[S_f(v) |Q(v)|^2 + \sigma_N^2 \right] D(v) = S_f(v) Q(v). \tag{40}$$

Equation (23) follows directly from this.

10.6 Existence, uniqueness, and stability

Given that we know the kernel K in equation (1) we should consider the questions of existence, uniqueness, and stability. By that I mean the following: given a measurement of g,

1. Does a function f exist that satisfies equation (1)?

2. Is there a *unique* function f that *must* have produced g, or are there several different functions that *might* have produced it?

3. Will a small error in the measurement of g produce only a small error in the estimated value if f?

If we take the position that equation (1) represents some physical process, then, *if there is no noise in the measurement of g*, the corresponding f must exist. However, when there is noise present, it is not clear that there must be a solution f corresponding to any possible measurement of $g' = g + n$. Whether or not there is depends on the specifics of a given problem.

10.6.1 Uniqueness

The possible solutions exist in a Hilbert space $\mathcal{H}$. The problem of uniqueness can be formulated as follows: if the solution f is not unique, then there must be some other function that also satisfies equation (1). Let $\mathcal{N}$ be the set of all functions that satisfy equation (1) for the case $g(x)$ identically equal to zero; *i.e.*,

$$\mathcal{N} = \{\phi(y): K\phi = 0\} \qquad (41)$$

[remember that the notation Kf is defined by equation (2), the limits a and b being understood]. $\mathcal{N}$ is called the *null space* of the kernel K. Note that, if ϕ_1 and ϕ_2 are both elements of $\mathcal{N}$, then any linear combination of ϕ_1 and ϕ_2 is also an element. If f is a solution to equation (1), then $f + \phi$ is also a solution for any $\phi \in \mathcal{N}$ (if there is one):

$$K(f + \phi) = g(x) + 0 = g(x). \qquad (42)$$

The solution f is unique if and only if $\mathcal{N} = \{\zeta\}$, where ζ is the function that maps everything to zero. What other elements $\mathcal{N}$ might contain depends on the kernel K.

In addition, we must realize that if $K\phi = \psi$ and $\psi(x) < \varepsilon$ for all x, for ε some small positive number, then $K\phi$ may be indistinguishable from the noise in the measurement of g. So, if $Kf = g$, then also $K(f + \phi) = g + \varepsilon = g$ to within the accuracy of the measurement and ϕ might just as well be an element of $\mathcal{N}$ for all practical purposes.

So far in this section, I have been considering that f and g are continuous functions, and that I can measure g at every value of x. In reality, however, we can only make a finite number of measurements, so that we can only measure $g(x_i)$ for a finite set $\mathcal{X} = \{x_i\}$. In this case, we are guaranteed that, if f exists, it is not unique (except, perhaps, in some degenerate cases). The reason for this nonuniqueness is that, since we have measured g at only a finite number of points, our measurements are consistent with any of the functions $g(x) + h(x)$, where $h(x)$ is only restricted by the condition that

$$h(x_i) = 0. \tag{43}$$

The values of $h(x)$ for any $x \notin \mathscr{H}$ are otherwise completely arbitrary.

10.6.2 Stability

The last question is that of stability, which means that small changes in g' should not produce large changes in $\hat{f}$. More formally, let $\varepsilon(x)$ be a function of x, and let $\hat{f}(y)$ be the solution corresponding to some measured function g'. Let $\hat{f}_\varepsilon(y)$ be the solution corresponding to $g' + \varepsilon$. Let $\delta(y)$ be the change in $\hat{f}(y)$ due to adding ε to g'; i.e.,

$$\delta(y) = \hat{f}(y) - \hat{f}_\varepsilon(y). \tag{44}$$

The solution of equation (1) is unstable if small values of ε produce large values of δ. Very roughly, this will happen—the solution will be unstable—when the noise-amplification factor α is large. How large it can be will depend, of course, on the particulars of the experiment at hand.

10.6.3 Comparison with matrix method

Let us take equation (22) one step further. We Fourier transform DG' to obtain $\hat{f}(y_j)$ for some set $\mathscr{Y} = \{y_j\}$. That is,

$$\hat{f}(y_j) = \int_{-\infty}^{\infty} D(v) G'(v) e^{2\pi i y_j v} \, dv$$

$$= \int_{-\infty}^{\infty} D(v) \left[\int_{-\infty}^{\infty} g'(t) e^{-2\pi i t v} \, dt \right] e^{2\pi i y_j v} \, dv. \tag{45}$$

Interchanging the order of integration,

$$\hat{f}(y_j) = \int_{-\infty}^{\infty} g'(t) \left[\int_{-\infty}^{\infty} D(v) \, e^{2\pi i v(y_j - t)} \, dv \right] dt. \tag{46}$$

Since we measure g' at only a finite number of values of x_i, the outer integral will, in actuality, be replaced with a sum. Also, we will only determine $\hat{f}$ at a finite set of points

$$\hat{f}(y_j) = \sum_i g'(x_i) w(x_i) \left[\int_{-\infty}^{\infty} D(v) \, e^{2\pi i v (y_j - x_i)} \, dv \right]. \qquad (47)$$

Here, $w(x_i) = w_i$ are suitable weights for estimating the integral with a numerical quadrature formula. Now if we define

$$\hat{f}_j = \hat{f}(y_j)$$

$$g_i = g'(x_i)$$

and

$$R_{ji} = w(x_i) \int_{-\infty}^{\infty} D(v) \, e^{2\pi i v (y_j - x_i)} \, dv, \qquad (48)$$

we get the simple matrix equation

$$\hat{f}_j = \sum_i R_{ji} g_i'. \qquad (49)$$

This shows the similarity of the matrix and integral-equation approaches to solving these problems. In each case, we are given a function that is the result of an operator acting on an unknown function, and we are asked to figure out what that original function was.

10.7 Eigenfunction expansion

Just as matrices have eigenvectors that we can use as the basis of a vector space, kernels of integral equations also have *eigenfunctions* that we can use as a basis of a Hilbert space. Furthermore, we can represent the kernel as a linear combination of these eigenfunctions.

A real kernel $K(x,y)$ is *symmetric* if $K(x,y) = K(y,x)$. If K is complex, the equivalent property is *Hermitian symmetry*, where $K(x,y) = K^*(y,x)$. Symmetric kernels have certain properties that make it possible to solve equations like equation (1).

10.7.1 Symmetric kernel

Let us return to the kernel K of equation (1). If K is symmetric (or Hermitian, if it is complex),[12] then there are functions ϕ_i, called *eigenfunctions* of K, that have the property that

$$\int_a^b K(x,y)\phi_i(y)\,dy \;=\; \lambda_i\phi_i(x). \tag{50}$$

(Compare this with the definition of eigenvectors in Chapter 5.) The scalars λ_i are called *eigenvalues* (just as with matrices). If ϕ_j is an eigenfunction of K, then for any scalar a, $a\phi_j$ is also an eigenfunction.

Exercise 10-2: Describe in detail the similarities between eigenfunctions of the kernel K and the eigenvectors of a square matrix $\mathbf{M}$.

Some properties of the eigenfunctions of *symmetric* kernels are

1. If K is not identically zero, there is at least one eigenvalue.

2. Every eigenvalue is real.

3. There is a maximum eigenvalue λ_1 and the eigenvalues $\lambda_n \to 0$ as n increases. In general, there is an infinite number of eigenvectors with the eigenvalue = 0.

Eigenfunctions corresponding to different eigenvalues are *orthogonal* and can be normalized so that

$$\left(\phi_i,\phi_j\right) \;=\; \int_a^b \phi_i(x)\,\phi_j(x)\,dx \;=\; \delta_{ij}, \tag{51}$$

where

$$\delta_{ij} \;=\; \begin{cases} 1 & i=j \\ 0 & i\neq j \end{cases}. \tag{52}$$

There may be several eigenfunctions that belong to the same eigenvalue λ. If ϕ_i and ϕ_j are two of these, then for any scalars a and b, $a\phi_i + b\phi_j$ is also an eigenfunction. (Please note that this is not true if ϕ_i and ϕ_j belong to different eigenvalues.) It is possible to find a set of n eigenfunctions belonging to the eigenvalue λ that are orthonormal and any other eigenfunction with eigenvalue λ is a linear combination of these n eigenfunctions. In this case, the ei-

---

[12] See, for example, S. G. Mikhlin, *Integral Equations,* MacMillan Co, New York, 341 pp (1964).

genvalue has *multiplicity n*. Note that the eigenfunctions depend both on the nature of the kernel K and the limits of integration a and b in equation (1).

A symmetric kernel is *degenerate* or has *finite rank* if it can be written in the form

$$K(x,y) = \sum_{k=1}^{m} c_k \, \phi_k(x)\phi_k(y), \tag{53}$$

where c_k are constants and the limit m is finite. We always denote the largest eigenvalue by λ_1 and succeeding eigenvalues denoted in non-increasing order. If the kernel is *not* degenerate, there is an infinite number of eigenvalues that get smaller and smaller as n increases. If it is degenerate, however, then there is a smallest nonzero eigenvalue. We shall see that the solution of an integral equation is unstable if its kernel is not degenerate.

Define the *adjoint* of the kernel K in the following way. Let $K^{\dagger}$ be the kernel that has the property that

$$\int_a^b \left[\int_a^b K(x,y)f(y)dy \right] g(x)dx \;=\; \int_a^b \left[\int_a^b K^{\dagger}(x,y)g(x)dx \right] f(y)dy \tag{54}$$

for any pair of functions f and g. To make this notation more compact, define

$$(Kf,g) \;=\; \int_a^b \left[\int_a^b K(x,y)f(y)dy \right] g(x)dx \tag{55}$$

and

$$(f,K^{\dagger}g) \;=\; \int_a^b \left[\int_a^b K(x,y)g(x)dx \right] f(y)dy. \tag{56}$$

In this notation,

$$(Kf,g) = (f,K^{\dagger}g). \tag{57}$$

This defines the adjoint of K. While, in general, K and $K^{\dagger}$ are different, in the case we treat below, where K is symmetric and all of the functions we consider are real, and $K = K^{\dagger}$. We say in this case that K is *self-adjoint*. In the more general case where complex functions are involved, the corresponding notion is Hermitian symmetry.

10.7.2 Asymmetric kernel

Even when K is not self-adjoint, we can find functions that serve the same purpose as the eigenfunctions of K did; they are eigenfunctions of $K^\dagger K$. Consider the following operator:

$$K^\dagger K f(y) = \int_a^b K^\dagger(x,y) \left[\int_a^b K(x,y') f(y') dy' \right] dx. \tag{58}$$

Let the functions $\phi_n(y)$ be eigenfunctions of $K^\dagger K$ associated with the eigenvalues λ_n^2:

$$K^\dagger K \phi_n = \lambda_n^2 \phi_n. \tag{59}$$

Also, let

$$\psi_n = \frac{1}{\lambda_n} K \phi_n. \tag{60}$$

Then

$$\phi_n = \lambda_n K^\dagger \psi_n \tag{61}$$

and

$$\psi_n = \lambda_n K^\dagger \phi_n. \tag{62}$$

This is easy to show using equation (59).[13]

10.7.3 Eigenfunction solution

We can solve equation (1) by expanding the function f as the weighted sum of the eigenfunctions ϕ_i [as in equation (4)]. We have already done this in the matrix formulation of the problem. The advantage of this method is that it is straightforward and we can include the effects of measurement noise quite easily. The drawback is that, in order to use this method, we have to know what the eigenfunctions of K are.

We shall represent f as a sum of the eigenfunctions ϕ_j of $K^\dagger K$ by[14]

---

[13] See Groetsch (1984).

[14] Do not confuse b_i in this equation, or a_i in subsequent equations, with the limits of integration a and b. They have nothing to do with each other.

$$f(y) = \sum_i b_i \phi_i(y). \tag{63}$$

The variance of $f(y)$ is

$$\left\langle \int_{-\infty}^{\infty} |f(y)|^2 \, dy \right\rangle = \sum_i \left\langle b_i^2 \right\rangle. \tag{64}$$

Exercise 10-3: Use the orthonormal property of the eigenfunctions to prove this. Show that

$$b_i = \int_{-\infty}^{\infty} f(y)\phi_i(y)dy. \tag{65}$$

Substitute equation (63) into equation (1):

$$g(x) = \sum_i b_i \int K(x,y) \phi_i(y)dy = \sum_i \lambda_i b_i \psi_i(x) \tag{65}$$

(note that we expand g in terms of ψ_i, and f, in terms of ϕ_i because, in general, g and f are elements of different Hilbert spaces).

If there is no noise in the measurement of $g(x)$,[15] we can easily find $f(y)$. Expand g in terms of $\{\psi_i\}$:

$$g(x) = \sum_i a_i \psi_i(x). \tag{66}$$

We estimate the value of a_i from[16]

$$\hat{a}_i = \int_a^b g(x)\psi_i(x)dx = \lambda_i b_i. \tag{67}$$

Then we would estimate $f(y)$ from (remember, this only works in the absence of noise)

$$\hat{f}(y) = \sum_i \frac{\hat{a}_i}{\lambda_i}\phi_i(y), \tag{68}$$

or $\hat{b}_i = \hat{a}_i / \lambda_i$.

---

[15] Unlikely as this might be.

[16] Remember that we have not included noise yet.

Of course, this method will only work in the absence of noise. By this time, I will be disappointed with the reader who fails to see the flaw in this method of estimating $f(y)$. The problem, of course, is that, when λ_i is small, any noise present in the estimate of a_i will be divided by a small number λ_i. To quantify this, we have to consider the effect of noise on $\hat{a}_i$. Let $n(x)$ be a random variable that represents the noise in the measurement of $g(x)$. We are really estimating a_i from

$$\hat{a}_i = \int_a^b [g(x) + n(x)]\psi_i(x)\,dx = a_i + \theta_i, \tag{69}$$

where θ_i is a random variable defined by

$$\theta_i = \int_a^b n(x)\psi_i(x)\,dx. \tag{70}$$

The variance of $\hat{a}_i$ is (remember that its mean value is zero) is

$$\text{var}(\hat{a}_i) = \left\langle \hat{a}_i^2 \right\rangle = \lambda_i^2 \left\langle b_i^2 \right\rangle + \sigma_N^2, \tag{71}$$

where

$$\sigma_N^2 = \left\langle \theta_i^2 \right\rangle = \left\langle \int_a^b \int_a^b \psi_i(x)\psi_i(x')n(x)n(x')\,dx\,dx' \right\rangle. \tag{72}$$

Since $n(x)$ is a random variable,

$$\left\langle n(x)n^*(x') \right\rangle = \sigma_N^2 \delta(x - x'). \tag{73}$$

Therefore, if we use equation (68) to invert equation (1), we will find that the variance of $\hat{f}(y)$ will be

$$\text{var}\left[\hat{f}(y)\right] = \int_a^b \left\langle \left| \sum_i \frac{(\hat{a}_i + \theta_i)}{\lambda_i}\phi_i(y) \right|^2 \right\rangle dx$$

$$= \int_a^b \left\langle \left| \sum_i \frac{(\lambda_i b_i + \theta_i)}{\lambda_i}\phi_i(y) \right|^2 \right\rangle dx = \sum_i \frac{\lambda_i^2 \left\langle b_i^2 \right\rangle + \sigma_N^2}{\lambda_i^2}.$$

$$= \sum_i \langle b_i^2 \rangle + \sigma_N^2 \sum_i \frac{1}{\lambda_i^2}. \tag{74}$$

Here, I have used the property that the θ_i's are independent random variables and are also uncorrelated with the coefficients b_i. Obviously, for the smaller λ_i's, the relative contribution of σ_N^2 is larger. Basically, the problem arises because, regardless of the magnitude of λ_i, the noise variance is the same in every component. If we know that there is some number h such that, for every $i > h$, the coefficient b_i is negligible, then the series can be truncated at $i = h$, and that will limit the noise. The total noise contribution to the variance of $\hat{f}(y)$ will be

$$\Sigma_N^2 = \sigma_N^2 \sum_{i=1}^{h} \frac{1}{\lambda_i^2}; \tag{75}$$

Note that, in most cases, this series does not converge as $h \to \infty$.

So how should we estimate $f(y)$? Let $\{\gamma_i\}$ be a set of coefficients to be determined, and let

$$\hat{f}(y) = \sum_i \gamma_i \hat{a}_i \phi_i(y), \tag{76}$$

where $\hat{a}_i = \lambda_i b_i$. The error is

$$\chi^2 = \int_a^b \left| f(y) - \hat{f}(y) \right|^2 dy = \int_a^b \left| \sum_i [b_i - \gamma_i \hat{a}_i] \phi_i(y) \right|^2 dy. \tag{77}$$

Using once again the orthonormality of the eigenfunctions,

$$\chi^2 = \sum_i \left\langle [b_i - \gamma_i \hat{a}_i]^2 \right\rangle = \sum_i \langle b_i^2 - 2b_i \gamma_i \hat{a}_i + \gamma_i^2 \hat{a}_i^2 \rangle. \tag{78}$$

Now

$$\langle b_i \hat{a}_i \rangle = \langle b_i (b_i \lambda_i + \theta_i) \rangle = \lambda_i \langle b_i^2 \rangle, \tag{79}$$

and

$$\langle \hat{a}_i^2 \rangle = \lambda_i^2 \langle b_i^2 \rangle + \sigma_N^2 \tag{80}$$

since b_i is independent of θ_j. So

$$\chi^2 = \sum_i \left[\langle b_i^2 \rangle - 2\lambda_i \gamma_i \langle b_i^2 \rangle + \gamma_i^2 \left(\lambda_i^2 \langle b_i^2 \rangle + \sigma_N^2 \right) \right]. \tag{81}$$

The error is minimum when

$$\frac{d\chi^2}{d\gamma_i} = -2\lambda_i \langle b_i^2 \rangle + 2\gamma_i \left(\lambda_i^2 \langle b_i^2 \rangle + \sigma_N^2 \right) = 0. \tag{82}$$

Then the coefficients that minimize the error are

$$\gamma_i = \frac{\lambda_i \langle b_i^2 \rangle}{\lambda_i^2 \langle b_i^2 \rangle + \sigma_N^2}. \tag{83}$$

Note two limiting cases:

$$\lim_{\sigma_N^2 \to 0} \gamma_i = \frac{1}{\lambda_i} \tag{84}$$

and

$$\lim_{\sigma_N^2 \to \infty} \gamma_i = 0, \tag{85}$$

as we should expect. Using these values of γ_i,

$$\chi^2 = \sum_i \frac{\langle b_i^2 \rangle \sigma_N^2}{\lambda_i^2 \langle b_i^2 \rangle + \sigma_N^2}. \tag{86}$$

Exercise 10-4: Show that any kernel can be written in terms of its eigenfunctions:

$$K(x,y) = \sum_{i=1} c_i \psi_i(x) \phi_i(y), \tag{87}$$

where the c_i are constants. What is the representation of $K^\dagger K$? of $KK^\dagger$?

10.7.4 Finding the eigenfunctions

How can we find the eigenvalues and eigenfunctions of a given kernel? There is no general method, even when the kernel is degenerate. The problem can be solved if $K(x-y)$ is the Fourier transform of a rational function,[17] although this may not be very helpful in practice.

I discuss a method of finding the largest eigenvalues of a matrix by iteration in §14.11. It involves repeated application of the matrix to an initial test vector, each application being followed by a renormalization. A similar scheme would work for finding the largest eigenvalues of a symmetric kernel, although the integration would have to be performed numerically. But a numerical integration is the equivalent of a finite summation, using an appropriate quadrature formula, so the problem of finding the eigenfunctions of a kernel is formally the same as that of finding the eigenvectors of a matrix, so there is nothing new involved.

We can get some more insight into the question of eigenfunctions by looking at the Fourier series expansion of the kernel. Suppose that $K(x,y)$ depends only on $|x - y|$ and that it is defined on the interval $(-1,1)$. We can extend $K(x,y)$ so that it is periodic with period 2. The we can expand it in a Fourier series. The Fourier cosine transform is

$$K(t) = \alpha_0 + \sum_{n=1}^{\infty} \alpha_n \cos(2\pi nt) \tag{88}$$

or, equivalently,

$$K(x,y) = K(x - y) = \sum_{n=0}^{\infty} \alpha_n \cos[2\pi n(x - y)]. \tag{89}$$

Now, since $\cos[2\pi n(x - y)] = \cos(2\pi nx)\cos(2\pi ny) + \sin(2\pi nx)\sin(2\pi ny)$, for $m \geq 0$,

$$\int_{-\frac{1}{2}}^{\frac{1}{2}} K(x - y)\cos(2\pi mx)dx = \int_{-\frac{1}{2}}^{\frac{1}{2}} \cos(2\pi mx)\sum_{n=0}^{\infty} \alpha_n \cos[2\pi n(x - y)]dx$$

$$= \alpha_m \int_{-\frac{1}{2}}^{\frac{1}{2}} \cos(2\pi mx)[\cos(2\pi mx)\cos(2\pi my) + \sin(2\pi mx)\sin(2\pi my)]dx$$

$$= \frac{1}{2}\alpha_m \cos(2\pi my). \tag{90}$$

This shows that $\cos(2\pi mx)$ is an eigenfunction of K with eigenvalue $\frac{1}{2}\alpha_m$.

---

[17] See Davenport and Root (1958, pp 375 *ff*)

Exercise 10-5: I specified that $K(x,y)$ depends only on $|x - y|$. An example of such a kernel might be a Gaussian $\exp\{-2\pi a^2(x - y)^2\}$. Why are there no sine components in K? Is this a realistic example?

Exercise 10-6: The spectrum—the set of eigenvalues—of any kernel is uniquely determined by that kernel. We know that there are other sets of orthogonal functions that we might have used to expand K in a Fourier series. Does this mean that there must be other eigenfunctions that have the same eigenvalues as given by equation (90), or that somehow sines and cosines play a special role in the scheme of things?

10.7.5 Information content

Let us look at a simple example of an integral equation and ask how many separate—*i.e.*, statistically independent—pieces of information can be retrieved from a set of observations. In order to keep things simple, I will take the kernel to be a normalized Gaussian

$$K(x,y) = \frac{1}{\sqrt{2\pi}w}e^{-\frac{(x-y)^2}{2w^2}}, \qquad (91)$$

where w is the width. The integral equation is defined on the interval $[-1,1]$ to be

$$g(x) = \frac{1}{\sqrt{2\pi}w}\int_{-1}^{1}e^{-\frac{(x-y)^2}{2w^2}}f(y)dy. \qquad (92)$$

To specify the problem further, suppose that we measure $g(x)$ at $N+1$ equally spaced points, including the endpoints. I denote these points by

$$x_j = \frac{2j}{N} - 1 \quad j = 0, 1, \ldots, N; \qquad (93)$$

we wish to infer the value of f at the n points y_k, specified similarly.

For now, suppose that the autocorrelation function of f is $\sigma_f^2\,\delta(y - y')$; *i.e.*, the values of f are uncorrelated at different values of y. Then the autocorrelation function of $g(x_i)$ is

$$C_g(\tau) = \frac{1}{2\pi w^2}\left\langle \int_{-1}^{1}\int_{-1}^{1}e^{-\frac{(x-y)^2}{2w^2}}e^{-\frac{(x+\tau-y')^2}{2w^2}}f(y)f(y')dy\,dy' \right\rangle_x$$

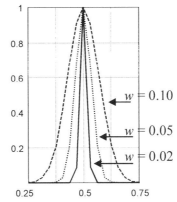

Figure 10-3. Widths of the kernels.

$$= \left\langle \frac{\sigma_f^2}{2\pi w^2} \int_{-1}^{1}\int_{-1}^{1} e^{-\frac{(x-y)^2}{2w^2}} e^{-\frac{(x+\tau-y')^2}{2w^2}} \delta(y-y') \, dy \, dy' \right\rangle_x$$

$$= \left\langle \frac{\sigma_f^2}{2\pi w^2} e^{-\frac{\tau^2}{4w^2}} \int_{-1}^{1} e^{-\frac{(y-x-\tau/2)^2}{w^2}} \, dy \right\rangle_x$$

$$\approx \frac{\sigma_f^2}{2\pi w^2} e^{-\frac{\tau^2}{4w^2}} \int_{-\infty}^{\infty} e^{-\frac{2(y-x-\tau/2)^2}{2w^2}} \, dy = \frac{\sigma_f^2}{2\sqrt{\pi}w} e^{-\frac{\tau^2}{4w^2}}. \qquad (94)$$

Here, $\langle \cdot \rangle_x$ denotes an average over x. In this last step, I have assumed that w is small enough that we can expand the limits of integration from $[-1,1]$ to $[-\infty, \infty]$ without making a serious error.

Since we make measurements of g at the coördinates x_j, we can also write the autocorrelation function $C_g(\tau)$ in terms of a covariance matrix

$$\mathbf{S}_g = \sigma_f^2 \begin{pmatrix} P_{11} & P_{12} & \cdots & P_{1n} \\ P_{21} & P_{22} & & \\ \vdots & & \ddots & \\ P_{n1} & & & P_{nn} \end{pmatrix} \qquad (95)$$

where $x_j - x_k = 2(j-k)/N_g$ and

$$P_{jk} \equiv \frac{1}{2\sqrt{\pi}w} e^{-\frac{(j-k)^2}{w^2 N_g^2}}. \qquad (96)$$

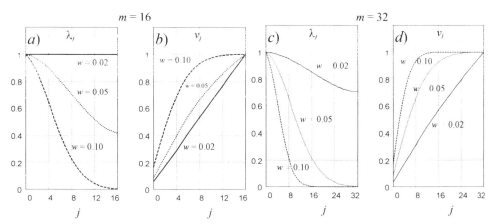

Figure 10-4. Eigenvalues (λ_j) and fraction of variance (v_j) explained for $m = 16$ [panels a) and b)] and for $m = 32$ [panels c) and d)].

Note that there is no noise involved yet; this is just the part of the covariance due to the variance of f. In the notation of Chapter 7, where we would have written in vector notation $\mathbf{g} = \mathbf{Af}$, the covariance matrix of $\mathbf{g}$ is $\mathbf{S_g} = \mathbf{AS_fA^t}$, where $\mathbf{S_f}$ is the covariance matrix of f.

I did some sample calculations for $N_g = 16$, and for $w = 0.02$, 0.05, and 0.1 (for simplicity, I have let $\sigma_f^2 = 1$). The kernels are shown in Figure 10-3; I have tabulated the results in Table 10-1. When $w = 0.02$, there is very little overlap in the kernels, so they are almost completely independent: therefore, the eigenvalues are almost equal. When $w = 0.1$, there is a lot of overlap of adjacent kernels. The eigenvalues quickly fall from 7.9 ($N = 0$) to 2.4 ($N = 7$). The first eight eigenvalues explain about 89% of the variance. The results for $w = 0.05$ are intermediate. We can see that we can retrieve about 17 independent pieces of information when the kernels are narrow, so the overlap is small, but only about eight pieces when the overlap is larger. *No matter how we process the data, we cannot get more pieces of information than this.*

I have shown these results graphically in Figure 10-4. Panel a) shows the eigenvalues, normalized to $\lambda_0 = 1$; panel b) shows the fraction of the variance explained, which ranges from 0 to 1. Panels c) and d) show the same information for $n_g = 32$. This shows that, when the kernel is wide ($w = 0.10$), making more measurements does not increase the number of independent pieces of information; the number only increases when the kernel becomes narrower.

We can derive a rough measure of the correlation length of g by considering the covariance of two measurements made a distance $\Delta x = 2/N_g$ units apart. Note that, for $j = k$, $P_{j,j} = 1/(2w\pi^{1/2})$. Therefore, the correlation coefficient is

$$\rho_{j,j+1} \equiv \frac{P_{j,j+1}}{P_{j,j}} \equiv e^{-\frac{1}{w^2 N_g^2}}. \tag{97}$$

Just consider two measurements, made at x_j and x_{j+1}, and assume that $P_{jj} = P_{j+1,j+1}$. Then the covariance matrix is proportional to

$$\mathbf{R}_1 = \begin{pmatrix} \rho_{j,j} & \rho_{j,j+1} \\ \rho_{j,j+1} & \rho_{j+1,j+1} \end{pmatrix} = \begin{pmatrix} 1 & e^{-1/w^2 n^2} \\ e^{-1/w^2 n^2} & 1 \end{pmatrix}. \tag{98}$$

It is not hard to show that the eigenvalues of this matrix are $\lambda_0 = 1 + \exp\{-1/w^2 N_g^2\}$ and $\lambda_1 = 1 - \exp\{-1/w^2 N_g^2\}$. So as $w \to 0$, $\lambda_1 \to 1$ and $\lambda_2 \to 1$. Conversely, as $w \to \infty$, $\lambda_0 \to 2$ and $\lambda_1 \to 0$. Arbitrarily, we can define the correlation length τ_g to be the value of w where $\exp\{-1/w^2 N_g^2\} = \frac{1}{2}$, or

$$N_g = \frac{1}{w\sqrt{\ln 2}}. \tag{99}$$

Roughly speaking, there are N_g possible independent measurements of g.

Note two aspects of this problem. The first is that, since we can diagonalize $\mathbf{S}_g$, we can treat each component separately. Second, since we assumed that the noise covariance matrix was $\mathbf{N} = \sigma_N^2 \mathbf{I}$ in the original basis, it is the same in any basis. Therefore we can write each component as $g_i = \lambda_i f_i + n_i$, where n_i is a noise variable with variance σ_N^2. But we saw in §1.7 that the solution to this problem is to multiply g_i by

$$\beta_i = \frac{\lambda_i \sigma_f^2}{\sigma_N^2 + \lambda_i^2 \sigma_f^2}. \tag{100}$$

We also saw that the error variance is

$$\chi_i^2 = \mathrm{var}\left(f_i - \hat{f}_i\right)^2 = \frac{\sigma_f^2 \sigma_N^2}{\lambda_i^2 \sigma_f^2 + \sigma_N^2}; \tag{101}$$

the fraction of the variance explained is $v_i \equiv 1 - \chi_i^2/\sigma_f^2$, or

$$v_i = \frac{\lambda_i^2 \sigma_f^2}{\sigma_N^2 + \lambda_i^2 \sigma_f^2}. \tag{102}$$

We can say, somewhat arbitrarily, that we can retrieve an independent component f_i if we can explain more than half its variance; i.e., $v_i > \frac{1}{2}$ or $\lambda_i^2 \sigma_f^2 > \sigma_N^2$.

There is a third matter to consider. In most applications, $f(y)$ is not δ-correlated: there is a correlation length τ_f such that, if $|y_1 - y_2| < \tau_f$, $f(y_1)$ is correlated with $f(y_2)$. So, roughly speaking, we can completely describe $f(y)$ with $n_f = 2/\tau_f$ measurements. Now if there are only n_f independent values of $f(y)$,

	w = 0.02		w = 0.05		w = 0.10	
j	λj	vj	λj	vj	λj	vj
0	14.11	0.0588	7.992	0.0833	7.853	0.1638
1	14.11	0.1177	7.881	0.1655	7.429	0.3187
2	14.11	0.1765	7.702	0.2458	6.772	0.4599
3	14.11	0.2353	7.459	0.3236	5.949	0.5839
4	14.11	0.2941	7.160	0.3982	5.035	0.6889
5	14.11	0.3530	6.815	0.4693	4.106	0.7746
6	14.11	0.4118	6.436	0.5364	3.225	0.8418
7	14.11	0.4706	6.034	0.5993	2.440	0.8927
8	14.10	0.5294	5.623	0.6579	1.776	0.9297
9	14.10	0.5883	5.213	0.7123	1.244	0.9557
10	14.10	0.6471	4.818	0.7625	0.838	0.9731
11	14.10	0.7059	4.450	0.8089	0.542	0.9844
12	14.10	0.7647	4.119	0.8519	0.335	0.9914
13	14.10	0.8235	3.835	0.8918	0.199	0.9956
14	14.10	0.8824	3.605	0.9294	0.119	0.9979
15	14.10	0.9412	3.436	0.9652	0.063	0.9992
16	14.10	1.0000	3.333	1.0000	0.037	1.0000

Table 10-1 First seventeen eigenvalues (λ) and the corresponding fraction of the variance explained (f) for w = 0.01 (λ_1, f_1); w = .033 (λ_2, f_2); and w = 0.1 (λ_3, f_3).

there can be no more than n_f independent values of $g(x)$ and we obviously cannot retrieve more than n_f independent values of $\hat{f}(y)$.

Therefore there are four different factors that restrict the number of independent measurements we need to make, and how many independent values of $f(y)$ we can retrieve: the correlation length of $f(y)$, the width of the kernel $K(x-y)$, and the number n_g of points where we measure $g(x)$. The correlation length of $f(y)$ is determined by nature; the width of the kernel, by the physics of the remote-sensing problem and the design of the sensor; while the number of measurements we make is under our control. Finally, the signal-to-noise ratio σ_g^2/σ_N^2 also affects how many independent values of $f(y)$ we can retrieve. We need to consider all of these factors to determine what fraction of the variance of $f(y)$ we can retrieve.

Exercise 10-7: Derive an expression for the covariance matrix of $g(x)$ when the autocorrelation function of $f(x)$ is a Gaussian width τ_f.

Exercise 10-8: Suppose the kernels were all Gaussians centered at $y = 0$, but with different widths. That is,

$$K(x,y) = \frac{1}{\sqrt{2\pi x}} e^{-\frac{y^2}{2x^2}}.$$ (103)

For $f(y)$ δ-correlated, what is the covariance matrix of $g(x)$? Can we retrieve more or fewer independent estimates of $f(y)$ than with the kernels that have a constant width and shift in y?

10.8 Discussion

The results of this chapter may seem a bit disappointing. We have just seen that, despite claims to the contrary, an iterated solution to an integral equation will not, in general, converge to the correct answer unless it is regularized. The eigenfunction approach of §10.7 also needs to be regularized. It is clear that neither approach can do what the other cannot. This anticipates a result of the next chapter, that, for matrix methods, iteration and matrix inversion give exactly the same results.

The only really useful result has been the one concerning integral equations where the kernel $K(x,y)$ depends only on the quantity $(x - y)$, so that $\int K(x,y)\phi(x)dx$ is a convolution.

A more fundamental problem is that, as far as I know, no one has found the eigenfunctions of any kernel that is germane to remote sensing. Unlike the case with matrices, there is no obvious way to find these eigenfunctions except by numerical quadrature, which is time-consuming and prone to roundoff error, or by calculating the Fourier series, for periodic kernels that depend only on $|x - y|$. However, because we are guaranteed that $\lambda_n \to 0$ as $n \to \infty$, we know that concern with the effects of having small eigenvalues is a necessary part of the problem. Even if the integral-equation approach provides no useful algorithms *per se*, its study should provide valuable insight into the inversion problem.

Furthermore, we must realize that not all inversion problems can be solved by making the noise sufficiently small: some kernels produce naturally unstable solutions, and there may not be much we can do about it. Delves and Mohamed[18] give some examples of simple-looking integral equations where roundoff error in the computation, rather than noise, makes the solutions unstable (like the logistic equation from §11.2.1).

As I mentioned before, the solution of a Fredholm equation of the first kind is unstable unless the kernel is degenerate (has only a finite number of non-zero eigenvalues). It is not clear how, in most physical situations, we can decide whether or not the kernel is degenerate, partly because we rarely know the analytic form of the kernel exactly (not to mention the problem that we have no simple method for finding the eigenfunctions). We can look at this another way. Suppose the kernel K is a mapping from $\mathcal{H}_1$ to $\mathcal{H}_2$, which are both Hilbert spaces, and suppose it has rank m (*i.e.*, m eigenvalues, including mul-

---

[18] (1985) pp 299 *ff.*

tiplicities). Then the eigenfunctions ϕ_i, $i = 1, \ldots, m$, span a subspace $\mathcal{H}_1' \subset \mathcal{H}_1$ and K maps $\mathcal{H}_1'$ to a subspace of $\mathcal{H}_2$ that is denoted by $R(\mathcal{H}_1')$. K maps the rest of $\mathcal{H}_1$ to zero. We can only invert K to find the components that are in $R(\mathcal{H}_1')$; all others are lost. So even with a degenerate kernel, we can only retrieve a function f if it lies in $\mathcal{H}_1'$.

As a parting thought, I hope that the reader will be wary of using iteration to solve integral equations unless the method is well regularized.

References

Davenport, W. B., and R. L. Root, *An Introduction to the Theory of Random Signals and Noise*, McGraw-Hill, New York (1958).

Delves, L. M., and J. L. Mohamed, *Computational Methods for Integral Equations*, Cambridge University Press, Cambridge (1985).

Foster, M., 'An application of the Wiener-Kolmogorov smoothing theory to matrix inversion,' *J. Society Indust. Appl. Math.*, **9**, 387-392 (1961).

Gardner, W. A., *Introduction to Random Processes*, McGraw-Hill Publishing Co., New York, 456 pp (1990).

Groetsch, C. W., *The theory of Tikhonov Regularization for Fredholm Equations of the First Kind*, Pitman Advanced Publishing Program, Boston (1984).

Landweber, L., 'An iteration formula for Fredholm integral equations of the first kind,' Amer. J. Math, **73**, 615-624 (1951).

Smithies, F., *Integral Equations*, Cambridge University Press, London (1958).

Strand, O. N. and E. R. Westwater, 'Minimum-rms estimation of the numerical solution of a Fredholm integral equation of the first kind,' *SIAM J. Numerical Analysis*, **2**, 287-295 (1968).

Twomey, S., 'On the numerical solution of Fredholm integral equations of the first kind by the inversion of the linear system produced by quadrature,' *J. Assoc. Comput. Mech.*, **10**, 97-101 (1963).

Twomey, S., 'Indirect measurements of atmospheric temperature profiles from satellites: II. Mathematical aspects of the inversion problem,' *Monthly Weather Review*, **94**, 363-366 (1966).

Westwater, E. R., and O. N. Strand, 'Inversion Techniques,' in *Remote Sensing of the Troposphere*, ed. V. E. Derr, Wave Propagation Laboratory, ERL, Boulder, CO, National Oceanic and Atmospheric Administration, Chapter 16 (1972).

Wiener, N., *Extrapolation, Interpolation, and Smoothing of Stationary Time Series, with Engineering Applications*, New York: Technology Press and John Wiley & Sons (1949).

11. Iteration

There is another way to solve the kinds of equations we have been considering here. Up to now, we have used a single application of some process—*e.g.*, multiplication by a matrix—that produces the desired answer. There is another possibility, however, that we repeatedly apply a process to some initial guess at the answer, each time getting closer to the truth. Consider once more the equation

$$\mathbf{y} = \mathbf{A}\mathbf{x}. \tag{1}$$

Suppose that $\mathbf{A}$ is either not square or that it is singular. In either case, $\mathbf{A}^{-1}$ does not exist in the usual sense. Although I have shown in the preceding chapters how different inverse matrices can be constructed, we can also approach the problem from a different angle. Multiplying $\mathbf{y}$ by $\mathbf{A}^{-1}$, when $\mathbf{A}$ is singular, is the logical equivalent of trying to divide by zero. But maybe we can *multiply* $\mathbf{y}$ by something else that will give us a reasonable answer. I shall show that this approach to solving equation (1) can work. I shall also show that it has no advantage over the methods derived in Chapter 7, for linear problems, but it will shed some light on the retrieval process in general, and we will need iterative techniques for nonlinear problems. I shall discuss the treatment of one nonlinear problem, that of inferring vertical temperature profiles in the earth's atmosphere from measurements of infrared radiances emitted by the atmosphere, in §11.6.

As before, let $\mathbf{n}$ be a random vector—representing the noise in the measurement of $\mathbf{y}$—and let $\mathbf{y}' = \mathbf{y} + \mathbf{n}$. One way to solve equation (1) is to *guess* the proper value of $\mathbf{x}$ and then use a systematic procedure for correcting the initial approximation until we discover an $\hat{\mathbf{x}}$ that satisfies equation (1) to within some desired accuracy. That is,

$$\left|\mathbf{y}' - \mathbf{A}\hat{\mathbf{x}}\right|^2 < \varepsilon \tag{2}$$

for some $\varepsilon > 0$. Clearly, it would be very inefficient simply to guess randomly at different values of $\hat{x}$ until we found the right one. Rather, we should use an algorithm that provides a method for correcting the guess $\hat{x}$ and, at each stage, getting a better approximation.

This process of obtaining a solution by successive approximations is called *iteration*. Let $\mathbf{x}^{(0)}$ be the initial guess at the solution; let $\mathbf{x}^{(1)}$, $\mathbf{x}^{(2)}$, ..., be successive approximations. In this and the following sections, superscripts will denote steps in an iteration process, not exponents. The iteration involves repeating the step

$$\mathbf{x}^{(k+1)} = \mathbf{B}\mathbf{x}^{(k)} + \mathbf{C}\mathbf{y} \tag{3}$$

using some matrices $\mathbf{B}$ and $\mathbf{C}$. Just as we found a way in Chapter 7 to regularize the matrix solution of equation (1), we will find that we can use iteration to do the same thing. Unfortunately, having done this, we shall also see that there is no particular advantage to using this iterative solution for matrix equations. The reader should not be disappointed, however, since this topic is worth our attention for other reasons.

I already discussed integral equations in Chapter 10: iteration is a standard method for solving these equations. Furthermore, iteration is the only way to solve many nonlinear equations.

11.1 One-dimensional examples

Prologue: As you read the following sections, it may help to remember a simple algebraic formula. For any scalar x,

$$\left(1 + x + x^2 + \cdots + x^{k-1}\right)(1 - x) = \left(1 + x + x^2 + \cdots + x^{k-1}\right)$$

$$- \left(x + x^2 + x^3 + \cdots + x^k\right)$$

$$= 1 - x^k. \tag{4}$$

Let $u = 1 - x$. Then

$$\left[1 + (1 - u) + (1 - u)^2 + \cdots + (1 - u)^{k-1}\right]u = 1 - (1 - u)^k. \tag{5}$$

11.1.1 Newton's method

As an example of iteration, let us consider the problem of finding a solution to

$$y = f(x), \tag{6}$$

where x and y are real numbers, and $f(x)$ is a known, continuous, real-valued, nonlinear function of x with a continuous first derivative. For any given y, there may be many—even infinitely many—x's that satisfy equation (6). But for the moment, assume that we can *a priori* restrict x to some closed interval $[a,b]$ where there is but one solution. Also, assume that $f'(x) \neq 0$ in that interval.

Take $x^{(0)}$ as an initial approximation: it might be $<x>$, or there may be a physical reason to pick some other value. Expand equation (6) in a Taylor series, keeping only the first-order term:

$$f(x) = f(x^{(0)}) + f'(x^{(0)})(x - x^{(0)}) + \cdots. \tag{7}$$

The next approximation for x comes from

$$y \approx f(x^{(0)}) + (x - x^{(0)}) f'(x^{(0)}), \tag{8}$$

or

$$x^{(1)} = x^{(0)} + \frac{y - f(x^{(0)})}{f'(x^{(0)})}. \tag{9}$$

Again, compare y with $f(x^{(1)})$ and decide whether or not $x^{(1)}$ is a satisfactory solution to equation (6). If it is not, we continue to refine our estimate using the equation

$$x^{(n+1)} = x^{(n)} + \frac{y - f(x^{(n)})}{f'(x^{(n)})}. \tag{10}$$

We stop when

$$\left| f(x^{(n)}) - y \right| < \varepsilon. \tag{11}$$

This method can be used, for instance, for solving polynomial equations. It can also be valuable when we cannot find an analytic form for $f^{-1}(y)$.

In general, for arbitrary functions $f(x)$, three questions should come to mind:

1. Does this procedure converge? Will equation (11) be satisfied after a finite number of steps, or might the iteration go on forever? Note that, if f is linear, then this process will converge after one or two steps.
2. Is the solution unique? Could there be many x's in the interval $[a,b]$ that satisfy equation (11)? Do we know, *a priori*, what interval will contain the desired solution and no other solutions?

3. Is the solution stable? Will a small change in the measured value of y' = $y + n$, where n is the noise—or uncertainty—in the measurement of y, cause a large change in the value of x that this scheme converges to?

It is not possible to answer any of these questions in a general manner. Nor is it always possible, as we did in Chapter 7, to calculate the error variance of the solution in a closed form. We could, of course, try to evaluate the error by generating a random set of x's; calculating the corresponding y's; adding random noise; and using the iterative method to find each estimated value $\hat{x}$. From this experiment, we could then estimate the error

$$\chi^2 = \frac{1}{N} \sum_{k=1}^{N} (x_k - \hat{x}_k)^2. \tag{12}$$

Doing this might produce a useful numerical result, but it would not produce much additional insight.

11.1.2 Division by multiplication

As another example, consider the problem

$$y = ax, \tag{13}$$

for x and y scalars, and a is close to zero. We measure $y' = y + n$, where n represents the noise. If we just estimate $\hat{x} = y' / a$, the noise variance σ_N^2 / a^2 might be uncomfortably large. Since we don't want to divide by zero, we might proceed as follows.

Remember that

$$\frac{1}{1+z} = 1 - z + z^2 - \cdots + (-1)^k z^k + \cdots \tag{14}$$

for $|z| < 1$. Suppose we take the first k terms in this series. Since

$$\left[1 - z + z^2 - \cdots + (-1)^k z^k\right](1 + z) = 1 + (-1)^k z^{k+1}, \tag{15}$$

$$1 - z + z^2 - \cdots + (-1)^k z^k = \frac{1 + (-1)^k z^{k+1}}{1 + z}. \tag{16}$$

Let $t = 1 + z$; then

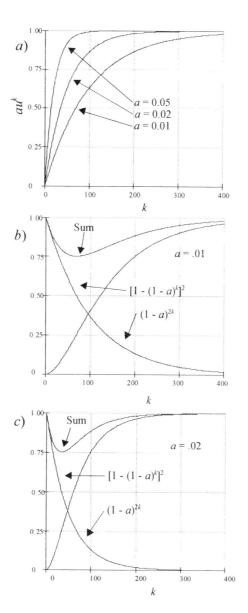

Figure 11-1. *a*) $au^{(k)}$ for $a = 0.01, 0.02$, and 0.05—see equation (19). Error for $a = 0.01$. *c*) Error for $a = 0.02$.

$$1 + (1 - t) + (1 - t)^2 + \cdots + (1 - t)^k = \frac{1 - (1 - t)^{k+1}}{t}. \tag{17}$$

Now, instead of dividing y' by a, for a small, we can perform the following iteration: Let $u^{(0)} = 0$ and

$$u^{(k+1)} = y' + (1-a)u^{(k)}. \tag{18}$$

Since $u^{(1)} = y'$; $u^{(2)} = y' + (1-a)y'$; $u^{(3)} = y' + (1-a)y' + (1-a)^2y'$; ...; we can see that

$$u^{(k)} = y'\frac{1-(1-a)^k}{a}. \tag{19}$$

So in one sense,[1] $u^{(k)} \to y'/a$ as $k \to \infty$. Figure 11-1a shows the first 400 values for $au^{(k)}/y'$ for $a = 0.01, 0.02$, and 0.05. Note the rate of convergence: as a becomes larger, the same fraction $au^{(k)}$ is reached with fewer iterations (*i.e.*, smaller k). Letting this processes continue long enough is equivalent to dividing by a. But we may find it advantageous to stop before k reaches ∞ so that we can limit the noise amplification. One way to regularize an iterative solution is to pick k in advance and only iterate k times, rather than continuing until some convergence criterion is met.

How can we use this to solve the equation $y = ax$ for x? Suppose that we measure $y' = ax_0 + n$, where x_0 is the true value of x, and n a random noise variable. Then

$$u^{(k)} - x_0 = (ax_0 + n)\frac{1-(1-a)^k}{a} - x_0$$

$$u^{(k)} - x_0 = -(1-a)^k x_0 + \frac{1-(1-a)^k}{a}n. \tag{20}$$

We see that, if $<x_0> = <y> = 0$ (our usual hypothesis), then $<u^{(k)} - x_0> = 0$, and the variance of $u^{(k)} - x_0$ is

$$\chi^2_{(k)} = \left\langle \left(u^{(k)} - x_0\right)^2 \right\rangle = (1-a)^{2k}\sigma^2_x + \frac{\left[1-(1-a)^k\right]^2}{a^2}\sigma^2_N, \tag{21}$$

where σ_x^2 and σ_N^2 are the variance of x_0 and of the noise, respectively. We see that the error, $\chi_{(k)}^2$, is the sum of two terms: one represents noise amplification and the other, how close we are getting to $\hat{x} = y/a$. If $k = 0$, the first term contributes σ_x^2 to the variance; the second, zero. This corresponds to an algorithm that always sets $\hat{x} = 0$ and ignores the data y'. At the other extreme, if $k \to \infty$, the first term $\to 0$ while the second term $\to \sigma_N^2/a^2$.

---

[1] Here, y' denotes $y+n$, not a derivative.

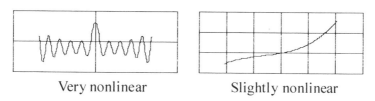

Very nonlinear   Slightly nonlinear

Figure 11-2. Different degrees of nonlinearity.

What is interesting about this is that there is a value of k where $\chi_{(k)}^2$ is minimum. Rather than stopping at $k = 0$ (before we've fairly begun) or iterating to $k = \infty$, we can minimize the error by iterating only k_0 times, where k_0 minimizes equation (21). Suppose that the variance of x is σ_x^2; then the variance of $y = ax$ is $a^2\sigma_x^2$. As a simple example, we can consider the case where the two variances in equation (21) are equal: $\sigma_N^2 = a^2\sigma_x^2$. For simplicity, we can take $\sigma_N^2 = 1$. Figure 11-1b shows the values of these two terms, and their sums, for $a = 0.01$; Figure 11-1c shows the same information for $a = 0.02$. We can see that the error is minimum if we use $k \approx 80$ terms or so when $a = 0.01$; the minimum is $k \approx 40$ for $a = 0.02$.

We will pick up this train of thought again in §11.5.2.

Exercise 11-1: We saw in Chapter 1 that the minimum-variance solution to y' $= ax$ is $\hat{x} = \beta y'$, where $\beta = a\sigma_x^2 / (\sigma_N^2 + a^2\sigma_x^2)$. If we choose the value of k in the iteration scheme above, what is the value of k when $[1 - (1 - a)^k]^2 / a^2 +$ $a^2(1-a)^{2k}$ is minimum? How is $\hat{x}$ related to x_0, the true value of x? What is the error variance $\chi_{(k)}^2$? How does it compare with the minimum-variance estimate for $\hat{x}$ from Chapter 1?

Exercise 11-2: Prove that there is a value of k, $0 < k < \infty$ where equation (21) is minimum. How does the value of k depend on a?

While iteration does not seem to be very practical for this one-dimensional problem—multiplication by β is much more straightforward—it might have some appeal where matrices are concerned.

11.1.3 Convergence

When we consider using an iterative solution to a problem, we must always consider how fast it will converge—and if it will converge at all. This depends, in part, on how nonlinear the problem is. On the one hand, if $f(x)$ is truly linear, iteration will converge very quickly. But if it is very nonlinear, iteration might not converge at all. To my knowledge, the equations that arise in remote sensing are, for the most part, only mildly nonlinear; I illustrate this

Table 11-1. $r = 2.500$			Table 11-2. $r = 2.750$		
	x^0	$(1+10^{-8})x^0$		x^0	$(1+10^{-8})x^0$
1	1.12500000000	1.12500000500	1	1.18750000000	1.18750000500
2	0.77343750000	0.77343748938	2	0.57519531250	0.57519529859
3	1.21151733398	1.21151733789	3	1.24714589119	1.24714588303
4	0.57087504258	0.57087503261	4	0.39952168871	0.39952171406
5	1.18331686343	1.18331685699	5	1.05925798834	1.05925802771
6	0.64101202381	0.64101203938	6	0.88664187015	0.88664178842
7	1.21630104666	1.21630105125	7	1.16303904683	1.16303913890
8	0.55858307304	0.55858306119	8	0.64158190836	0.64158166469
9	1.17500313192	1.17500312354	9	1.27395695723	1.27395690330
10	0.66093006166	0.66093008157	10	0.31418118521	0.31418136083
...			...		
51	1.22499616904	1.22499616904	51	0.34822758320	0.39342131897
52	0.53594755622	0.53594755622	52	0.97238170038	1.04968402703
53	1.15771698923	1.15771698923	53	1.04623440562	0.90626457072
54	0.70123789443	0.70123789443	54	0.91321133445	1.13987459181
55	1.22499616904	1.22499616904	55	1.13116641543	0.70141598538
56	0.53594755622	0.53594755622	56	0.72314604453	1.27735288767
57	1.15771698923	1.15771698923	57	1.27371211226	0.30308972975
58	0.70123789443	0.70123789443	58	0.31497842246	0.88396217979
59	1.22499616904	1.22499616904	59	0.90833771602	1.16603805214
60	0.53594755622	0.53594755622	60	1.13730356762	0.63361966318
...			...		
91	1.22499616904	1.22499616904	91	1.25529053424	1.27403341487
92	0.53594755622	0.53594755622	92	0.37401510868	0.31393216470
93	1.15771698923	1.15771698923	93	1.01786657837	0.90622375653
94	0.70123789443	0.70123789443	94	0.96785564765	1.13992497052
95	1.22499616904	1.22499616904	95	1.05341115330	0.70128905881
96	0.53594755622	0.53594755622	96	0.89868541565	1.27736652452
97	1.15771698923	1.15771698923	97	1.14907274886	0.30304506255
98	0.70123789443	0.70123789443	98	0.67801030726	0.88386913224
99	1.22499616904	1.22499616904	99	1.27836921616	1.16614147785

concept in Figure 11-2. Suppose that we know *a priori* that the solution to equation (6) lies in some interval [*a,b*]. Without trying to quantify the term "mildly nonlinear" exactly, we could say that it means, in some sense, that the first derivative of *f* does not change too much in the interval [*a,b*], but that we cannot ignore that change, either.

11.2 Two perverse examples

Iteration may engender some computational problems that the reader should be aware of. Sometimes numerical calculations that seem simple turn out to have unexpected complications. Computations that seem straightforward may get completely out of hand. Here are two examples.

11.2.1 Logistic equation

Consider the innocuous-looking equation

Table 11-3. $r = 3.000$

	x^0	$(1+10^{-8})x^0$
1	1.25000000000	1.25000000500
2	0.312500000000	0.31249998250
3	0.957031250000	0.95703121281
4	1.08039855957	1.08039862436
5	0.819811095716	0.81981093489
6	1.26297368489	1.26297383267
7	0.266587153399	0.26658662467
8	0.853142482524	0.85314121331
9	1.22901364363	1.22901506369
10	0.384630965819	0.38462617443
...		
24	0.012058394943	0.03393567585
25	0.047797365106	0.13228781313
26	0.184335696093	0.47665105600
27	0.635403837810	1.22501553644
28	1.33040123993	0.39807295220
29	0.011702582102	1.11690558297
30	0.044.63994771	0.72518808807
31	0.179139174069	1.32305906305
32	0.620284165218	0.04078039926
33	1.32687932401	0.15813247415

$$x^{(n+1)} = x^{(n)} + rx^{(n)}\left(1 - x^{(n)}\right). \qquad (22)$$

Here, r is a real number, $0 \le r \le 4$. The notation here $x^{(n)}$ still means the nth iterated value of x. This equation is called the *logistic equation*. Given an initial value, it should be possible to iterate equation (22) for any r in the given interval and calculate $x^{(n)}$ for any n. Is it?

The logistic equation has been used to model the way the population of a species changes with time. Suppose that x represents the number of fish in a pond. If there are only a few fish, the food supply (proportional to $1 - x$) will be relatively abundant. So the population can increase rapidly. But when there are too many fish, the food supply is smaller, and the population declines.

Peitgen *et al.* (1992) point out that very small errors in the iteration of equation (22) can lead to huge changes in the subsequent values of $x^{(n)}$ that are calculated. One would think that a small change in the starting value, $x^{(0)}$, would produce only a very small change in the values of $x^{(1)}$, $x^{(2)}$, ..., but in fact, the differences can become very large. How large they are will depend on the value of r. As an example, I have performed the iteration in equation (22) using, as starting values, $x^{(0)}$ and $(1 + 10^{-8})x^{(0)}$. Tables 1 to 3 show results for an initial value $x^{(0)} = \frac{1}{2}$.

Table 11-1 shows some of the values for $r = 2.5$. The results are the same to about six decimal places, even to the 99th iteration—and beyond—as shown in Table 11-2. The situation changes, however, when $r = 2.75$. The first ten iterations are very close, but by the 50th iteration, the values differ by 10 – 30%. Table 11-3 shows that, for $r = 3$, the numbers become completely unrelated by 25 iterations.

Tables 11-1 to 11-3 show that, depending on the value of r, a change in the initial value of one part in one hundred million, may make no appreciable difference (as we would normally expect), or it may make an enormous difference. There are very, very few physical measurements that can approach a precision anywhere near as small as 10^{-8}: if the initial guess depends on some measurement, the result of iterating equation (22) may be meaningless. This is a simple example of *chaos*, the situation that arises when infinitesimal changes in a dynamical system cause large changes in the way it evolves with time.[2]

11.2.2 Recursion

As another example of an iteration scheme that looks simple but may sometimes go awry, let us look briefly at some difference equations that are often used to evaluate certain kinds of functions by recursion.[3] There are many families of functions that are used frequently in mathematical physics because they are solutions to commonly encountered differential equations. One such family are the Bessel functions of the first and second kinds, $J_v(x)$ and $Y_v(x)$, which are defined for all real values of v. They are the two independent solutions to the second-order differential equation[4]

$$x^2 \frac{d^2 w(x)}{dx^2} + x \frac{dw(x)}{dx} + (x^2 - v^2) w(x) = 0. \tag{23}$$

They have the series representations for $v \neq n$, where n is an integer:

$$J_v(x) = (\tfrac{1}{2}x)^v \sum_{k=0}^{\infty} \frac{(-1)^v (\tfrac{1}{2}x)^{2v}}{k! \, \Gamma(v+k+1)} \tag{24}$$

and

$$Y_v(x) = \frac{J_v(x) \cos v\pi - J_{-v}(x)}{\sin v\pi}, \tag{25}$$

for any $x > 0$. When $v = n$ is an integer,

---

[2] See Delves and Mohamed (1985), pp 299-301 for a discussion of similar effects that arise in the solution of certain integral equations that are ill-conditioned.

[3] See also Wimp (1984).

[4] See Abramowitz and Stegun (1968).

$$J_n(x) = \sum_{k=0}^{\infty} \frac{(-1)^n (\tfrac{1}{2}x)^{n+2k}}{k!(n+k)!}. \tag{26}$$

For integral values of $v = n$, Y_n is defined by the value of Y_v in the limit as $v \to n$ in equation (25).

Although they are formally convergent, these representations are not very useful for actually calculating values of J_n or Y_n; equation (24) converges only slowly for $x \gg 1$, and equation (25) is defined only as a limit when v is an integer (since $\sin n\pi = 0$ if n is an integer).

Because of these difficulties, we often use recursion relations to find the values of Bessel functions. The values of $J_n(x)$ are related for different values of n by

$$J_{n+1}(x) = \frac{2n}{x} J_n(x) - J_{n-1}(x); \tag{27}$$

the $Y_n(x)$ obey the same relation. (I discussed the first-order recursion equation for the exponential integrals in §4.3) Since equation (27) is a second-order difference equation, there are two linearly independent solutions.

Each solution can have one of three properties:

1. A solution[5] can increase exponentially with increasing values of n;

2. A solution can increase or decrease with increasing values of n, but less than exponentially; or

3. A solution can increase exponentially with decreasing values of n.

At most one solution can increase exponentially with increasing n, and at most one solution can increase exponentially with decreasing n.

Let U_n and D_n be the two solutions to equation (27), with the former increasing exponentially with increasing n, and the latter, with decreasing n. Suppose that D_n is the one we want. So we start by evaluating D_0 and D_1, and then calculating D_2, D_3, ..., from equation (27). However, any calculation contains some roundoff error, so what we really calculate is not exactly D_2, but $D_2 + \varepsilon_2 U_2$, and then $D_3 + \varepsilon_3 U_3$, ... Because the U_n increase exponentially with increasing n, the small errors ε_2, ε_3, ..., grow and quickly swamp D_n, the desired solution. Calculating D_n by forward recursion—in order of increasing n—is unstable under these circumstances.

The way around this problem is to calculate them by backward recursion. If the U_n increase exponentially with increasing n, then they decrease exponentially with decreasing n. If the recursion is unstable for calculating the D_n in the direction of increasing n, it will be stable in the other direction. There-

---

[5] A *solution* is a set of J_n, $n = 0$, 1, ..., or other functions, that obey the recursion relation.

fore, pick a large integer N, and set $D'_{N+1} = 0$ and $D'_N = 1$. Then use equation (27) to evaluate $D'_{N-1}, ..., D'_0$. If a is any scalar, the set of aD_n is a solution to equation (27) whenever the D_n are. We have evaluated $D'_n = aD_n$; we find the value of a by calculating the value of D_0 and

$$a = \frac{D'_n}{D_0}. \tag{28}$$

If we have chosen N large enough, the error made by assuming that $D'_{N+1} = 0$ and $D'_N = 1$ will be negligible.

This topic is discussed more fully by Abramowitz and Stegun (1968), and by Press *et al*(1992, pp 172 *ff*). The latter authors give several theorems about recursion. They also give a simple method for telling whether recursion in a given direction is stable. Given a recursion relation like equation (27), test it in the forward direction by taking $D_0 = 0$, $D_1 = 1$, and then calculating the first 30 or so of the D_n. If they all remain about order unity, then the recursion is stable; if they grow wildly, it is not.

11.3 Matrix iteration...

We can turn now to iterative solutions of linear equations. For the present, let us assume that the problem is linear; I will discuss nonlinear problems later in this chapter.

A meeting at NASA's Langley Research Center in December, 1976, brought together many of the leading investigators in remote sensing. The printed proceedings of this meeting are not—as far as I know—readily available. Among other topics, there was a discussion of iterative and matrix inverse methods of retrieving geophysical parameters from satellite data. The late Henry Fleming (1976) of NOAA showed that these two methods will give statistically similar results. His iteration method regularized the solution by only iterating for a preset number of terms (*cf.* §11.4.2). I shall describe the matrix iteration method in this section, and prove a slightly stronger result than Fleming did, that matrix-inverse and iterative methods can give exactly the same results. Therefore, there can be no advantage to iteration, which takes far more computation than multiplying a vector by a matrix.

First, let me recapitulate the unregularized matrix-inverse solution (we will regularize it later) to equation (1) as a point of reference. I will then take up iterated solutions. As with matrix methods, regularization is needed to ensure that the iteration will converge to the correct solution. If there were a unique vector $\hat{\mathbf{x}}$ that satisfied equation (1), in the sense that $|\mathbf{R}\hat{\mathbf{x}} - \mathbf{y}'| \leq \sigma_N^2$, there would be no problem. However, if the solution is not unique, there is an infinite number of solutions. Which one will an iteration scheme converge to? Without regularization, we will get one of these solutions chosen at random.

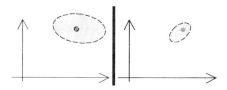

Figure 11-3. The range of solutions for the unregularized solution (*left*), and the regularized one (*right*).

Regularization reduces the range of the solutions that we will get. I show this schematically in Figure 11-3. The dark circle in each panel is the correct solution; the iterative solution will converge to some point in the lightly shaded region. Regularization reduces the size of this region, although it also moves its center of gravity toward the origin. So let us look at the unregularized solution first, and then two regularized solutions.

First, let me describe the following unregularized matrix-inverse method. I bring it up here because it will illustrate the iterative method that will follow, not because I think it will work in real applications. To solve equation (1) for $\mathbf{x}$, multiply on the left by $\mathbf{A}^t$:

$$\mathbf{A}^t \mathbf{y} = \mathbf{A}^t \mathbf{A} \mathbf{x}. \tag{29}$$

If $\mathbf{A}$ is an $n \times m$ matrix, then $\mathbf{A}^t \mathbf{A}$ is an $m \times m$ matrix. If it is nonsingular, we can solve for $\mathbf{x}$ by

$$\hat{\mathbf{x}} = \left(\mathbf{A}^t \mathbf{A}\right)^{-1} \mathbf{A}^t \mathbf{y}. \tag{30}$$

But now consider the noise in the measurement $\mathbf{y}$. Suppose we measured

$$\mathbf{y}' = \mathbf{A}\mathbf{x} + \mathbf{n}, \tag{31}$$

where $\mathbf{n}$ is a random noise vector. Then, if we made many measurements under identical conditions, so that only $\mathbf{n}$ varied each time, the error variance of $\hat{\mathbf{x}}$ would be

$$\chi^2 = \left(\mathbf{A}^t \mathbf{A}\right)^{-1} \mathbf{A}^t \langle \mathbf{n}\mathbf{n}^t \rangle \mathbf{A} \left(\mathbf{A}^t \mathbf{A}\right)^{-1} = \left(\mathbf{A}^t \mathbf{A}\right)^{-1} \mathbf{A}^t \mathbf{N} \mathbf{A} \left(\mathbf{A}^t \mathbf{A}\right)^{-1} \tag{32}$$

where $\langle \mathbf{n}\mathbf{n}^t \rangle = \mathbf{N}$ is the noise covariance matrix. If, as is usually the case, $\mathbf{N} = \sigma_N^2 \mathbf{I}$, then

$$\chi^2 = \sigma_N^2 \left(\mathbf{A}^t \mathbf{A}\right)^{-1}. \tag{33}$$

As we saw in Chapter 7, amplifying the noise by $(\mathbf{A}^t \mathbf{A})^{-1}$ is usually not acceptable; that is why more appropriate methods were developed. I provided this digression because $(\mathbf{A}^t \mathbf{A})^{-1} \mathbf{A}^t$ will play a major role in the iteration methods we will turn to next.

11.3.1 Outline of the iteration process

In order to derive an inverse to equation (1), we need to use two identities.

Lemma 1. For any square matrix $\mathbf{B}$,[6]

$$\left[\mathbf{I} + (\mathbf{I} - \mathbf{B}) + \cdots + (\mathbf{I} - \mathbf{B})^{k-1}\right]\mathbf{B} = \mathbf{I} - (\mathbf{I} - \mathbf{B})^{k}, \tag{34}$$

where $\mathbf{I}$ is the identity matrix. We can prove equation (34) by induction (see §14.13). First, if $k = 1$, the result is obvious. Therefore, assume that it is true for all $n \le k$. For $n = k + 1$,

$$\mathbf{I} - (\mathbf{I} - \mathbf{B})^{k+1} = \mathbf{I} - (\mathbf{I} - \mathbf{B})^{k}(\mathbf{I} - \mathbf{B})$$

$$= \mathbf{I} - (\mathbf{I} - \mathbf{B})^{k} + (\mathbf{I} - \mathbf{B})^{k}\mathbf{B}. \tag{35}$$

Using the induction hypothesis, this becomes

$$\mathbf{I} - (\mathbf{I} - \mathbf{B})^{k+1} = \left[\mathbf{I} + (\mathbf{I} - \mathbf{B}) + \cdots + (\mathbf{I} - \mathbf{B})^{k-1}\right]\mathbf{B} + (\mathbf{I} - \mathbf{B})^{k}\mathbf{B}$$

$$= \left[\mathbf{I} + (\mathbf{I} - \mathbf{B}) + \cdots + (\mathbf{I} - \mathbf{B})^{k}\right]\mathbf{B}. \tag{36}$$

This shows that, if equation (34) is true for $n = k$, it must also be true for $n = k + 1$. Therefore it is true for *every* k. Furthermore, if $\mathbf{B}$ is nonsingular,

$$\mathbf{I} + (\mathbf{I} - \mathbf{B}) + \cdots + (\mathbf{I} - \mathbf{B})^{k-1} = \left[\mathbf{I} - (\mathbf{I} - \mathbf{B})^{k}\right]\mathbf{B}^{-1}. \tag{37}$$

◆

Corollary 1. If $\|\mathbf{M}\|$ is small enough, we can use the property that $\|\mathbf{M}^{k}\| = \|\mathbf{M}\|^{k}$ and show that, for any vector $\mathbf{v}$, $\|\mathbf{M}\|^{k}\mathbf{v} \to 0$ as $k \to \infty$. Therefore, if $\|(\mathbf{I} - \mathbf{B})^{k}\| \to 0$ as $k \to \infty$, then[7]

$$\mathbf{B}^{-1} = \sum_{k=0}^{\infty}(\mathbf{I} - \mathbf{B})^{k}. \tag{38}$$

◆

Lemma 2. If $\mathbf{A}$ is an $m \times n$ matrix and $\mathbf{I}_{m}$ is the $m \times m$ identity matrix, then

---

[6] $\mathbf{B}$ is not necessarily symmetric.

[7] I discuss the matrix norm in §14.4.

$$\left(\mathbf{I}_n - \gamma \mathbf{A}^t \mathbf{A}\right)^k \mathbf{A}^t \; = \; \mathbf{A}^t \left(\mathbf{I}_m - \gamma \mathbf{A} \mathbf{A}^t\right)^k. \tag{39}$$

Note that two different identity matrices are involved in this equation: $\mathbf{A}^t \mathbf{A}$ is an $n \times n$ matrix, while $\mathbf{A}\mathbf{A}^t$ is $m \times m$. Be careful not to confuse them. We can prove this identity by induction also. First, for $k = 1$, the result is obvious. Assume that it is true for every $n \leq k$; then, for $n = k + 1$,

$$\left(\mathbf{I}_n - \gamma \mathbf{A}^t \mathbf{A}\right)^{k+1} \mathbf{A}^t \; = \; \left(\mathbf{I}_n - \gamma \mathbf{A}^t \mathbf{A}\right)^k \left(\mathbf{I}_n - \gamma \mathbf{A}^t \mathbf{A}\right) \mathbf{A}^t$$

$$= \; \left(\mathbf{I}_n - \gamma \mathbf{A}^t \mathbf{A}\right)^k \mathbf{A}^t \; - \gamma \left(\mathbf{I}_n - \gamma \mathbf{A}^t \mathbf{A}\right)^k \mathbf{A}^t \mathbf{A} \mathbf{A}^t$$

$$= \; \mathbf{A}^t \left(\mathbf{I}_m - \gamma \mathbf{A} \mathbf{A}^t\right)^k \; - \; \gamma \mathbf{A}^t \left(\mathbf{I}_m - \gamma \mathbf{A} \mathbf{A}^t\right)^k \mathbf{A} \mathbf{A}^t$$

$$= \; \left(\mathbf{I}_m - \gamma \mathbf{A}^t \mathbf{A}\right)^k \mathbf{A}^t \; - \; \gamma \left(\mathbf{I}_m - \gamma \mathbf{A}^t \mathbf{A}\right)^k \mathbf{A}^t \mathbf{A} \mathbf{A}^t \; = \; \left(\mathbf{I}_m - \gamma \mathbf{A}^t \mathbf{A}\right)^{k+1} \mathbf{A}^t \tag{40}$$

which proves the desired result. Here, I have used equation (39) to get from the third line to the fourth line above. ◆

Corollary 2. By taking the transpose of both sides[8] of equation (39), we can show that

$$\mathbf{A}\left(\mathbf{I}_n - \gamma \mathbf{A}^t \mathbf{A}\right)^k \; = \; \left(\mathbf{I}_m - \gamma \mathbf{A} \mathbf{A}^t\right)^k \mathbf{A}. \tag{41}$$

Corollary 3.

$$\mathbf{A}^t \left(\mathbf{A} \mathbf{A}^t + \alpha \mathbf{I}\right)^{-1} \; = \; \left(\mathbf{A}^t \mathbf{A} + \alpha \mathbf{I}\right)^{-1} \mathbf{A}^t. \tag{42}$$

Use equation (39) with $k = 1$ and substitute α for $-1 / \gamma$. This becomes

$$\left(\alpha \mathbf{I}_n + \mathbf{A}^t \mathbf{A}\right)\mathbf{A}^t \; = \; \mathbf{A}^t \left(\alpha \mathbf{I}_m + \mathbf{A} \mathbf{A}^t\right). \tag{43}$$

Multiply both sides of this equation first, from the left by $(\mathbf{A}\mathbf{A}^t + \alpha \mathbf{I})^{-1}$, and then from the right by $(\mathbf{A}^t\mathbf{A} + \alpha \mathbf{I})^{-1}$. So

$$\left(\alpha \mathbf{I} + \mathbf{A}^t \mathbf{A}\right)^{-1}\left(\alpha \mathbf{I} + \mathbf{A}^t \mathbf{A}\right)\mathbf{A}^t\left(\alpha \mathbf{I} + \mathbf{A} \mathbf{A}^t\right)^{-1}$$

---

[8] Remember that $(\mathbf{A}\mathbf{B})^t = \mathbf{B}^t\mathbf{A}^t$.

$$= \left(\alpha \mathbf{I} + \mathbf{A}^{t}\mathbf{A}\right)^{-1}\mathbf{A}^{t}\left(\alpha \mathbf{I} + \mathbf{A}\mathbf{A}^{t}\right)\left(\alpha \mathbf{I} + \mathbf{A}\mathbf{A}^{t}\right)^{-1}. \tag{44}$$

The desired result follows immediately.◆

Corollary 4. Setting $\alpha = 0$, we see that

$$\mathbf{A}^{t}\left(\mathbf{A}\mathbf{A}^{t}\right)^{-1} = \left(\mathbf{A}^{t}\mathbf{A}\right)^{-1}\mathbf{A}^{t}. \tag{45}$$

11.3.2 An expansion theorem

To begin, take a square matrix $\mathbf{B}$ and vector $\mathbf{v}$; define the vectors $\mathbf{u}^{(k)}$ by

$$\mathbf{u}^{(k)} = \mathbf{u}^{(k-1)} + \left(\mathbf{v} - \mathbf{B}\mathbf{u}^{(k-1)}\right) = \mathbf{v} + \left(\mathbf{I} - \mathbf{B}\right)\mathbf{u}^{(k-1)}. \tag{46}$$

I shall show that this converges to $\mathbf{u}^{(k)} = \mathbf{v}$ as $k \to \infty$. Write the first few vectors:

$$\mathbf{u}^{(1)} = \mathbf{v} + \left(\mathbf{I} - \mathbf{B}\right)\mathbf{u}^{(0)};$$

$$\mathbf{u}^{(2)} = \mathbf{v} + \left(\mathbf{I} - \mathbf{B}\right)\mathbf{u}^{(1)} = \mathbf{v} + \left(\mathbf{I} - \mathbf{B}\right)\mathbf{v} + \left(\mathbf{I} - \mathbf{B}\right)^{2}\mathbf{u}^{(0)}; \tag{47}$$

and so forth. At the kth step,

$$\mathbf{u}^{(k)} = \left[\mathbf{I} + \left(\mathbf{I} - \mathbf{B}\right) + \left(\mathbf{I} - \mathbf{B}\right)^{2} + \cdots + \left(\mathbf{I} - \mathbf{B}\right)^{k-1}\right]\mathbf{v} + \left(\mathbf{I} - \mathbf{B}\right)^{k}\mathbf{u}^{(0)}. \tag{48}$$

Multiply each side by B and use equation (34) to get

$$\mathbf{B}\mathbf{u}^{(k)} = \left[\mathbf{I} + \left(\mathbf{I} - \mathbf{B}\right) + \left(\mathbf{I} - \mathbf{B}\right)^{2} + \cdots + \left(\mathbf{I} - \mathbf{B}\right)^{k-1}\right]\mathbf{B}\mathbf{v} + \left(\mathbf{I} - \mathbf{B}\right)^{k}\mathbf{B}\mathbf{u}^{(0)}$$

$$= \left[\mathbf{I} - \left(\mathbf{I} - \mathbf{B}\right)^{k}\right]\mathbf{v} + \left(\mathbf{I} - \mathbf{B}\right)^{k}\mathbf{B}\mathbf{u}^{(0)} = \mathbf{v} + \left(\mathbf{I} - \mathbf{B}\right)^{k}\left(\mathbf{B}\mathbf{u}^{(0)} - \mathbf{v}\right). \tag{49}$$

Therefore, if

$$\lim_{k \to \infty} \left\|\left(\mathbf{I} - \mathbf{B}\right)^{k}\right\|^{2} = 0, \tag{50}$$

then $\mathbf{B}\mathbf{u}^{(k)} \to \mathbf{v}$ as $k \to \infty$, so

$$\mathop{Lim}_{k \to \infty} \mathbf{u}^{(k)} = \mathbf{B}^{-1}\mathbf{v}. \tag{51}$$

This establishes a very important point. *Any iterative method that can be written in the form of equation (46), if it converges, is exactly equivalent to an appropriate matrix-inverse solution.*

To ensure convergence, we can modify equation (46) to include a *convergence factor* γ, where γ is a positive scalar:

$$\mathbf{u}^{(k)} = \mathbf{u}^{(k-1)} + \gamma\left(\mathbf{v} - \mathbf{B}\mathbf{u}^{(k-1)}\right) = \gamma\mathbf{v} + \left(\mathbf{I} - \gamma\mathbf{B}\right)\mathbf{u}^{(k-1)}. \tag{52}$$

It is not hard to show that

$$\mathbf{u}^{(k)} = \left[\mathbf{I} + \left(\mathbf{I} - \gamma\mathbf{B}\right) + \left(\mathbf{I} - \gamma\mathbf{B}\right)^2 + \cdots + \left(\mathbf{I} - \gamma\mathbf{B}\right)^{k-1}\right]\gamma\mathbf{v} + \left(\mathbf{I} - \gamma\mathbf{B}\right)^k \mathbf{u}^{(0)}. \tag{53}$$

Therefore

$$\mathbf{B}\mathbf{u}^{(k)} = \left[\mathbf{I} - \left(\mathbf{I} - \gamma\mathbf{B}\right)^k\right]\mathbf{v} + \left(\mathbf{I} - \gamma\mathbf{B}\right)^k \mathbf{B}\mathbf{u}^{(0)} = \mathbf{v} + \left(\mathbf{I} - \gamma\mathbf{B}\right)^k\left(\mathbf{B}\mathbf{u}^{(0)} - \mathbf{v}\right). \tag{54}$$

Now $\mathbf{B}\mathbf{u}^{(k)} \to \mathbf{v}$ as $k \to \infty$ if

$$\|\mathbf{I} - \gamma\mathbf{B}\| < 1. \tag{55}$$

We can adjust γ to ensure that the iteration converges. Regardless of the value of γ, it converges to $\mathbf{B}^{-1}\mathbf{v}$. Therefore, the duality between iteration and matrix-inverse solutions is complete: there is an equivalent iterative solution corresponding to any matrix-inverse solution.

11.3.3 Back to iteration

Here, I want to examine an iterative solution to equation (1) and compare it to the matrix-inverse solution. To start the iteration process, take some vector $\mathbf{x}^{(0)}$ as an initial approximation.[9] Then compare $\mathbf{A}\mathbf{x}^{(0)}$ with $\mathbf{y}'$ and use the difference between them to generate better approximation, $\mathbf{x}^{(1)}$. This continues until, at the kth step, $\mathbf{A}\mathbf{x}^{(k)}$ agrees with $\mathbf{y}'$ to within whatever precision will satisfy us. Then we stop.

One possible iteration procedure is the following. Let $\mathbf{x}^{(k)}$ be the kth approximate solution to equation (1). An improved solution is found from

$$\mathbf{x}^{(k+1)} = \mathbf{x}^{(k)} + \gamma\,\mathbf{A}^{t}\left(\mathbf{y} - \mathbf{A}\mathbf{x}^{(k)}\right), \tag{56}$$

---

[9] A superscript here on a vector shows its position in a series; a superscript on a matrix will denote a power of that matrix.

where γ is a scalar convergence factor that is chosen to ensure that the solution will converge. Obviously, if $\gamma = 0$, the convergence will be infinitely slow. However, if γ is too large, $\mathbf{x}^{(k)}$ will either oscillate around the desired solution or diverge. We shall see that, if this process converges, it converges to $\hat{\mathbf{x}} = (\mathbf{A}^t\mathbf{A})^{-1}\mathbf{A}\mathbf{y}$. This will not be very useful in practice, but it illustrates the method—we will derive a regularized method presently.

Rewrite equation (56) as

$$\mathbf{x}^{(k)} = \gamma\mathbf{A}^t\mathbf{y} + (\mathbf{I} - \gamma\mathbf{A}^t\mathbf{A})\mathbf{x}^{(k-1)}. \tag{57}$$

We can write this in terms of $\mathbf{x}^{(k-2)}$:

$$\mathbf{x}^{(k)} = \gamma\mathbf{A}^t\mathbf{y} + \left(\mathbf{I} - \gamma\mathbf{A}^t\mathbf{A}\right)\left[\gamma\mathbf{A}^t\mathbf{y} + \left(\mathbf{I} - \gamma\mathbf{A}^t\mathbf{A}\right)\mathbf{x}^{(k-2)}\right]$$

$$= \gamma\mathbf{A}^t\mathbf{y} + \left(\mathbf{I} - \gamma\mathbf{A}^t\mathbf{A}\right)\gamma\mathbf{A}^t\mathbf{y} + \left(\mathbf{I} - \gamma\mathbf{A}^t\mathbf{A}\right)^2\mathbf{x}^{(k-2)}$$

$$= \gamma\left[\mathbf{I} + \left(\mathbf{I} - \gamma\mathbf{A}^t\mathbf{A}\right)\right]\mathbf{A}^t\mathbf{y} + \left(\mathbf{I} - \gamma\mathbf{A}^t\mathbf{A}\right)^2\mathbf{x}^{(k-2)}. \tag{58}$$

We can continue this process, using equation (57) to write $\mathbf{x}^{(k-2)}$ in terms of $\mathbf{x}^{(k-3)}$, and so forth; this eventually yields

$$\mathbf{x}^{(k)} = \gamma\left[\mathbf{I} + \left(\mathbf{I} - \gamma\mathbf{A}^t\mathbf{A}\right) + \cdots + \left(\mathbf{I} - \gamma\mathbf{A}^t\mathbf{A}\right)^{k-1}\right]\mathbf{A}^t\mathbf{y} + \left(\mathbf{I} - \gamma\mathbf{A}^t\mathbf{A}\right)^k\mathbf{x}^{(0)}. \tag{59}$$

Using Lemma 1, with $\mathbf{B} = \gamma\mathbf{A}^t\mathbf{A}$

$$\mathbf{x}^{(k)} = \gamma\left[\mathbf{I} - \left(\mathbf{I} - \gamma\mathbf{A}^t\mathbf{A}\right)^k\right]\left[\gamma\mathbf{A}^t\mathbf{A}\right]^{-1}\mathbf{A}^t\mathbf{y} + \left(\mathbf{I} - \gamma\mathbf{A}^t\mathbf{A}\right)^k\mathbf{x}^{(0)}. \tag{60}$$

Note that this only makes sense if $(\mathbf{A}\mathbf{A}^t)^{-1}$ exists. Then

$$\mathbf{x}^{(k)} = \left[\mathbf{A}\mathbf{A}^t\right]^{-1}\mathbf{A}^t\mathbf{y} + \left(\mathbf{I} - \gamma\mathbf{A}^t\mathbf{A}\right)^k\left(\mathbf{x}^{(0)} - \left[\mathbf{A}^t\mathbf{A}\right]^{-1}\mathbf{A}^t\mathbf{y}\right)$$

$$= \hat{\mathbf{x}} + \left(\mathbf{I} - \gamma\mathbf{A}^t\mathbf{A}\right)^k\left(\mathbf{x}^{(0)} - \hat{\mathbf{x}}\right) \tag{61}$$

where

$$\hat{\mathbf{x}} \equiv \left[\mathbf{A}^t\mathbf{A}\right]^{-1}\mathbf{A}^t\mathbf{y}. \tag{62}$$

This will converge to $\hat{\mathbf{x}}$ if the constant γ is chosen so that

$$\left|\left(\mathbf{I} - \gamma\mathbf{A}^{\mathsf{t}}\mathbf{A}\right)\left(\mathbf{x}^{(0)} - \hat{\mathbf{x}}\right)\right| \;<\; \left|\left(\mathbf{x}^{(0)} - \hat{\mathbf{x}}\right)\right|. \tag{63}$$

This can happen, in turn, if

$$\left|(\mathbf{I} - \gamma\mathbf{A}^{\mathsf{t}}\mathbf{A})\mathbf{x}\right| \;\le\; \left\|(\mathbf{I} - \gamma\mathbf{A}^{\mathsf{t}}\mathbf{A})\right\| \cdot |\mathbf{x}| \;\le\; |\mathbf{x}|. \tag{64}$$

since $|\mathbf{Mx}| \le |\mathbf{x}| \cdot \|\mathbf{M}\|$ for any $m \times m$ matrix $\mathbf{M}$ and vector $\mathbf{x}$ of dimension m. Therefore, the iteration will converge if

$$\left\|\mathbf{I} - \gamma\mathbf{A}^{\mathsf{t}}\mathbf{A}\right\|^2 \;<\; 1, \tag{65}$$

Since $\mathbf{A}^{\mathsf{t}}\mathbf{A}$ is non-negative definite—see §5.7 —this is equivalent to the condition that

$$0 \;<\; \gamma \left\|\mathbf{A}^{\mathsf{t}}\mathbf{A}\right\|^2 \;< 2. \tag{66}$$

As we can see, if γ is too small, the iterative algorithm will converge very slowly, while if γ is too large, the process will not converge at all. However, we should note that when it converges,

1. The value it converges to does not depend on γ; and
2. It does not depend on $\mathbf{x}^{(0)}$, either. (However, as we have seen, this unregularized solution may amplify the noise to an unacceptable level; any regularized solution will depend on the choice of the initial approximation $\mathbf{x}^{(0)}$.)

Therefore, this iteration scheme converges to a solution that is the same as the matrix-inverse solution in equation (30). Note that, as long as equation (66) is satisfied, it converges to the same value of $\hat{\mathbf{x}}$ regardless of the initial approximation $\mathbf{x}^{(0)}$. However, it will converge only for certain values of γ; for others, it will diverge.

This proves a point, perhaps, but it is not particularly relevant to anything practical since we would not use the unregularized solution of $\mathbf{y} = \mathbf{Ax}$ anyway. However, I shall show in the next section that there is also a duality between the regularized matrix inverse solution and an iterative solution. Note that γ here controls how fast the iteration converges; it does not affect the value that it converges to. One way to regularize this iteration process will be to select a number k and iterate only k times. Another regularization method introduces a new constant α that I shall deal with presently.

11.4 ….Is the same as the matrix-inverse solution

We shall see that matrix-inverse solutions and iteration can produce exactly the same results. Fleming (1976) compared the result of using what I have called "truncated iteration" (see §11.4.2) to results as the Twomey-Phillips solution that I discussed in Chapter 7. He developed a principle he called *virtual duality* that showed that they produce the same results, in that they are statistically equivalent. Specifically, let $\hat{\mathbf{x}}_m$ and $\hat{\mathbf{x}}_i$ be matrix-inverse and iterated solutions, respectively: he showed that average properties of these vectors will satisfy

$$\left\langle \left| \mathbf{A}\hat{\mathbf{x}}_m - \mathbf{A}\hat{\mathbf{x}}_i \right|^2 \right\rangle \le \sigma_N^2. \tag{67}$$

That is, the error variance will be the same whichever method is used. I shall prove his result, as well as a stronger version, here.

There are two ways we can regularize the iterative method under discussion. We have seen that, for any matrix-inverse method, there is an iterative method that converges to the same solution. So we can use the results of Chapter 7 to find the proper iteration formula. The other way to do it, which is not so obvious, is to choose a suitable number k and iterate only k times, rather than persevering until we have obtained convergence. I shall discuss the first way briefly, and the second at some length.

11.4.1 Shifted eigenvalues

First, consider the regularized matrix-inverse solution we derived in Chapter 7, in the form

$$\mathbf{R} = (\mathbf{A}^t\mathbf{A} + \alpha\mathbf{I})^{-1}\mathbf{A}^t. \tag{68}$$

We shall derive an iterated solution that converges to this one. If we identify $\mathbf{B}$ with $\mathbf{A}^t\mathbf{A} + \alpha\mathbf{I}$, then $\mathbf{I} - \mathbf{B} = (1 - \alpha)\mathbf{I} - \mathbf{A}^t\mathbf{A}$, and the iteration equation that corresponds to equation (46) is (setting $\mathbf{v} = \mathbf{A}^t\mathbf{y}$)

$$\mathbf{u}^{(k)} = \gamma\mathbf{A}^t\mathbf{y} + \left[(1 - \alpha\gamma)\mathbf{I} - \gamma\mathbf{A}^t\mathbf{A}\right]\mathbf{u}^{(k-1)}. \tag{69}$$

This converges to $\mathbf{R}\mathbf{y} = (\mathbf{A}^t\mathbf{A}+\alpha\mathbf{I})^{-1}\mathbf{A}^t\mathbf{y}$ if γ is chosen appropriately. This is a regularized iterative solution, although it is not very interesting since it provides nothing new. However, note that the solution it converges to is independent of the initial approximation $\mathbf{x}^{(0)}$. (Note that it is *unregularized* in the sense of the following section.)

By introducing the scalar α in equation (68), we have shifted all of the eigenvalues: each $\lambda_i \rightarrow \lambda_i + \alpha$. Therefore (and as we have seen before), the noise in the solution is amplified by a factor $1/(\lambda_i + \alpha)$, instead of $1/\lambda_i$.

11.4.2 Truncated iteration

Alternatively, we can regularize the iterative solution [equation (56)] by limiting ahead of time the number of times we iterate. I discussed this method of solving a one-dimensional problem in §11.1.2.

Since we have adopted the convention that all $<\mathbf{x}> = \mathbf{0}$, we can see that $\mathbf{x} = \mathbf{0}$ is as good an initial guess as any. Although this regularized solution depends to some (hopefully small) extent on $\mathbf{x}^{(0)}$, I propose that we neglect that dependence here. If it were important, the whole method would be useless. By looking at equation (60), we can see that the choice $\mathbf{x}^{(0)} = \mathbf{0}$ minimizes the magnitude of $\mathbf{x}^{(k)}$. So I shall pursue the analysis of this regularization method only for this special case. Now there is a *true* value of $\mathbf{x}$, call it $\mathbf{u}$. The measured value is $\mathbf{y}' = \mathbf{Au} + \mathbf{n}$. Putting this into equation (60), and letting $\mathbf{B} = \mathbf{A}^t\mathbf{A}$,

$$\mathbf{x}^{(k)} = \left[\mathbf{I} - \left(\mathbf{I} - \gamma\mathbf{A}^t\mathbf{A}\right)^k\right]\left[\mathbf{A}^t\mathbf{A}\right]^{-1}\mathbf{A}^t\left(\mathbf{Au} + \mathbf{n}\right); \tag{70}$$

therefore

$$\mathbf{x}^{(k)} - \mathbf{u} = -\left(\mathbf{I} - \gamma\mathbf{A}^t\mathbf{A}\right)^k\mathbf{u} + \left[\mathbf{I} - \left(\mathbf{I} - \gamma\mathbf{A}^t\mathbf{A}\right)^k\right]\left[\mathbf{A}^t\mathbf{A}\right]^{-1}\mathbf{A}^t\mathbf{n}. \tag{71}$$

Remember that the covariance matrix of a vector $\mathbf{w}$ was defined to be $<\mathbf{ww}^t>$. A little matrix algebra shows that the covariance matrix of $\mathbf{Aw}$ is $<\mathbf{Aww}^t\mathbf{A}^t>$. In the special case that $<\mathbf{ww}^t>$ is of the form $q\mathbf{I}$ for some scalar q, $<\mathbf{Aww}^t\mathbf{A}^t> = q\mathbf{AA}^t$. Now let us assume that the covariance matrix of $\mathbf{u}$ is

$$\mathbf{S}_u = \sigma_u^2\mathbf{I}. \tag{72}$$

Neither this assumption nor the one that $\mathbf{x}^{(0)} = \mathbf{0}$ is necessary, but a more general analysis would be more complicated than this subject warrants. As usual, we also assume that the noise covariance matrix is $\mathbf{S}_N = \sigma_N^2\mathbf{I}$. Then the error covariance matrix of $\mathbf{x}^{(k)} - \mathbf{u}$ is

$$\chi_{(k)}^2 = \left\langle\left(\mathbf{x}^{(k)} - \mathbf{u}\right)\left(\mathbf{x}^{(k)} - \mathbf{u}\right)^t\right\rangle$$

$$= \left(\mathbf{I} - \gamma\mathbf{AA}^t\right)^k\left\langle\mathbf{uu}^t\right\rangle\left(\mathbf{I} - \gamma\mathbf{AA}^t\right)^k$$

$$+ \left[\mathbf{I} - \left(\mathbf{I} - \gamma\mathbf{A}^t\mathbf{A}\right)^k\right]\left[\mathbf{A}^t\mathbf{A}\right]^{-1}\mathbf{A}^t\left\langle\mathbf{nn}^t\right\rangle\mathbf{A}\left[\mathbf{A}^t\mathbf{A}\right]^{-1}\left[\mathbf{I} - \left(\mathbf{I} - \gamma\mathbf{A}^t\mathbf{A}\right)^k\right]^t$$

$$= \sigma_u^2\left(\mathbf{I} - \gamma\mathbf{A}^t\mathbf{A}\right)^{2k} + \sigma_N^2\left[\mathbf{I} - \left(\mathbf{I} - \gamma\mathbf{A}^t\mathbf{A}\right)^k\right]$$

$$\times \left[\mathbf{A}^{t}\mathbf{A}\right]^{-1} \mathbf{A}^{t}\mathbf{A}\left[\mathbf{A}^{t}\mathbf{A}\right]^{-1}\left[\mathbf{I} - \left(\mathbf{I} - \gamma\mathbf{A}^{t}\mathbf{A}\right)^{k}\right]$$

$$= \sigma_{u}^{2}\left(\mathbf{I} - \gamma\mathbf{A}^{t}\mathbf{A}\right)^{2k} + \sigma_{N}^{2}\left[\mathbf{A}^{t}\mathbf{A}\right]^{-1}\left[\mathbf{I} - \left(\mathbf{I} - \gamma\mathbf{A}^{t}\mathbf{A}\right)^{k}\right]^{2}. \tag{73}$$

To make this last step, I used the property that $(\mathbf{A}^{t}\mathbf{A})^{-1}[\mathbf{I} - (\mathbf{I} - \gamma\mathbf{A}^{t}\mathbf{A})^{k}] = [\mathbf{I} - (\mathbf{I} - \gamma\mathbf{A}^{t}\mathbf{A})^{k}] (\mathbf{A}^{t}\mathbf{A})^{-1}$. Note that

$$\chi_{(0)}^{2} = \sigma_{u}^{2}; \tag{74}$$

$$\chi_{(1)}^{2} = \sigma_{u}^{2}\left(\mathbf{I} - \gamma\mathbf{A}^{t}\mathbf{A}\right)^{2} + \sigma_{N}^{2}\gamma^{2}\mathbf{A}^{t}\mathbf{A}; \tag{75}$$

and so forth. We can see that, if we iterate zero times (*i.e.,* always use $\hat{\mathbf{x}} = \mathbf{0}$ to estimate $\mathbf{x}$), the error variance is just the variance of $\mathbf{u}$. If we iterate k times, the error variance is the sum of two terms: one proportional to σ_{u}^{2}; the other, σ_{N}^{2}. Suppose we choose γ so that $\|\mathbf{I} - \gamma\mathbf{A}^{t}\mathbf{A}\| < 1$. Then the former term decreases as we iterate more times; the latter term increases. In the limit as $k \rightarrow \infty$, the error due to the first term vanishes, but the error due to the noise amplification increases to a value that, in general, may be unacceptably large: that is why we needed to regularize the solution in the first place.

We want to see how the error changes each time we perform another iteration. To do this, we need to use two identities:

$$\left(\mathbf{I} - \gamma\mathbf{A}^{t}\mathbf{A}\right)^{2} - \mathbf{I} = -\gamma\mathbf{A}^{t}\mathbf{A}\left(2\mathbf{I} - \gamma\mathbf{A}^{t}\mathbf{A}\right), \tag{76}$$

so

$$\left(\mathbf{I} - \gamma\mathbf{A}^{t}\mathbf{A}\right)^{2k+2} - \left(\mathbf{I} - \gamma\mathbf{A}^{t}\mathbf{A}\right)^{2k} = \left(\mathbf{I} - \gamma\mathbf{A}^{t}\mathbf{A}\right)^{2k}\left[\left(\mathbf{I} - \gamma\mathbf{A}^{t}\mathbf{A}\right)^{2} - \mathbf{I}\right]$$

$$= -\gamma\mathbf{A}^{t}\mathbf{A}\left(\mathbf{I} - \gamma\mathbf{A}^{t}\mathbf{A}\right)^{2k}\left(2\mathbf{I} - \gamma\mathbf{A}^{t}\mathbf{A}\right). \tag{77}$$

Also,

$$\left[\mathbf{I} - \left(\mathbf{I} - \gamma\mathbf{A}^{t}\mathbf{A}\right)^{k+1}\right]^{2} - \left[\mathbf{I} - \left(\mathbf{I} - \gamma\mathbf{A}^{t}\mathbf{A}\right)^{k}\right]^{2}$$

$$= \left[\mathbf{I} - \left(\mathbf{I} - \gamma\mathbf{A}^{t}\mathbf{A}\right)^{k} + \gamma\mathbf{A}^{t}\mathbf{A}\left(\mathbf{I} - \gamma\mathbf{A}^{t}\mathbf{A}\right)^{k}\right]^{2} - \left[\mathbf{I} - \left(\mathbf{I} - \gamma\mathbf{A}^{t}\mathbf{A}\right)^{k}\right]^{2}$$

$$= \gamma^2\left(\mathbf{A^tA}\right)^2\left(\mathbf{I}-\gamma\mathbf{A^tA}\right)^{2k} - 2\gamma\mathbf{A^tA}\left(\mathbf{I}-\gamma\mathbf{A^tA}\right)^{2k} + 2\gamma\mathbf{A^tA}\left(\mathbf{I}-\gamma\mathbf{A^tA}\right)^k$$

$$= 2\gamma\mathbf{A^tA}\left(\mathbf{I}-\gamma\mathbf{A^tA}\right)^k - \gamma\mathbf{A^tA}\left(\mathbf{I}-\gamma\mathbf{A^tA}\right)^{2k}\left(2\mathbf{I}-\gamma\mathbf{A^tA}\right). \tag{78}$$

Then it is not hard to show from equation (73) that

$$\chi^2_{(k+1)} - \chi^2_{(k)} = -\sigma_u^2\gamma\mathbf{A^tA}\left(\mathbf{I}-\gamma\mathbf{A^tA}\right)^{2k}\left(2\mathbf{I}-\gamma\mathbf{A^tA}\right)$$

$$+ \sigma_N^2\left(\mathbf{A^tA}\right)^{-1}\left[2\gamma\mathbf{A^tA}\left(\mathbf{I}-\gamma\mathbf{A^tA}\right)^k - \gamma\mathbf{A^tA}\left(\mathbf{I}-\gamma\mathbf{A^tA}\right)^{2k}\left(2\mathbf{I}-\gamma\mathbf{A^tA}\right)\right]$$

$$= \gamma\left(\mathbf{I}-\gamma\mathbf{A^tA}\right)^k$$

$$\times\left\{-\sigma_u^2\mathbf{A^tA}\left(\mathbf{I}-\gamma\mathbf{A^tA}\right)^k\left(2\mathbf{I}-\gamma\mathbf{A^tA}\right)+ \sigma_N^2\left[2\mathbf{I}-\gamma\left(\mathbf{I}-\gamma\mathbf{A^tA}\right)^k\left(2\mathbf{I}-\gamma\mathbf{A^tA}\right)\right]\right\}$$

$$= \gamma\left(\mathbf{I}-\gamma\mathbf{A^tA}\right)^k\left[2\sigma_N^2\mathbf{I}-\left(\sigma_u^2\mathbf{A^tA}+\sigma_N^2\right)\left(\mathbf{I}-\gamma\mathbf{A^tA}\right)^k\left(2\mathbf{I}-\gamma\mathbf{A^tA}\right)\right]. \tag{79}$$

The value of k for which $\chi^2_{(k)}$ is minimum occurs when $\chi^2_{(k+1)}\approx\chi^2_{(k)}$, or

$$\left(\mathbf{I}-\gamma\mathbf{A^tA}\right)^k \approx 2\sigma_N^2\left(2\mathbf{I}-\gamma\mathbf{A^tA}\right)^{-1}\left[\sigma_u^2\mathbf{A^tA}+\sigma_N^2\mathbf{I}\right]^{-1}. \tag{80}$$

It is inconvenient to pursue this further for general matrices. Instead, we can derive more insight by considering the case where we have chosen the basis vectors for $\mathbf{x}$ to be the eigenvectors of $\mathbf{A^tA}$, so that $\mathbf{A^tA}$ is diagonal; *i.e.*, we use the canonical form introduced in §5.11, where I showed that any matrix equation can, by a proper choice of basis vectors, be expressed in the form

$$\mathbf{y} = \mathbf{Dx}, \tag{81}$$

where $\mathbf{D}$ is a diagonal matrix, the components of $\mathbf{x}$ vary independently, and so do the components of $\mathbf{y}$. I can also write this relationship $y_i = \lambda_i x_i$. Also, remember that the singular values $\lambda_i \geq 0$ for all i.

Let λ_i^2 be i^{th} eigenvalue of $\mathbf{A^tA}$, then

$$\mathbf{A^tA} = \text{diag}\left(\lambda_1^2, \lambda_2^2, \cdots, \lambda_m^2\right). \tag{82}$$

In this case, $\chi^2_{(k)}$ is also diagonal, and

$$\chi^2_{i,(k)} = \sigma^2_u \left(1 - \gamma\lambda^2_i\right)^{2k} + \frac{\sigma^2_N}{\lambda^2_i}\left[1 - \left(1 - \gamma\lambda^2_i\right)^k\right]^2, \tag{83}$$

and

$$\chi^2_{i,(k+1)} - \chi^2_{i,(k)} = \gamma\left(1 - \gamma\lambda^2_i\right)^k \left\{2\sigma^2_N - \left(\sigma^2_u\lambda^2_i + \sigma^2_N\right)\left(2 - \gamma\lambda^2_i\right)\left(1 - \gamma\lambda^2_i\right)^k\right\}. \tag{84}$$

The error is minimum when

$$2\sigma^2_N \approx \left(\sigma^2_u\lambda^2_i + \sigma^2_N\right)\left(2 - \gamma\lambda^2_i\right)\left(1 - \gamma\lambda^2_i\right)^k, \tag{85}$$

or

$$\left(1 - \gamma\lambda^2_i\right)^k \approx \frac{2\sigma^2_N}{\left(\sigma^2_u\lambda^2_i + \sigma^2_N\right)\left(2 - \gamma\lambda^2_i\right)}. \tag{86}$$

Now when this obtains, we can substitute equation (86) into equation (21) to get

$$\chi^2_{i,(k)} = \sigma^2_u \left[\frac{2\sigma^2_N}{\left(2 - \gamma\lambda^2_i\right)\left(\sigma^2_u\lambda^2_i + \sigma^2_N\right)}\right]^2 + \sigma^2_N\lambda^2_i\left[\frac{2\sigma^2_u - \gamma\left(\sigma^2_u\lambda^2_i + \sigma^2_N\right)}{\left(2 - \gamma\lambda^2_i\right)\left(\sigma^2_u\lambda^2_i + \sigma^2_N\right)}\right]^2$$

$$= \frac{4\sigma^2_N\sigma^2_u\left(\lambda^2_i\sigma^2_u + \sigma^2_N\right) - \sigma^2_N\lambda^2_i\left[4\sigma^2_u\gamma\left(\sigma^2_u\lambda^2_i + \sigma^2_N\right) - \gamma^2\left(\sigma^2_u\lambda^2_i + \sigma^2_N\right)^2\right]}{\left[\left(2 - \gamma\lambda^2_i\right)\left(\sigma^2_u\lambda^2_i + \sigma^2_N\right)\right]^2}$$

$$= \frac{2\sigma^2_N\sigma^2_u\left(2 - \gamma\lambda^2_i\right)^2 + \gamma^2\sigma^4_N\lambda^2_i}{\left(2 - \gamma\lambda^2_i\right)^2\left(\sigma^2_u\lambda^2_i + \sigma^2_N\right)}. \tag{87}$$

The iteration will not converge unless $\gamma\lambda^2_i < 1$ [see equation (79)], so $(2 - \gamma\lambda^2_i)$ ≈ 1 and we can neglect $\gamma\lambda^2_i\sigma^2_N$ compared with σ^2_u. We can approximate the error by

$$\chi^2_{i,(k)} = \frac{2\sigma^2_N\sigma^2_u}{\left(\sigma^2_u\lambda^2_i + \sigma^2_N\right)}. \tag{88}$$

This is the same as the variance given in equation (22) of Chapter 1. Therefore, we have discovered two important facts:

1. If we pick the appropriate value of k, we get the same result (in a statistical sense) as we would have from the regularized matrix solution in Chapter 7; and

2. There cannot possibly be any advantage to using iteration instead of a matrix-inverse method for linear problems.

In addition, remember that we assumed that we could always start with the approximation $x^{(0)} = 0$. It should be clear, however, that we could have started with some other initial guess. In this case, the result would depend to some extent on what that initial guess is. If the dependence is too large, this dependence could be detrimental because it would bias the results. On the other hand, suppose that we had a sequence of measurements $\{y_i\}$ (*i.e.*, a sequence of measurements of the vector y) and had reason to believe that consecutive sets of measurements are correlated to some degree. Then using $\hat{x}_i$ as the initial approximation $x_{i+1}^{(0)}$ might have some benefits.

Exercise 11-3: Can you analyze this situation, where you allow each initial approximation to depend on the last result and the iterative solution is purposely allowed to depend on $x^{(0)}$? Might there be a problem with parasitic solutions that arise because each solution depends on the preceding one?

Exercise 11-4: Show that any regularized, iterated solution must depend on the initial approximation $x^{(0)}$. We would expect that choosing $x^{(0)} = 0$ would be optimum in many circumstances, since $<x> = 0$ by hypothesis. Are there circumstances where it would be better to pick some other value for $x^{(0)}$?

11.4.3 Summary

I have introduced two parameters in relation to iterative solutions of matrix equations, a convergence parameter γ and a regularization parameter α. One of them controls the convergence properties; the other, what solution the process converges to. Under very general conditions, an equation of the form $y = Ax$ can be solved either by finding an inverse matrix R and estimating $\hat{x} = Ry$ or by taking an initial approximation $x^{(0)}$ and using iteration to converge to a solution. Either process can produce the same results; there is no particular mathematical advantage to one approach or the other. However, since the former method requires but one matrix multiplication (assuming that A remains fixed and that the problem is truly linear), it would seem to be preferable. We will need the iterative process, however, when the equations to be solved are not linear.

We have also seen something that, at first, may seem peculiar: we can regularize an iterative solution to a matrix equation by iterating for a pre-set number of terms. We have also seen that it has no obvious benefit as opposed to using a matrix-inverse solution to a set of linear equations. So what have we learned so far?

First, we have, I hope, gained some insight into the nature of inverse problems. At every turn we see the prominent role played by σ_x^2 and σ_N^2. Note that we could write equation (88) in such a way that it is obvious that the error depends on the *ratio* of $\lambda^2 \sigma_x^2 / \sigma_N^2$, suggesting that what is important is the signal-to-noise ratio of the measurements. Second, if we need to solve nonlinear problems, we usually need to do it by iteration. This regularization technique can be applied in this situation also, and it may have some advantages over other methods.

Finally, we should note that we could modify this procedure slightly and, rather than iterating a fixed number of times, iterate until some other criterion is met. One that has been proposed is to iterate until $|\mathbf{x}^{(k)} - \mathbf{y}'|^2 \approx \sigma_N^2$, and then stop.

Exercise 11-5: Work out the details of how to implement this regularization method as it applies to multivariate problems, assuming that the number of measurements = number of independent variables (*i.e.*, $n = m$).

Exercise 11-6: Work out the details of the iterative process where one iterates not a fixed number of times, but until $|\mathbf{x}^{(k)} - \mathbf{y}'|^2 \approx \sigma_N^2$. How well will this work? Does it have any advantages or disadvantages?

11.5 Integral equations

I discussed integral equations in the last chapter. Here I resume the discussion of integral equations to take up iterated solutions. We are considering Fredholm equations of the first kind, where $g(x)$ is a measured function, $f(y)$ is the function we want to estimate, and $K(x,y)$ is a known kernel. In the last chapter I showed how to solve them using Fourier transforms, by converting them to equivalent matrix equations, or by studying the eigenfunctions of the kernel. Often these methods are not available to us; then we can use iteration. I shall first discuss the unregularized iterative solution, and then a regularized one.

11.5.1 Iterated solution of integral equations

We wish to solve

$$g(x) = \int_a^b f(y) K(x,y) \, dy \qquad (89)$$

by iteration.[10] Suppose we are given measurements of $g(x)$ and we know the kernel $K(x,y)$. Remember that the adjoint of the kernel K operates on $g(x)$ so that, by definition,

$$K^\dagger g \equiv \int_a^b K(x,y)g(x)\mathrm{d}x. \tag{90}$$

Furthermore,

$$\left(Kf,g\right) = \left(f,K^\dagger g\right). \tag{91}$$

Select some function $f^{(0)}(y)$ to be the first approximation to the solution to equation (89). Then calculate the functions $f^{(n)}$ defined by

$$f^{(n)} = f^{(n-1)} + K^\dagger\left(g - Kf^{(n-1)}\right) \tag{92}$$

(cf. §11.3.3). The functions $f^{(n)}$ approach a solution to equation (89) if $f^{(0)}$ is of the form $K^\dagger h$ for some function $h(x)$ and $K^\dagger K$ has eigenvalues λ^2 that lie in the range $(0,2)$. Westwater and Strand (1972) cite Landweber (1951) and Smithies[11] (1958) for a proof that this procedure converges and a discussion of other properties of the solution.

I defined the eigenfunctions of K in equation (50) of Chapter 10. We can examine the iterative solution in terms of the eigenfunctions of K and $K^\dagger$. In §10.7 I defined ϕ_i to be the eigenfunctions of $K^\dagger K$ associated with the eigenvalues λ_i^2. Here, I allow that g and f might belong to different spaces. If K is a mapping from $\mathscr{H}_1$ to $\mathscr{H}_2$, both Hilbert spaces, then a $K^\dagger$ is a mapping from $\mathscr{H}_2$ to $\mathscr{H}_1$. For any kernel K, the eigenfunctions ϕ_i belonging to different eigenvalues λ_i are orthogonal. The set of eigenfunctions $\{\phi_i\}$ that all belong to the same eigenfunction form a subspace that is spanned by n orthogonal eigenfunctions. Therefore, we can assume that the eigenfunctions ϕ_i, $i = 1, 2, \ldots,$ span $\mathscr{H}_1$ (since $K^\dagger K$ is a mapping from $\mathscr{H}_1$ to $\mathscr{H}_1$). Assume also that they are normalized.

Expand $f^{(n)}(y)$ in terms of ϕ_j:

$$f^{(n)}(y) = \sum_i b_i^{(n)}\phi_i(y). \tag{93}$$

Also, expand g in terms of ψ_i:

$$g(x) = \sum_j a_j\psi_j(x). \tag{94}$$

---

[10] E.g., see Westwater and Strand (1972) and references cited therein.

[11] I looked at Smithies (1958) recently and couldn't find the section on iteration, so this reference may not be germane here. Or I just didn't look hard enough.

Using equation (92),

$$K^\dagger\left[g - Kf^{(n)}(y)\right] = f^{(n+1)}(y) - f^{(n)}(y)$$

$$= K^\dagger\left(\sum a_j\psi_j(x) - K\sum_j b_j^{(n)}\phi_j(y)\right)$$

$$= K^\dagger\left(\sum a_j\psi_j(x) - \sum_j \lambda_j b_j^{(n)}\psi_j(x)\right)$$

$$= K^\dagger\sum\left(a_j - \lambda_j b_j^{(n)}\right)\psi_j(x)$$

$$= \sum_j \lambda_j\left(a_j - \lambda_j b_j^{(n)}\right)\phi_j(x)$$

$$= \sum_j\left(b_j^{(n+1)} - b_j^{(n)}\right)\phi_j(x). \tag{95}$$

Therefore

$$b_j^{(n+1)} = b_j^{(n)} + \lambda_i\left(a_j - \lambda_j b_j^{(n)}\right). \tag{96}$$

This converges when

$$b_j^{(n)} = \frac{a_j}{\lambda_j}. \tag{97}$$

Therefore, iteration of equation converges to

$$\hat{f}(y) = \sum_{j=1}\hat{b}_j\phi_j(y), \tag{98}$$

where

$$\hat{b}_j = \frac{a_j}{\lambda_j}. \tag{99}$$

Since $\lambda_i \to 0$ as $i \to \infty$, this solution is unstable. The exception is when K is degenerate (*i.e.*, is a finite sum and has a finite number of eigenvalues; see §10.7.1).

Returning to equation (96) for a moment, we can see that each eigenvector component separately obeys

$$b_j^{(n)} = b_j^{(n-1)}\left(1 - \lambda_j^2\right) + \lambda_j a_j. \tag{100}$$

Therefore, each successive estimate of b_j is multiplied by another factor $(1 - \lambda_j^2)$. This will diverge unless $\lambda_j^2 \leq 2$, since otherwise $|(1 - \lambda_j^2)| > 1$. In addition, it is clear that this process cannot converge unless $a_i / \lambda_i \to 0$ as $i \to \infty$.

Exercise 11-7: How can we add a convergence factor γ this method so that it does not diverge? How does it relate to the matrix solution?

Exercise 11-8: Does this solution differ from the eigenvalue solution in the previous section? Is there an advantage to using an iterative solution?

11.5.2 Regularized solutions

Just as we discuss regularized matrix solutions in the last chapter, there are regularized iterative solutions that apply in general to solutions of integral equations. The unregularized solution is given in equation (92). We can look at two regularized solutions here. Since the development follows that of the iterated matrix solutions closely, I shall be brief here.

Shifted Eigenvalues. Let $K(x,y)$ be the kernel; let α be a positive real number; and iterate the solution according to

$$f^{(k)} = K^\dagger g + \left(\alpha - K^\dagger K\right)f^{(k-1)}. \tag{101}$$

Let $f^{(0)}$ be the initial approximation to the solution; then

$$f^{(1)} = K^\dagger g + \left(\alpha - K^\dagger K\right)f^{(0)}; \tag{102}$$

$$f^{(2)} = K^\dagger g + \left(\alpha - K^\dagger K\right)f^{(1)}$$

$$= K^\dagger g + \left(\alpha - K^\dagger K\right)K^\dagger g + \left(\alpha - K^\dagger K\right)^2 f^{(0)}; \tag{103}$$

and so forth. The kth iterated term is

$$f^{(k)} = \left[1 + \left(\alpha - K^\dagger K\right) + \cdots + \left(\alpha - K^\dagger K\right)^{(k-1)}\right]$$

$$K^\dagger g + \left(\alpha - K^\dagger K\right)^{(k)} f^{(0)}. \tag{104}$$

It is easy to show that [*cf.* equation (15)]

$$\left[1 + \left(\alpha - K^\dagger K\right) + \cdots + \left(\alpha - K^\dagger K\right)^{(k-1)}\right]\left[1 - \left(\alpha - K^\dagger K\right)\right]$$

$$= 1 - \left(\alpha - K^\dagger K\right)^k, \tag{105}$$

which is a result that we have used many times already. Then

$$\left[1 - \left(\alpha - K^\dagger K\right)\right] f^{(k)}$$

$$= \left[1 - \left(\alpha - K^\dagger K\right)^k\right] K^\dagger g + \left[1 - \left(\alpha - K^\dagger K\right)\right]\left(\alpha - K^\dagger K\right)^{(k)} f^{(0)}. \tag{106}$$

Then [compare this step with equation (49)]

$$\left(1 - \alpha + K^\dagger K\right) f^{(k)}$$

$$= K^\dagger g + \left(\alpha - K^\dagger K\right)^k \left[\left(1 - \alpha + K^\dagger K\right) f^{(0)} - K^\dagger g\right]. \tag{107}$$

If $\| \alpha - K^\dagger K\| < 1$, this will converge to

$$\hat{f} = \left(1 - \alpha + K^\dagger K\right)^{-1} K^\dagger g \tag{108}$$

since the second term on the r.h.s. of equation (107) $\to 0$ as $k \to \infty$ (I shall explain what $(1 - \alpha + K^\dagger K)^{-1}$ means in a moment). If we represent $g(x)$ by (*cf.* §10.7.2 for definition of the eigenfunctions ϕ_i and ψ_i)

$$g(x) = \sum_i a_i \psi_i(x) \tag{109}$$

and

$$f(y) = \sum_{i=1} b_i \phi_i(y), \tag{110}$$

then

$$K^\dagger g = \sum_{i=1}^{n} a_i \lambda_i \phi_i(y). \tag{111}$$

Also,

$$(1 - \alpha + K^\dagger K)f = \sum_i (1 - \alpha + \lambda_i^2)b_i \phi_i(y). \tag{112}$$

We can define what we mean by $(1 - \alpha + K^\dagger K)^{-1}$ with:

$$(1 - \alpha + K^\dagger K)^{-1}\phi_j(y) = \frac{1}{1 - \alpha + \lambda_j^2}\phi_j(y). \tag{113}$$

Finally, we can estimate f from

$$\hat{f}(y) = \sum_i \frac{a_i \lambda_i}{\lambda_i^2 + 1 - \alpha}\phi_i(y). \tag{114}$$

Comparing this with §10.7.3, we see that, by shifting the eigenvalues up by $(1 - \alpha)$, we have removed the instability inherent in the unregularized result

$$\hat{f}(y) = \sum_i \frac{\hat{a}_i}{\lambda_i}\phi_i(y). \tag{115}$$

Truncated Iteration. The second approach is truncated iteration.[12] Instead of using equation (101), use

$$f^{(k)} = \alpha K^\dagger g + (1 - \alpha K^\dagger K)f^{(k-1)}, \tag{116}$$

where, as before, α is a positive real number—compare this with equation (57). Working as before,

$$K^\dagger K f^{(k)} = K^\dagger g + (1 - \alpha K^\dagger K)^k [K^\dagger K f^{(0)} - K^\dagger g]. \tag{117}$$

This converges to

$$f^{(k)} = (K^\dagger K)^{-1} K^\dagger g \tag{118}$$

when

---

[12] See Groetsch (1984, p 27).

$$0 \; < \; \|1 - \alpha K^\dagger K\| \; < \; 1. \tag{119}$$

However, if we iterated until it converged, we would get the same unregularized solution that we have seen before. As before, we can find an acceptable regularized solution if we iterate only for a predetermined number n steps. We can see that the solution will depend on $f^{(0)}$.

11.6 Nonlinear equations

In the previous section, the iteration scheme using equation (57) used a constant matrix $(\mathbf{I} - \gamma \mathbf{A}^t \mathbf{A})$. If, however, the problem is nonlinear, then $\gamma \mathbf{A}^t \mathbf{A}$ can change at each step. This makes questions of convergence more complicated. This section is necessarily only a cursory treatment of nonlinear inversion.

I am very wary of iterative solutions to linear equations, since there seems to be no advantage over matrix-inverse solutions. But there's not much we can do for nonlinear equations except to iterate. However, such enterprises are always filled with potential pitfalls, not all of which are obvious. In the introduction to a recent conference on inverse problems, Sabatier (1987) wrote the following about nonlinear inverse problems:

> Unfortunately, many applied physicists forget that non linear optimization problems were already deeply studied by mathematicians, who showed in particular that simple optimization techniques (e.g. generalized least squares methods) could be treacherous, and that the hard work is not to formulate the method but to circumvent secondary minima and, in specific problems, to guarantee convergence and stability. Of course, a physicist may feel allowed to replace mathematical proofs by a large number of well-chosen numerical experiments. But too many so-called "new" methods do not contain either justification, ignore difficulties, and are nothing but a cheap, unchecked variant of some known optimization technique.

Perhaps Sabatier is overly harsh, but one should not underestimate the difficulties.

For linear equations, it is possible to prove that a certain method is stable and is optimal, in relation to a given criterion. We cannot, in general, do this for nonlinear equations. Any inverse method involves minimizing some quantity, presumably a measure of how well the nth approximate solution agrees with the measurements. Now for a linear problem, we can prove that there is a single minimum: all we have to do is find it. But for nonlinear problems there can be many relative minima and we need to know which of them is the physically meaningful one that we want.

While there is an abundant literature on solving nonlinear equations, there is an infinite number of possible nonlinear problems, each of which is likely to have its special properties and require a different kind of solution. It is not without trepidation that I discuss the following nonlinear problem. It relates to remote sensing of the earth from space, and may not be applicable in other areas. On the other hand, it will introduce to the reader some ways of thinking about nonlinear problems.

I hope that the reader will keep three questions in mind. The first is, "Is this particular method optimal, or might there be a substantially better one somewhere (if only I could find it)?" There is no general answer to this question. The second question is, "What is the error variance when I use this method?" This may be hard to calculate, and such calculations might be tedious, but they are necessary. Lastly, if the method is to be applied to real data, "How can I compare the calculated results with the correct values? Is there a way to measure the accuracy independently from the inversion process?" *Caveat emptor.*

I discuss briefly the problem of inferring vertical temperature profiles in the atmosphere from infrared radiance measurements made from satellites. The nonlinearity arises here because the Planck radiation law (see §2.2.1) is highly nonlinear in the infrared part of the spectrum at temperatures that obtain in the earth's atmosphere. In addition, the temperature of the atmosphere—which is what we are trying to measure—affects the absorption coefficients. Much of the original work on this subject was done in the late 60s and the 70s by M. Chahine, W. Smith,[13] S. Twomey,[14] and other of the pioneering satellite meteorologists; it was published in various journal articles at the time. Liou[15] gives a concise description of the methods, which I recommend to the reader for more details, and which I shall follow here.

Assume that the equation we want to invert is a nonlinear Fredholm equation of the first kind. The measured function $g(x)$ is given by

$$g(x) = \int_a^b A[y, f(y)]K_f(x, y)\,\mathrm{d}y, \qquad (120)$$

where A is a known nonlinear function of $f(y)$; $K_f(x,y)$ is a known kernel; and a and b are the known limits of integration. I write K_f to indicate that the kernel depends on f. Here, if we are discussing the measurement of vertical temperature profiles in the atmosphere, $g(x)$ is the measurement of the radiance at wavelength x; y represents the height in the atmosphere; $f(y)$ is the temperature at that height; and $A[y, f(y)]$ represents the Planck function at this temperature. This equation is different from equation (1) of Chapter 10 because the kernel here is multiplied by a nonlinear function of $f(y)$, not just $f(y)$ itself.

Since any integral is reduced to a sum when it is calculated numerically, we can also express equation (120) as a matrix equation. For convenience,

---

[13] See Smith (1970). Also see the footnote at the beginning of §11.6.5.

[14] Be sure to read his book also: Twomey (1977).

[15] Liou, Kuo-Nan (1980) Chapter 7.

suppose that the vector **y** has components y_j that are evenly spaced with $y_{j+1} - y_j = \delta y$. Then equation (120) can also be represented as a sum by

$$g_i = g(x_i) = \delta y \sum_j A[y, f(y_j)] K_f(x_i, y_j). \qquad (121)$$

The kernel K_f has become a matrix, but it still depends on $f(y)$. Therefore, this too will require an iterative solution.

11.6.1 An example: the equation of radiative transfer

I discussed the equation of radiative transfer in Chapter 4. In the present situation, we can write[16]

$$I(\lambda_i) = \int_0^p B[\lambda_i, T(p')] \frac{dw(p')}{dp'} dp', \qquad (122)$$

where $I(\lambda_i)$ is the radiance at wavelength λ_i; $B[\lambda_i, T(p)]$ is the Planck function at wavelength λ_i and temperature $T(p)$; p is the pressure level in the atmosphere; and $w(p)$ is the weighting function. Letting $\kappa(\lambda, T, p)$ be the absorption coefficient at level p. Then the weighting function is

$$w(p) = \exp\left\{-\frac{1}{g} \int_0^p \kappa[\lambda_i, T(p'), p'] dp'\right\}, \qquad (123)$$

where p is pressure, used as the independent variable, and g, the acceleration of gravity. I discussed this in Chapter 4. So

$$\frac{dw(p)}{dp} = -\frac{\kappa[\lambda_i, T(p'), p']}{g} \exp\left\{-\frac{1}{g} \int_0^p \kappa[\lambda_i, T(p'), p'] dp'\right\}. \qquad (124)$$

Then the integral equation is

$$I(\lambda_i) = \frac{1}{g} \int_0^p B[\lambda_i, T(p')] \kappa[\lambda_i, T(p'), p']$$

$$\times \exp\left\{-\frac{1}{g} \int_0^{p'} \kappa[\lambda_i, T(p''), p''] dp''\right\} dp'. \qquad (125)$$

---

[16] For simplicity, I assume that we are looking straight down, toward nadir.

We can see that this is not, strictly speaking, the kind of integral equation we have been considering, since we cannot write it as an integral of $T(p)K_f(\lambda,p)$. The Planck function B is a nonlinear function of T; the absorption coefficient κ depends on T as well as p. Therefore, at best, we have to replace equation (125) with an approximation to apply this theory of integral equations directly.

Having admitted this, I shall continue to describe this equation as having a kernel. What makes inverting infrared measurements so difficult is that the kernel itself depends on the factor (T) that we are trying to measure. If this were not the case—if the kernel K_f were independent of $f(y)$, in our original notation—we could simply solve a linear problem for $A[y,f(x)]$. Then we would operate on each value with A^{-1} to find the values of $f(y)$.

Although many remote-sensing problems are even more complicated than the theory in this chapter would suggest, they can usually be approximated by more tractable forms.

11.6.2 Inverse methods for nonlinear integral equations

Ignore the noise in any measurements for the moment. The iteration will proceed as follows. Given a measurement of the vector $\mathbf{y}$, select an initial approximation, $\mathbf{x}^{(0)}$. (Again, superscripts denote steps in the iteration.) We seek an operator $\mathbf{R}(\mathbf{x})$ such that, at each stage of the iteration,

$$f^{(k)}(\mathbf{x}) = \mathbf{R}\left[f^{(k-1)}(\mathbf{x})\right] \tag{126}$$

where the function $\mathbf{R}$ may depend at each step on k or on the value of $\mathbf{x}^{(k-1)}$.

It seems to me that the first question we should ask, given the circumstances, is whether or not this procedure converges. A *perfect* iteration scheme would have two properties:

1. Given a measurement g, the iteration would always converge to a solution $\hat{f}$ independent of the initial guess $f^{(0)}$. The choice of $f^{(0)}$ may affect the speed with which the iteration converges, but this is not important.

2. The method would be stable, in the sense that, for two measurements g_1 and g_2 that are close together, the corresponding solutions $\hat{f}_1$ and $\hat{f}_2$ would also be close together. The converse of this property is chaos, where a small perturbation of the input leads to an arbitrarily large change in the output. Clearly, if a small change in $\mathbf{y}$ causes a large change in $\hat{f}$, then a small amount of noise will cause a large error in the solution.

Convergence might be global, or it might be local. In the first case, the iteration would converge to the same point, no matter what the initial guess might be. In the second case, the iteration might converge to any one of sev-

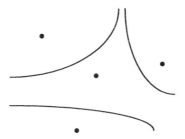

Figure 11-4. Basins of attraction.

eral points, depending on the initial guess. In this case, we can define a *basin of attraction* around each of these points. Let these points be $\mathbf{p}_1, \cdots, \mathbf{p}_m$, for some integer m, and let B_k be the basin of attraction around $\mathbf{p}_k$. Then, if at any step, $\mathbf{f}^{(k)}$ falls into B_k, the iteration will converge to $\mathbf{p}_k$. It will be attracted to that point, hence the name. See Figure 11-4, which shows four basins of attraction and the respective points that the iteration will converge to. Clearly, basins of attraction cannot overlap, but they do not have to have any particularly simple shapes, either.

How important the local nature of the convergence will be depends on how large the basins of attraction are and how far apart they are. If I need to estimate $f^{(0)}$ only to within 50 K of the right answer to be in the right basin of attraction for a temperature profile, say, then there is no problem. Temperatures in the atmosphere are not that variable. But if I need to be within 2 K to be in the right basin, I am in trouble.

We can find the solution as follows. Start with an initial approximation $f^{(0)}$. Calculate at each step

$$\delta g^{(k)} = g - \int_a^b A[y, f(y)]K_f(x,y)\mathrm{d}y. \tag{127}$$

Then use $\delta g^{(k)}$ to find an improved solution $f^{(k+1)}$. The usual practice is to keep iterating until $|\delta g^{(k)}| <$ some predetermined limit. I shall mention here two methods of solving nonlinear problems that are commonly used in remote sensing; see Liou[17] for details. First, I shall discuss an obvious approach, expanding A in a Taylor series, and then discuss Chahine's and Smith's methods.

11.6.3 Taylor series

Start with equation (120). Let $f^{(k)}(y)$ be the kth estimate of $f(y)$; and let $K_f^{(k)}(x,y)$ be the kth estimate of the kernel. I shall write $\mathrm{d}K_f/\mathrm{d}f$ to represent the

---

[17] Liou (1980) pp 263-266.

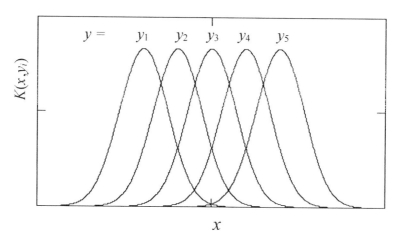

Figure 11-5. Plots of $K(x,y_i)$ for a kernel with well spaced peaks. An exceedingly simplified picture of the real world.

change in K_f due to a change in f at each point x. Assume that the kernel $K_f^{(k)}(x,y)$ has a single maximum in x, and that the value of x where $K_f^{(k)}(x,y)$ is maximum is different for each value of y (see Figure 11-5). Define

$$g^{(k)}(x) = \int_a^b A\left[y, f^{(k)}(y)\right] K_f^{(k)}(x,y) dx. \tag{128}$$

(To avoid confusion, note that the corresponding matrix equation would be $\mathbf{g} = \mathbf{Af}$, rather than $\mathbf{y} = \mathbf{Ax}$.) The error at the kth step is

$$g(x) - g^{(k)}(x) = \int_a^b \left\{ A[y, f(y)] K_f(x,y) - A\left[y, f^{(k)}(y)\right] K_f^{(k)}(x,y) \right\} dy. \tag{129}$$

Suppose that we measure $g(x)$ at a set of points, denoted x_i. Using the mean value theorem, there is a value of y, denoted y_i, such that

$$g(x_i) - g^{(k)}(x_i)(b-a)$$

$$= \left\{ A[y_i, f(y_i)] K_f(x_i, y_i) - A\left[y_i, f^{(k)}(y_i)\right] K_f^{(k)}(x_i, y_i) \right\}(b-a). \tag{130}$$

Assume that each x_i corresponds to a single peak in $K_f(x,y_i)$ and that, for $y_i \neq y_j$, $x_i \neq x_j$. I show an idealized case of this in Figure 11-5. When the peaks are pronounced, x_i will be in the center of the peak for the corresponding y_i.

Expand $A(f)K_f$ in a Taylor series:

$$A[f(y)] K_f(x,y) = A[f_0(y)] K_{f_0}(x,y)$$

$$+\left[f(y)-f_0(y)\right]\left[A\left[y,f_0(y)\right]\frac{\mathrm{d}K_{f_0}}{\mathrm{d}f}+K_{f_0}(x,y)\frac{\mathrm{d}A(y,f)}{\mathrm{d}f}\right]\Bigg|_{f_0(y)}+\cdots. \quad (131)$$

Denote $g(x_i)$ by g_i, and $f(y_i)$ by f_i. Combining these two equations,

$$g_i - g_i^{(k)} = \left[\hat{f}_i - f_i^{(k)}\right]$$

$$\times (b-a)\left[K_f(x_i,y_i)\frac{\mathrm{d}A(y,f)}{\mathrm{d}f}\bigg|_{f_i^{(k)}}+A(f)\frac{\mathrm{d}K}{\mathrm{d}f}\bigg|_{f_i^{(k)}}\right]. \quad (132)$$

Here, $\hat{f}$ denotes the true value of f. The iteration equation is (*cf.* §11.1)

$$f_i^{(k+1)} = f_i^{(k)}$$

$$+\frac{g_i - g_i^{(k)}}{(b-a)}\left[K(x_i,y_i)\frac{\mathrm{d}A(y,f)}{\mathrm{d}f}\bigg|_{f_i^{(k)}}+A(y,f)\frac{\mathrm{d}K_f}{\mathrm{d}f}\bigg|_{f_i^{(k)}}\right]^{-1}. \quad (133)$$

One continues on in this way, updating $\mathrm{d}A(y,f)/\mathrm{d}f$ and $\mathrm{d}K_f/\mathrm{d}f$ each time, until $|g_i - g_i^{(k)}| < \varepsilon$, where ε is some small number > 0, for all values of i.

This method depends on K_f having the property that, for each y_i, there is a different x_i where K_f is maximum. If these peaks are not pronounced, or there are multiple peaks, this method probably will not work, because we can no longer use equation (130).

This iterative solution involves the following steps:

1. Choose an initial approximation $\mathbf{f}^{(0)} = (f_1^{(0)}, \cdots, f_n^{(0)})$.

2. Use equation (128) to calculate each $g_i^{(k)}$.

3. Use equation (133) to calculate $\mathbf{f}^{(k+1)}$.

4. Stop when $|g_i^{(k)} - g_i| < \varepsilon$ for some preselected $\varepsilon > 0$; otherwise go back to step 2 and iterate once more.

This is very similar to the linear iteration scheme, except that here the counterpart of the matrix $\mathbf{B}$ in equation (46) changes at each step. This kind of solution can be computationally intensive. While the cost of computing seems to be falling by one-half every year, this still might be an important consideration in operational remote-sensing situations, where a day's worth of satellite measurements must be processed every day.

Because the "matrix" in equation (132) changes at each step, it is a little harder than it was before to evaluate the convergence of this process. It should be clear, however, that $\|AK_f\| < 1$, the method should converge.

11.6.4 Chahine's method

This method is similar to the Taylor series expansion. Suppose that dA/df is more important than dK_f/df. Go back to equation (128) and use the mean value theorem again: there is a value of x_i such that

$$g_i^{(k)} = A\left(f_i^{(k)}\right)K_f(x_i,y_i)\Delta x_i. \tag{134}$$

Remember that the measured values are at $\{y_1,\cdots, y_m\}$, and that $g_i^{(k)}$ is the kth estimate of g_i. Apply the mean value theorem to equation (120) also. Chahine (1970) now takes the ratio

$$\frac{g_i}{g_i^{(k)}} = \frac{A\left(f_i\right)}{A\left(f_i^{(k)}\right)}. \tag{135}$$

Note that the factors $K_f(x_i,y_i)\Delta x_i$ approximately cancel each other. This replaces equation (134) in the iterative method above:

$$A\left(f_i\right) = \frac{g_i}{g_i^{(k)}} A\left(f_i^{(k)}\right). \tag{136}$$

The advantage here is that, while equation (120) is not linear in f, it is linear in $A(f)$. Therefore, it may be sensible to solve for $A(f)$ instead of f. At each iteration step, then, we arrive at a new estimate $A(f^{(k)})$. However, we must solve for f because the kernel K depends on f; we must calculate a new kernel at each iteration step.

Twomey et al.[18] suggest an extension of this method. Consider, for each y_i, the x_i where $K_f(x,y_i)$ is maximum. Then let $K_{max}(y_i)$ be this maximum and define

$$\zeta_{ij} = \frac{K_f(x_j,y_i)}{K_{max}(y_i)}. \tag{137}$$

Then modify equation (136) to be

---

[18] Twomey et al. (1977).

$$A[f^{(k+1)}(x_j)] = \frac{g_i}{g_i^{(k)}} A[f^{(k)}(x_j)]\zeta_{ij} + A[f^{(k)}(x_j)](1 - \zeta_{ij}). \quad (138)$$

That is, at each step, take a single one of the y_j's and calculate f_j for each value of j, not just the one where $K_f(x_j, y_i)$ is the maximum. The sequence of steps becomes

1. Calculate $g_1^{(k)}$.

2. Use equation (136) to update *all* of the f_j's.

3. Calculate the next g_i^k; update all of the x_j's again.

4. When all of the g_i's have been gone through, evaluate the error

5. Either stop or repeat at steps 2 - 5.

Twomey *et al.* add one additional step. At each stage, check $f^{(k)}(y_j)$ to make sure that the value is physically reasonable. For instance, we would reject negative temperatures. Use the old value $f^{(k)}(y_j)$ if $f^{(k+1)}(y_j)$ is not physically reasonable for some j. If none of them is physically reasonable, the iteration is abandoned and declared not to have converged.

The benefit of this method is supposed to be that the requirement that $K(x, y_i)$ have a single peak that is different for each y_i can be relaxed.

11.6.5 Smith's method

Here, W. Smith[19] approximates equation (129) by

$$g_i - g_i^{(k)} = \left\{ A[f] - A[f^{(k)}] \right\} \int_a^b K_f(x, y_i) dx. \quad (139)$$

Define

$$K_i = \int_a^b K_f(x, y_i) dx \quad (140)$$

and $Y_i = y_i / K_i$. Also, assume that $A(f)$ has a known inverse, so that $f = A^{-1}(Y)$. Then

$$f_i^{(k+1)} = A^{-1}\left[Y_i - Y_i^{(k)} + A(f_i^{(k)}) \right] \quad (141)$$

---

[19] Smith (1970), as referenced in Liou, (1980, p 265), which is out of print as of this writing, and is not available in the local university library. See also Houghton *et al.* (1984).

where the notation $f_i^{(k)}$ denotes the value of f found from equation (139) using y_i and y_i^k. It is not necessarily identified with $f^{(k)}(y_i)$ for any particular y_i. However, since we have several estimates for f, each coming from a different peak value of K_f, we can find f from

$$f^{(k+1)}(x_j) = \frac{\sum_i f_i^{(k)} K_f(x_j, y_i)}{\sum_i K_f(x_j, y_i)}.$$

(142)

Exercise 11-9: Is this the same as Twomey's extension of Chahine's method I mentioned in the last section? How do these methods compare?

11.6.6 Expansion in terms of the kernel

Twomey[20] suggests another method, which I shall describe briefly here. To make the notation easier, define[21]

$$K_i(y) = K(x_i, y).$$

(143)

He notes that a solution to equation (120) must be of the form

$$\hat{f}(y) = \sum_j \alpha_j K_j(y) + \phi(y),$$

(144)

where the α_j's are constants that depend on the x_i's and $\phi(y)$ is a function that is orthogonal to every one of the $K_j(x)$, in the sense that

$$\int \phi(y) K_j(y) \, dy = 0$$

(145)

for each value of j. Substitute equation (144) into (120) and define

$$Q_{ij} = \int K_i(y) K_j(y) \, dx.$$

(146)

Then we get a simple matrix equation

$$\mathbf{y} = \mathbf{Q}\alpha.$$

(147)

---

[20] 1977, p 158 *ff.*

[21] Having made the point by now, I shall write simply K instead of K_f.

Since, however, $\mathbf{Q}$ depends on α, this still must be solved by iteration. Note that the component $\phi(y)$, being orthogonal to the kernel at each y_j, does not affect the values of g_i. Twomey remarks that an advantage of this method is that $\phi(y)$ does not enter the solution explicitly. Clearly, if $\hat{f}$ satisfies equation (120), then $\hat{f} + \gamma\phi$ satisfies it for any real number γ. The most parsimonious solution is the one that sets $\gamma = 0$. He also remarks that the matrix $\mathbf{Q}$ will be just as ill-conditioned as the inverse problems we have discussed earlier.

Since $\mathbf{Q}$ must be updated at each step of the iteration, it seems natural to use one of the regularization procedures outlined in the last section. Exactly how to do this is, however, left to the reader.

11.7 Discussion

I think that it is fair to say that, as far as linear problems are concerned, we have shown that iteration and matrix-inverse methods are equivalent, in that they will give the same results, so there is no advantage of using one or the other, except that the matrix-inverse method is easier to apply and to analyze.

Given a more general, but linear, integral equation, we can solve it either by finding the eigenvalues and eigenfunctions of the kernel, or by iteration. In general, it is not possible to determine the eigenvalues and eigenfunctions analytically, nor is it possible to calculate them with much precision. So our discussion of the eigenfunction-expansion method I have described here may be enlightening, but may not be practical to use. If we approximate the integral equation by an quadrature formula, we get back a matrix problem, and we already know how to solve them. We can solve linear integral equations by iteration, but, if we are going to use a quadrature formula, this reduces to the equivalent matrix-iteration method.

There are straightforward ways to solve linear integral equations. There is no general method for solving nonlinear equations, but I have described some methods that we can use if the equations are only mildly nonlinear.

References

Abramowitz, M., and I. Stegun, *Handbook of Mathematical Functions*, Dover, New York, (1968).

Chahine, M. T., 'Inverse problems in radiative transfer: determination of atmospheric parameters,' *J. of the Atmospheric Sciences*, **27**, 960-967 (1970).

Defrise, C., and C. de Mol, 'A note on stopping rules for iterative regularization methods and filtered SVD,' in *Inverse Problems: An Interdisciplinary Study*, P. C. Sabatier, (Ed.), Academic Press, London, pp. 261-268 (1987).

Delves, L. M., and J. L. Mohamed, *Computational Methods for Integral Equations*, Cambridge University Press, Cambridge, 376 pp (1985).

Fleming, H. E., 'Comparison of linear inversion methods by examination of the duality between iterative and inverse matrix methods,' in *Inversion Methods in Atmospheric Remote Sensing, a workshop held at Langley Research Center, Hampton Virginia, December 15-17, 1976*, NASA Conference Publication CP-004 (1976).

Groetsch, C. W., *The Ttheory of Tikhonov Regularization for Fredholm Equations of the First Kind*, Pitman Advanced Publishing Program, Boston, 104 pp (1984).

Houghton, J. T., F. W. Taylor, and C. D. Rogers, *Remote Sounding of Atmospheres*, Cambridge University Press, London (1984).

Liou, Kuo-Nan, *An Introduction to Atmospheric Radiation,* Academic Press, New York (1980).

Peitgen, H.-O., H. Jürgens, and D. Saupe, *Chaos and Fractals: New Frontiers of Science*, Springer-Verlag, New York (1992).

Press, W. H., S. A. Teukolsky, W. T. Vettering, and B. R. Flannery, *Numerical Recipes in Fortran, the Art of Scientific Computing*, Second Edition, Cambridge University Press, Cambridge (1992).

Sabatier, P. C., 'Introduction and a few questions,' in P.C. Sabatier (Ed.), *Inverse Problems: An Interdisciplinary Study*, Academic Press, London (1987).

Smith, W., 'Iterative solution of the radiation transfer equation for the temperature and absorbing gas profile of an atmosphere,' *Applied Optics*, **9** (1993).

Smithies, F., *Integral Equations*, Cambridge University Press, London (1958).

Twomey, S., B. Herman, and R. Rabinoff, 'An extension to the Chahine method of inverting the radiative transfer equation,' *J. of the Atmospheric Sciences*, **34**, 1085-1090 (1977).

Twomey, S., *Introduction to the Mathematics of Inversion in Remote Sensing and Indirect Measurements*, Elsevier Scientific Pub. Co., Amsterdam (1977).

Westwater, E. R., and O. N. Strand, 'Inversion Techniques,' in *Remote Sensing of the Troposphere*, ed. V. E. Derr, Wave Propagation Laboratory, ERL, Boulder, CO, National Oceanic and Atmospheric Administration, Chapter 16 (1972).

Wimp, J., *Computation with Recurrence Relations*, Pitman Advanced Publishing Program, Boston, 310 pp (1984).

12. Resolution and Noise

I had a feeling once about Mathematics—that I saw it all. Depth beyond depth was revealed to me—the Byss and Abyss. I saw—as one might see the transit of Venus or even the Lord Mayor's Show—a quantity passing through infinity and changing its sign from plus to minus. I saw exactly why it happened and why the tergiversation was inevitable but it was after dinner and I let it go.

Winston Spencer Churchill (1874-1965)

I derived matrix and integral-equation methods for solving remote sensing problemsn Chapters 7 and 10. I have stressed that the solution of the inversion problem depends on what quantity we wish to minimize. Both of these methods employed the device of minimizing the error in the estimated values of the parameters being retrieved. Although minimizing the error is an obvious choice, it is not the only reasonable one. In this chapter, I want to investigate another possibility, that of trading resolution for the amplification of the noise in the retrieval. I shall quantify both of these notions presently.

Instead of minimizing the r.m.s. error in the retrieval, we can consider a quantity called the *spread* of the solution. Suppose we want to retrieve $f(y)$ from measurements of a related quantity $g(x)$. We would like the estimated value of f, denoted by $\hat{f}$ to have the property that, to the extent possible, the errors at different values of y to be independent: *i.e.*, $\hat{f}(y_i) - <f(y_i)>$ to be independent of $\hat{f}(y_j) - <f(y_j)>$ for $i \neq j$. The spread is the minimum distance between y_i and y_j that is needed to ensure this independence. It is also called the *resolution* of the inversion.

For some applications, minimizing the spread may be more appropriate than minimizing the r.m.s. error. How this can be done is the subject of this chapter.

There is an important, and quite general, result that is derived in §12.1.5. We must consider both the spread in the retrieval and the effects of noise. We shall see that there is always a tradeoff between the spread and the noise amplification: making the spread smaller necessarily increases the effects of noise, and *vice versa*. There is no way to minimize both terms simultaneously. As you might expect, everything has its cost, even in remote sensing. I start here with an explanation of the method and show how it can be generalized. As is commonly the case, we can work either with matrix equations or integral equations. I shall treat both of them in turn.

12.1 Matrix approach

Backus and Gilbert, in a series of three articles,[1] discuss the following method of solving inverse problems that arise in studies of the structure of the earth. Their articles are quite long and involve some rather advanced mathematics. I suspect that relatively few people have read them, although it would be well worth the effort. The first one cited starts, "Any single number which describes some property of the whole Earth will be called a gross datum of the Earth...." They look like heavy reading: it is tempting to rely on secondary sources in cases like these. However, having read the original papers in preparation for writing this chapter, I have come to see that the secondary sources that I am familiar with do not really do them justice. While Backus and Gilbert do describe another way to invert remote-sensing data, they also provide some important mathematical insight into necessary limitations of any inversion method. The secondary authors that I know of who describe Backus and Gilbert's method either do not give the solution, or they simply state the result without proof, claiming that it is obvious. While I would encourage the reader to consult the original papers for themselves, I will explain the method, demonstrate the solution, and discuss some of its implications.

Backus and Gilbert treat the same integral equation that I discussed in Chapter 10. There is a known kernel $K(x,y)$ that relates the functions f and g by

$$g(x) = \int_a^b K(x,y) f(y)\, dy. \tag{1}$$

As before, I will first discuss an ideal, noiseless case before investigating the effects of noise, in order to make the exposition easier to follow.

To specify the problem a bit further, suppose that there is a (finite) number of different values of x, x_k, $k = 1,...,n$, where g is measured. Denote the measured values by

$$g_k = g(x_k), \quad k = 1, \cdots, n. \tag{2}$$

We want to infer the value of $f(y_i)$ for some set $\{y_i\}$ of values of y. One way to do this is to use a suitable linear combination of the g_k's. Let the coefficients be r_{jk}; then the estimated value of f is

$$\hat{f}(y_j) = \sum_{k=1}^n r_{jk} g_k = \sum_{k=1}^n r_{jk} \int_a^b K(x_k,y) f(y)\, dy. \tag{3}$$

---

[1] Backus and Gilbert (1967, 1968, and 1970).

The subscript j on the coefficients r_{jk} shows explicitly that the coefficients depend on the value of y_j. Since K is a known function, we can define a function $A_j(y)$ to be

$$A_j(y) = \sum_{k=1}^{n} r_{jk} K(x_k, y).$$ (4)

Combining these equations, we see that the estimated value of $f(y)$ is related to the true value by

$$\hat{f}(y_j) = \int_a^b A_j(y) f(y) \, dy.$$ (5)

We will require for now that the functions $K(x_k, y)$, $k = 1, ..., n$, be linearly independent. That is, we assume that there is no j such that it is possible to write $K(x_j, y)$ as a linear combination of the other $n - 1$ functions. If it were possible, then we would not need to measure $g(x_j)$; we could have inferred its value from the other measurements. We need these functions to be linearly independent so that the matrix $\mathbf{Q}$ to be introduced presently will be nonsingular. We can relax this assumption when we consider explicitly the noise in the measurements of g.

12.1.1 The spread of A(y)

Obviously, there is an infinite number of possible sets of coefficients r_{jk}, or equivalently, sets of functions $A_j(y)$, that we could use. Which one of them is best? Backus and Gilbert suggest that we minimize the *spread* of $A_j(y)$. This means that, given two adjacent values of y, say y_j and y_{j+1}, we want the estimates $\hat{f}(y_j)$ and $\hat{f}(y_{j+1})$ to be as little correlated as possible. Backus and Gilbert define the spread of $A_j(y)$ in several alternative ways; the one that is usually discussed is

$$s_j = 12 \int_a^b (y - y_j)^2 A_j^2(y) \, dy.$$ (6)

Here, the factor 12 is used so that if A is a rectangle function with width w_j and unit area (*i.e.*, height $1/w_j$), centered on y_j, the spread $s_j = w_j$. This is purely a matter of convenience. In general, just as the coefficients r_{jk} are different for the different y_j's, the spread is different also.

The spread s_j is a measure of the *resolution* of the retrieval in the vicinity of y_j. Let us assume for the moment that all of the values of s_j are approximately equal and denote their common value by s. Then we can retrieve $f(y_j)$ and $f(y_{j+1})$ independently if and only if

$$\left| y_j - y_{j+1} \right| > s. \tag{7}$$

If $s = b - a$, we can retrieve, at best, only one average value of f. Otherwise, if the spread is a constant $s < b - a$, we can retrieve, roughly,

$$m = \frac{b-a}{s} \tag{8}$$

independent values of f. By minimizing the spread, we maximize the number of independent pieces of information that we can retrieve.

     If it were possible to choose coefficients so that every A_j was a δ-function, the retrieval would be perfect: we would have

$$\hat{f}(y_j) = \int_a^b \delta(y - y_j) f(y) \, dy = f(y_j) \tag{9}$$

exactly. Needless to say, this will never be the case. Now a δ-function has a spread of zero—it is the epitome of narrowness—so presumably the function that is most like a δ-function will be the most desirable one. We can find the coefficients r_{jk} minimize the spread at y_j as follows.

12.1.2 Coefficients that minimize the spread

First, there is a constraint on the permissible functions $A_j(y)$: it must be *unimodular*. That is, we require that

$$\int_a^b A_j(y) \, dy = 1. \tag{10}$$

This obviously is a necessary condition if the estimate $\hat{f}$ is to make any sense: $A_j(y) = 0$ would minimize s_j, but it would not be a useful solution. To implement this constraint, let

$$u_k = \int_a^b K(x_k, y) \, dy. \tag{11}$$

Then let the vectors $\mathbf{u} = \mathrm{col}(u_1, \ldots, u_n)$ and $\mathbf{r}_j = \mathrm{col}(r_{j1}, \ldots, r_{jn})$. For A_j to be unimodular, it is necessary that[2]

$$\mathbf{r}_j \cdot \mathbf{u} = 1. \tag{12}$$

---

[2] If you have not read Chapter 14 yet, please do so now, so that the following equations will look familiar.

Now define the matrix $\mathbf{Q}_j$ by its components

$$q_{ik} = 12 \int_a^b (y - y_j)^2 \, K(x_i, y) \, K(x_k, y) \, \mathrm{d}y. \tag{13}$$

Although I do not show it explicitly, the coefficients q_{ik} clearly depend on j. Substituting equation (4) into equation (6),

$$A_j^2(y) = \left[\sum_i r_{ji} K(x_i, y) \right]\left[\sum_k r_{jk} K(x_k, y) \right]$$

$$= \sum_{i,k} r_{ji} \, r_{jk} \, K(x_i, y) \, K(x_k, y). \tag{14}$$

Some straightforward algebra then shows that

$$s_j = 12 \int_{-\infty}^{\infty} (y - y_j)^2 \sum_{k,m=1}^n r_{jk} r_{jm} K(x_k, y) K(x_m, y) \, \mathrm{d}y = \sum_{k,m=1}^n r_{ij} \, q_{ik} \, r_{kj}, \tag{15}$$

which we can write in matrix notation as

$$s_j = \mathbf{r}_j \cdot \mathbf{Q}_j \mathbf{r}_j. \tag{16}$$

I have already shown (see §14.3) that the vector $\mathbf{r}_j$ that minimizes s_j, subject to the constraint that $\mathbf{r}_j \cdot \mathbf{u} = 1$, is

$$\mathbf{r}_j = \frac{\mathbf{Q}_j^{-1} \mathbf{u}}{\mathbf{u} \cdot \mathbf{Q}_j^{-1} \mathbf{u}}. \tag{17}$$

Remember that for the inverse of $\mathbf{Q}_j$ to exist, its rows (and columns) must be linearly independent. It was necessary to assume that the functions $A_j(y)$ are linearly independent to ensure that $\mathbf{Q}_j^{-1}$ will exist; we can relax this requirement later. Finally, the minimum value of s_j is

$$s_j = \frac{1}{\mathbf{u} \cdot \mathbf{Q}_j^{-1} \mathbf{u}}. \tag{18}$$

Therefore, we can calculate s_j in a straightforward manner from our knowledge of K and the values of x where we measure g.

12.1.3 Effects of noise

But by now the reader should be getting impatient: "But what about the *noise?*" How does the value of s_j affect the contribution that measurement noise makes to $\hat{f}$? It turns out that we can answer that question quite easily. Let n_k be the random noise component of the measurement of $g(x_k)$. Then the measured values will be

$$g_k = g(x_k) + n_k. \tag{19}$$

Assume that the different noise components are independent with zero mean, and have equal variance σ_N^2. Now equation (3) becomes

$$\hat{f}(y_j) = \sum_k r_{jk}\left[\int_a^b K(x_k,y)f(y)\,dy + n_k\right]$$

$$= \left[\int_a^b A_j(y)f(y)\,dy + \sum_k r_{jk}n_k\right]. \tag{20}$$

If we make many measurements of $g(x)$, keeping each $f(y_j)$ fixed but letting the noise n_j vary randomly each time, the variance of $\hat{f}(y_j)$ due to noise will be

$$\mathrm{var}\left[\hat{f}(y_j)\right] \equiv \chi_f^2 = \left\langle \hat{f}^2(y_j)\right\rangle - \left\langle \hat{f}(y_j)\right\rangle^2$$

$$= \left\langle\left[\int_a^b A_j(y)f(y)\,dy\right]^2\right\rangle + \left\langle \sigma_N^2 \sum_k r_{jk}^2\right\rangle_k - \left\langle\int_a^b A_j(y)f(y)\,dy\right\rangle^2. \tag{21}$$

Note I have made use of the property of the noise that it is uncorrelated with *f*. The first and third terms on the r.h.s. of this equation are constant (since we have fixed *f* for the moment), so the error variance in $\hat{f}$ due to noise is

$$\chi_f^2 = \sigma_N^2 \sum_k r_{jk}^2 = \sigma_N^2\, \mathbf{r}_j \cdot \mathbf{r}_j. \tag{22}$$

I call

$$\alpha_j \equiv \mathbf{r}_j \cdot \mathbf{r}_j \tag{23}$$

noise amplification factors. Because this is an awkwardly long phrase, I shall simply denote it by α in the text. Since the spread s_j is different for different values of y_j; α_j may also be different. In general, when all of the components

of $\mathbf{r}_j$ are positive, α_j will be relatively small. However, when some are negative and the others positive, the α_j can be quite large.

(Backus and Gilbert consider the more general case where n_j is correlated
with every other n_k. Then the noise covariance matrix Σ is not diagonal, so
that they would write

$$\chi_f^2 = \langle \mathbf{n} \cdot \Sigma \mathbf{n} \rangle. \tag{24}$$

This makes only a minor difference in the formalism.)

But now we see that different retrieval vectors $\mathbf{r}_j$ will produce different
amounts of noise in $\hat{f}$. Unfortunately, using the vector $\mathbf{r}_j$ that minimizes s_j will
maximize χ_f^2. (I will return to this presently; just take it on faith for the moment.) Choosing a different $\mathbf{r}_j$ that reduces χ_f^2 will increase s_j. Since they represent different physical quantities, there is no single best way to choose $\mathbf{r}_j$.
There is no single quantity that we can minimize to arrive at an optimal solution. However, we can parameterize the problem and examine a *family* of solutions.

To that end, define a generalized width w_j

$$w_j \equiv p\,\mathbf{r}_j \cdot \mathbf{Q}_j\,\mathbf{r}_j + (1-p)\,\sigma_N^2\,h\,\mathbf{r}_j \cdot \mathbf{I}\,\mathbf{r}_j$$

$$= p\,\mathbf{r}_j \cdot \mathbf{Q}_j\,\mathbf{r}_j + (1-p)\,\alpha_j\,\sigma_N'^2, \tag{25}$$

where p is a scalar, $0 \le p \le 1$, and h is a constant that is introduced so that the
units will match. $\mathbf{I}$ is the identity matrix, and note that $\mathbf{r}_j \cdot \mathbf{I}\,\mathbf{r}_j = \alpha_j$. The scalar
$\sigma_N'^2$ incorporates the factor h. Define a matrix

$$\mathbf{T}_j(p) \equiv p\,\mathbf{Q}_j + (1-p)\,\sigma_N'^2\,\mathbf{I}. \tag{26}$$

Then

$$w_j = \mathbf{r}_j \cdot \mathbf{T}_j(p)\mathbf{r}_j, \tag{27}$$

and the constraint $\mathbf{r}_j \cdot \mathbf{u} = 1$ still applies. We already know the solution:

$$w_j(p) = \frac{1}{\mathbf{u} \cdot \mathbf{T}_j^{-1}(p)\mathbf{u}}, \tag{28}$$

and

$$\mathbf{r}_j(p) = \frac{\mathbf{T}_j^{-1}(p)\mathbf{u}}{\mathbf{u} \cdot \mathbf{T}_j^{-1}(p)\mathbf{u}}. \tag{29}$$

I write the solution as $\mathbf{r}_j(p)$ to emphasize the way $\mathbf{r}_j$ depends on p. Note that the matrix $\mathbf{T}_j$ is not singular unless $p = 1$. The spread depends on p in the following manner:

$$s_j(p) = \mathbf{r}_j \cdot \mathbf{Q}_j \mathbf{r}_j = \frac{\mathbf{u} \cdot \mathbf{Q}\left[p\mathbf{Q} + (1-p)\sigma_N'^2 \mathbf{I}\right]^{-2} \mathbf{u}}{\left\{\mathbf{u} \cdot \left[p\mathbf{Q} + (1-p)\sigma_N'^2 \mathbf{I}\right]^{-1} \mathbf{u}\right\}^2}. \tag{30}$$

The upshot of all this is that we can select the value of p that seems to us to be the best tradeoff between good resolution and low noise. Remember that $\mathbf{r}_j(p)$ also depends on which y_j we are centered on. Simply put, when p is large, we will be minimizing the spread s_j; when p is small, we will be minimizing the noise amplification factor α_j. This quantifies the tradeoff between these two aspects of the retrieval.

12.1.4 Tradeoff between resolution and noise

To my way of thinking, the real importance of Backus and Gilbert's work is that they prove that there is indeed a tradeoff between noise and resolution. It certainly seems reasonable that, in order to achieve better resolution, we must pay a penalty in the form of increased noise. But is it true? How can we be sure that there is no vector $\mathbf{r}_j$ that somehow gives us the best of both worlds? Backus and Gilbert proved that there is, in fact, always a tradeoff. I shall treat these questions in this section and the next one.

Suppose we consider different possible values of the spread s_j. There is an infinite number of possible vectors $\mathbf{r}_j$ that will produce a given spread s_j, subject to $\mathbf{u} \cdot \mathbf{r}_j = 1$, provided that s_j is greater than the minimum value found in the last section. We have already seen that the error due to noise is proportional to $\alpha_j = \mathbf{r}_j \cdot \mathbf{r}_j$. Denote the minimum α_j for that spread by $\alpha(s_j)$. Backus and Gilbert have proved that, if $s_j < s_j'$, then the corresponding minimum α_j must satisfy

$$\alpha(s_j) \geq \alpha(s_j'). \tag{31}$$

This means that $\chi_f^2(s_j) \geq \chi_f^2(s_j')$. This theorem is important because it applies quite generally in remote sensing.

We can quantify this notion as follows for the special—but most common—case that the noise covariance matrix $\mathbf{N}$ is $\sigma_N^2 \mathbf{I}$. Equation (25) shows that there is a quantity, w_j, that incorporates both the effects of the spread s_j and noise amplification α. Let us pursue this thought a bit farther; we would like to see if there is a way to analyze the tradeoff between spread and noise. Combine equations (26) and (27):

$$w_j = \frac{1}{\mathbf{u} \cdot \left[p\mathbf{Q}_j + (1-p)\alpha \sigma_N'^2 \mathbf{I} \right]^{-1} \mathbf{u}}. \tag{32}$$

This shows that w_j depends on two terms, where $\mathbf{u} \cdot \mathbf{Q}_j^{-1}\mathbf{u}$ represents the spread s_j and $(1 - p)\alpha\sigma_N'^2$ represents the noise variance. The quantity p represents the relative weight we give to each quantity.

Exercise 12-1: Derive a similar result when $\mathbf{N}$ is not diagonal.

There is a point that I have glossed over: remember, we required that the matrix $\mathbf{Q}_j$ be nonsingular, so that equation (17) will make sense. But suppose, after we have determined the retrieval vector $\mathbf{r}_j$ for a given y_j, we decide to add another measurement at some x_m that is very close to one of the x_j's that we have already used. This added measurement would make $\mathbf{Q}_j$ singular, or nearly so: would this invalidate the method? Certainly, we would expect that adding an additional measurement, even if it is redundant, would still improve the retrieval a little bit, since it would reduce the effect of noise, at least. As we have seen, matrices really span a continuum, from the truly singular, to the grossly ill-conditioned, to moderately ill-conditioned, and finally to well conditioned matrices. Given a matrix based on some physical measurements, it would be surprising to find it to be truly singular, in the strict mathematical sense—even though, for all computational purposes, it might as well be. So: does an additional measurement make $\mathbf{Q}_j$ more ill-conditioned and make the retrieval worse?

Remember that the effect of inverting ill-conditioned matrices is that, generally, the noise amplification increases as the smallest eigenvalue $\to 0$. All of the eigenvalues of $\mathbf{Q}_j$ are non-negative because $s_j = (\mathbf{r}_j \cdot \mathbf{Q}\mathbf{r}_j)^{-1} > 0$. But we derived equation (17) under the assumption that there was no noise. Naturally, in this ideal situation the noise amplification is immaterial. Therefore, adding a redundant measurement has no effect on the quality of the retrieval when there is no noise present.

In the real world, where there is noise, equation (29) is used instead of equation (17). But, as we have seen before, $\mathbf{T}_j$ is always nonsingular (in the loose, computational sense of not being ill-conditioned) as long as it is the sum of a positive scalar times the identity matrix and any symmetric real matrix. Therefore, when we use equation (29) it does not matter whether or not $\mathbf{Q}_j$ is singular. Adding a redundant measurement will only improve the retrieval (as we know it should).

It will be worthwhile to prove that $\mathbf{T}_j$ is nonsingular, although it repeats material covered in Chapter 5. To start, note that $\mathbf{Q}_j$ is a real, symmetric matrix. Therefore, there is a unitary matrix $\mathbf{V}$ that diagonalizes $\mathbf{Q}_j$:

$$\mathbf{V}^t \mathbf{Q}_j \mathbf{V} = \mathbf{D}_j, \tag{33}$$

where $\mathbf{D}_j$ is a diagonal matrix. This corresponds to changing the basis of the vector space to be the eigenvectors of $\mathbf{Q}_j$. Let λ_{ji} be the elements of $\mathbf{D}_j$ that lie on the diagonal (*i.e.* they are the eigenvalues of $\mathbf{Q}_j$). As usual, we assume that the noise covariance matrix is a scalar times the identity matrix, $\sigma_N^2 \mathbf{I}$. Therefore, the sum of the noise variance and the spread [see equation (26)] is

$$\mathbf{T}_j = p \begin{pmatrix} \lambda_{j1} & & 0 \\ & \ddots & \\ 0 & & \lambda_{jn} \end{pmatrix} + (1-p)\sigma_N'^2 \begin{pmatrix} 1 & & 0 \\ & \ddots & \\ 0 & & 1 \end{pmatrix}. \tag{34}$$

In other words, $\mathbf{T}_j$ is diagonal; the non-zero elements of $\mathbf{T}_j$ are[3]

$$t_{ji} = p\lambda_{ji} + (1-p)\sigma_N'^2, \tag{35}$$

while the eigenvalues of $\mathbf{Q}_j$, λ_j can (and do) approach 0, $t_{ji} > 0$. Therefore, $\mathbf{T}_j^{-1}$ exists, is diagonal, and its diagonal elements are

$$\tau_{ji} = \frac{1}{t_{ji}} = \frac{1}{p\lambda_{ji} + (1-p)\sigma_N'^2}. \tag{36}$$

Let me make two comments here:

1. We can pick any value of $s_j \geq s_{min}$ to get the best tradeoff between noise and resolution. Where this point is depends on the nature of each specific problem; there is no general method for finding it.

2. We have not said anything about how we should pick the values of x_j where we will measure $g(x)$, or even how many points there should be. Since measurements always cost money—sometimes a great deal of it—it is important to determine what the smallest number of measurements is that will get the job done.

12.1.5 Resolution vs. noise

So far, we have seen that we can find an inverse to the equation $\mathbf{y} = \mathbf{Ax}$ by minimizing a weighted sum of the spread s and the noise-amplification factor α: we can select the combination that best suits our purposes.[4] The foregoing discussion suggests that there is a tradeoff between making the spread smaller and making the noise-amplification larger, and *vice-versa*, but it doesn't prove

---

[3] We could use a different value of p for each y_j if we wished, but let us just use a constant value here.

[4] For the balance of this section, I shall drop the explicit dependence on j.

it. We need to know that there is no stratagem that will minimize both quantities simultaneously. I shall prove it in this section.

Clearly, when $p = 1$, the spread is minimum, since we have minimized s subject only to the constraint $\mathbf{u} \cdot \mathbf{r} = 1$. Call this value s_{min}; let $\mathbf{r}_0$ be the value found from (17). Remember that this applies only for one particular value of y, which I have heretofore denoted y_j, but it will be less confusing if I drop the explicit dependence on j. Also, remember that $\mathbf{u}$ is fixed by the physics of the problem.

Suppose we pick a different value of $\mathbf{r}$; the spread will be $s = \mathbf{r} \cdot \mathbf{Q} \mathbf{r} > s_{min}$. Introducing the Lagrange multipliers λ and μ, the minimum value of α obtains when[5]

$$\frac{d}{d\mathbf{r}}\left\{\mathbf{r} \cdot \mathbf{r} + \lambda(\mathbf{r} \cdot \mathbf{Q}\mathbf{r} - s) - 2\mu(\mathbf{u} \cdot \mathbf{r} - 1)\right\} = 0. \tag{37}$$

The result is

$$(\mathbf{I} + \lambda\mathbf{Q})\mathbf{r} = \mu\mathbf{u}. \tag{38}$$

Therefore

$$\mathbf{r} = \mu(\mathbf{I} + \lambda\mathbf{Q})^{-1}\mathbf{u}. \tag{39}$$

We evaluate μ by taking the dot product with $\mathbf{u}$ (remember that $\mathbf{u} \cdot \mathbf{r} = 1$)

$$\mu = \frac{1}{\mathbf{u} \cdot (\mathbf{I} + \lambda\mathbf{Q})^{-1}\mathbf{u}} \tag{40}$$

and

$$\mathbf{r} = \frac{(\mathbf{I} + \lambda\mathbf{Q})^{-1}\mathbf{u}}{\mathbf{u} \cdot (\mathbf{I} + \lambda\mathbf{Q})^{-1}\mathbf{u}}. \tag{41}$$

Finally, taking the dot product of $\mathbf{r}$ with each side of equation (38),

$$\alpha = \frac{1}{\mathbf{u} \cdot (\mathbf{I} + \lambda\mathbf{Q})^{-1}\mathbf{u}} - \lambda s. \tag{42}$$

So far, we have eliminated μ. In principle, we could eliminate λ also. However, we cannot do it in the context of a general discussion, but we see that a different spread s corresponds to each value of λ—at least for values of

---

[5] You may want to consult Chapter 14 about Lagrange multipliers and vector derivatives.

λ where s is defined. Therefore, we can consider λ to be a parameter that we can use to select different values of s, and also of α.

This shows that, for each value of $s > s_{min}$, there is a vector $\mathbf{r}$ that minimizes α; we have parameterized this with the parameter $\lambda \geq 0$. We cannot evaluate λ explicitly in closed form, but we can evaluate $s = \mathbf{r \cdot Q}$ for different values of λ and, if necessary, use an iterative procedure to find the λ that corresponds to a given spread s. Compare this with equation (26), where the parameter p is used to give more or less weight to the spread. The value $\lambda = 0$ corresponds to $p = 0$; $p = 1$ corresponds to $\lambda = \infty$. (Note that I have implicitly set the value of the noise to $\sigma'_N{}^2 = 1$. I can always do this by adjusting the units of g appropriately.) We can also use equation (39) to see that

$$\alpha = \frac{\mathbf{u} \cdot (\mathbf{I} + \lambda \mathbf{Q})^{-2} \mathbf{u}}{\left[\mathbf{u} \cdot (\mathbf{I} + \lambda \mathbf{Q})^{-1} \mathbf{u} \right]^2} \tag{43}$$

and

$$s = \frac{\mathbf{u} \cdot \mathbf{Q} (\mathbf{I} + \lambda \mathbf{Q})^{-2} \mathbf{u}}{\left[\mathbf{u} \cdot (\mathbf{I} + \lambda \mathbf{Q})^{-1} \mathbf{u} \right]^2}. \tag{44}$$

When $\lambda = 0$, $\alpha = 1 / |\mathbf{u}|^2$; this is the minimum possible value. We have already derived the value for $\lambda \to \infty$.

We can find the relationship between α and s as follows. First, by combining equations (40) and (42), we get the relationship

$$\mu = \alpha + \lambda s. \tag{45}$$

Second, note that, for any square, nonsingular matrix $\mathbf{M}$,

$$0 = \frac{d}{d\lambda} \mathbf{I} = \frac{d}{d\lambda} \mathbf{M} \mathbf{M}^{-1} = \mathbf{M}^{-1} \frac{d\mathbf{M}}{d\lambda} + \mathbf{M} \frac{d\mathbf{M}^{-1}}{d\lambda}, \tag{46}$$

so

$$\frac{d\mathbf{M}^{-1}}{d\lambda} = -\mathbf{M}^{-2} \frac{d\mathbf{M}}{d\lambda}. \tag{47}$$

Similar (and familiar) rules apply to derivatives of $\mathbf{M}^{-2}$, and so forth. Therefore,

$$\frac{d}{d\lambda} \left[\mathbf{u} \cdot (\mathbf{I} + \lambda \mathbf{Q})^{-1} \mathbf{u} \right] = -\mathbf{u} \cdot \mathbf{Q} (\mathbf{I} + \lambda \mathbf{Q})^{-2} \mathbf{u}. \tag{48}$$

Now using equations (39) and (45), $\mathbf{r} = (\alpha + \lambda s)(\mathbf{I} + \lambda \mathbf{Q})^{-1}\mathbf{u}$, so

$$1 = (\alpha + \lambda s)\mathbf{u} \cdot (\mathbf{I} + \lambda \mathbf{Q})^{-1}\mathbf{u} \qquad (49)$$

so

$$0 = \left(s + \frac{d\alpha}{d\lambda} + \lambda \frac{ds}{d\lambda}\right)\mathbf{u} \cdot (\mathbf{I} + \lambda \mathbf{Q})^{-1}\mathbf{u} - (\alpha + \lambda s)\mathbf{u} \cdot \mathbf{Q}(\mathbf{I} + \lambda \mathbf{Q})^{-2}\mathbf{u}$$

$$= \left(s + \frac{d\alpha}{d\lambda} + \lambda \frac{ds}{d\lambda}\right)\frac{1}{\alpha + \lambda s} - \frac{s}{\alpha + \lambda s}, \qquad (50)$$

Since $s = (\alpha + \lambda s)^2 \mathbf{u} \cdot \mathbf{Q}(\mathbf{I} + \lambda \mathbf{Q})^{-2}\mathbf{u}$ and $(\alpha + \lambda s)^{-1} = \mathbf{u} \cdot (\mathbf{I} + \lambda \mathbf{Q})^{-1}\mathbf{u}$. Finally, we get the simple result that

$$\frac{d\alpha}{d\lambda} = -\lambda \frac{ds}{d\lambda}. \qquad (51)$$

Therefore, making s smaller must make α larger, and *vice versa*.

This is an important result. It says that, for a wide class of problems where there is a tradeoff between resolution s and noise amplification α, the minimum value of s corresponds to the maximum value of α, and that decreasing α necessarily increases s. There is no way that both quantities can be minimized simultaneously.

12.1.6 Application to other formulations

So far, we have been concentrating on the spread s. However, we should note that there is nothing special about the spread as we defined it; s could equally well stand for some other quantity. For example, suppose s represents the fit between f and $\hat{f}$. We might define it to be

$$s = \left\langle \int_{-\infty}^{\infty} \left[\sum_k r_k g(x_k) - f(y)\right]^2 dy \right\rangle, \qquad (52)$$

where the average is taken over realizations of f. We can see that s represents the error we would have had if there were no noise in the measurement of g. It is clear that the error variance of $\hat{f}$, denoted χ_f^2, is the sum of this term and a noise term $\alpha \sigma_N'^2$. In Chapter 7, I discussed inversion from the point of view of minimizing χ_f^2, which is certainly a reasonable choice, but it is not the only choice. We see here that there is a family of possible inverses that give different weight to the noise amplification α and the fit s.

Just as there was a necessary tradeoff between α and s in the last section, there is a tradeoff here. There cannot be an inverse that simultaneously minimizes both quantities.

12.2 Integral-equation approach

The method described in the last section uses a matrix formulation because it envisages that we measure $g(x)$ at only a finite number of different points. While this is certainly not an unreasonable assumption, we can also gain insight into the retrieval problem by looking upon $g(x)$ as a continuous function. One reason for doing this is, as I said before, that it is not clear how the result depends on which x's are in the set $\{x_i\}$ of points where we make measurements of g.

To begin with, consider again the situation where there is no noise in the measurement of $g(x)$. Let $R(x,y)$ be the function we use to estimate the value of f from

$$\hat{f}(y) = \int_a^b R(x,y) g(x) \, dx = \int_a^b \int_a^b R(x,y) K(x,y') f(y') \, dy' \, dx. \quad (53)$$

That is, the sum in equation (3) has been replaced by an integral, representing the continuous limit; define

$$A(y,y') \equiv \int_a^b R(x,y) K(x,y') \, dx, \quad (54)$$

and we note that equation (5) still applies. Here, y denotes the point where we are estimating f, while y' is a variable of integration. As before, A must be unimodular:

$$\int_a^b A(y,y') \, dy' = 1. \quad (55)$$

Define

$$Q(x,x';y) = 12 \int_a^b (y - y')^2 K(x',y') K(x,y') \, dy'. \quad (56)$$

Note that Q is symmetric in x and x', and that it depends (explicitly) on y. The spread is

$$s(y) = 12 \int_a^b (y-y')^2\, A^2(y,y')\mathrm{d}\, y'. \tag{57}$$

Using equation (54), this is

$$s(y) = \int_a^b \int_a^b R(x,y)\, R(x',y)\, Q(x,x';y)\mathrm{d}\, x\, \mathrm{d}\, x'. \tag{58}$$

How should we pick R? Again, a good approach is to choose the function that minimizes the spread s. The method for finding this R is similar to taking a derivative of an ordinary function and setting it equal to zero.

If R is the function that minimizes s, then the derivative of s with respect to an infinitesimal change in R must be zero. To do this, replace $R(x,y)$ with $R(x,y) + \varepsilon h(x)$ where ε is any small positive number and h is any function of x other than $h = 0$ everywhere. There could be a different h for each value of y, but it is not denoted explicitly. As we did in the §14.1, we shall use a Lagrange multiplier -2λ and write

$$s(y) = \int_a^b \int_a^b [R(x,y) + \varepsilon h(x)][R(x',y) + \varepsilon h(x')]Q(x,x';y)\,\mathrm{d}x\mathrm{d}x'$$

$$- 2\lambda \left[\int_a^b \int_a^b [R(x,y) + \varepsilon h(x)]K(x,y')\,\mathrm{d}\, x\, \mathrm{d}\, y' - 1 \right]. \tag{59}$$

In this equation, the term in square brackets will always be zero if the constraint (that A be unimodular) is fulfilled.

In order for $A(y,y')$ to remain unimodular, it is necessary that

$$\int_a^b h(x)\, K(x,y)\mathrm{d}\, x = 0. \tag{60}$$

If this were not true, then we could assume, without loss of generality, that $\int_a^b h(x)\, K(x,y)\mathrm{d}\, x = 1$. Then we would retain unimodularity by substituting $\varepsilon(h - R)$ for εh. The result would be the same.

Take the derivative with respect to ε (we neglect terms proportional to ε^2):

$$\left. \frac{\partial s(y)}{\partial \varepsilon} \right|_{\varepsilon=0} = \int_a^b \int_a^b Q(x,x';y)\big[R(x,y)\, h(x') + R(x',y)h(x)\big]\mathrm{d}\, x\, \mathrm{d}\, x'$$

$$- 2\lambda \int_a^b h(x)u(x)\,\mathrm{d}x \;=\; 0, \tag{61}$$

where, by analogy with equation (11),

$$u(x) \;=\; \int_a^b K(x,y')\,\mathrm{d}\,y'. \tag{62}$$

Since Q is symmetric in x and x', the two terms inside the square brackets in equation (61) are the same. Then

$$\left. \frac{\partial s(y)}{\partial \varepsilon} \right|_{\varepsilon=0} \;=\; 2\int_a^b \left[\int_a^b Q(x,x';y)R(x',y)\,\mathrm{d}\,x' + \lambda u(x) \right] h(x)\,\mathrm{d}\,x \;=\; 0. \tag{63}$$

Since h is an arbitrary function, this equation is zero for all possible functions h if and only if

$$\int_a^b Q(x,x';y)R(x',y)\,\mathrm{d}\,x' \;=\; \lambda u(x). \tag{64}$$

As we know, this is subject to the constraint that

$$\int_a^b R(x,y)u(x)\,\mathrm{d}\,x \;=\; 1. \tag{65}$$

So far, so good. But how do we perform the analogue of the matrix inverse to find R?

We can use the method from Chapter 10 for solving integral equations; equation (64) is a Fredholm equation of the first kind. Since Q is symmetric in x and x', it has an (orthonormal) set of eigenvectors $v_i(x',y)$:[6]

$$\int_a^b Q(x,x';y)v_i(x',y)\,\mathrm{d}\,x' \;=\; \mu_i v_i(x,y), \tag{66}$$

where the eigenvalues μ_i are non-negative definite. Expand $u(x)$ in terms of these eigenvectors:

$$u_y(x) \;=\; \sum_i \phi_i v_i(x,y) \tag{67}$$

where

---

[6] See §14.8.

$$\phi_i \;=\; \int_a^b u_y(x)\,v_i(x,y)\,\mathrm{d}x. \tag{68}$$

Similarly, expand R as[7]

$$R(x,y) \;=\; \sum_i \alpha_i v_i(x,y). \tag{69}$$

Then equation (64) becomes

$$\int_a^b Q(x,x';y)\sum_i \alpha_i v_i(x',y)\,\mathrm{d}x' \;=\; \sum_i \alpha_i \mu_i v_i(x,y) \;=\; \lambda \sum_i \phi_i v_i. \tag{70}$$

Consequently, equating the coefficients of each v_i,

$$\alpha_i \;=\; \lambda\frac{\phi_i}{\mu_i}. \tag{71}$$

We can use equation (65) to find the value of λ:

$$\int_a^b \left(\sum_i \alpha_i v_i(x,y)\right)\left(\sum_j \phi_j v_j(x,y)\right)\mathrm{d}x \;=\; \sum_{i,j}\alpha_i \phi_j \delta_{ij} \;=\; 1. \tag{72}$$

Consequently, multiply each side of equation (71) by ϕ_i and sum over i:

$$\lambda \;=\; \frac{1}{\displaystyle\sum_j \frac{\phi_j^2}{\mu_j}}. \tag{73}$$

Therefore,

$$\alpha_i \;=\; \frac{\phi_i}{\displaystyle\sum_j\left(\phi_j^2/\mu_j^2\right)}. \tag{74}$$

To evaluate the spread, use equation (58):

$$s(y) \;=\; \sum_{i,j}\alpha_i \alpha_j \int_a^b\!\!\int_a^b v_i(x,y)\,v_j(x',y)\,Q(x,x';y)\,\mathrm{d}x\,\mathrm{d}x'$$

---

[7] Do not confuse the coefficients α_i here with the noise amplification $\alpha = \mathbf{r}\cdot\mathbf{r}$.

$$= \sum_{i,j} \alpha_i \, \alpha_j \, \mu_j \int_a^b v_i(x,y) v_j(x,y) \, dx. \qquad (75)$$

Using the orthogonality property of the eigenvectors,

$$s(y) = \sum_i \alpha_i^2 \, \mu_i. \qquad (76)$$

We can add the effects of noise amplification easily, in the same manner as we did using the matrix method.

Exercise 12-2: Derive the corresponding result where noise is included.

Finally, the estimated value of f will be

$$\hat{f}(y) = \sum_i \alpha_i \int_a^b v_i(x,y) g(x) \, dx. \qquad (77)$$

Clearly, this integral must be approximated by a sum. In general, in order to evaluate the integral numerically, we must measure g at at least as many points as $v_i(x,y)$ has zeroes. That is, we need to evaluate g for at least one point between each pair of zeroes, more or less. Very often, $v_i(x,y)$ has approximately i zeroes. Therefore, the more terms we keep in the expansion for $\hat{f}$, the greater the number of points where we must measure g.

In addition, since the eigenfunctions are orthonormal, when we add an additional term to the expansion, we do not change the values of the previous coefficients. Therefore, each additional term adds additional information. This gives a method for analyzing how many terms to use, or how many points there are where we should measure g. The matrix method does not provide for an answer that is this simple. As I have said earlier, integral methods are often the best way of thinking about a problem, even though we need to resort to matrix methods in the end.

If $Q(x,x';y)$ depends only on $x-x'$ (and y, of course), we could use the ideas developed in §10.2. There, we found the solution that minimizes the r.m.s. difference between the estimated and true values of $f(y)$. However, we could also adopt the same formalism, but minimize the spread of the estimate instead.

Exercise 12-3: Do what I suggested in the last paragraph.

12.3 Discussion

In this discussion of the possible tradeoff between resolution and noise, we have considered both matrix and integral-equation approaches that both assume that the quantities we are retrieving are values of a single function f evaluated at different values of the independent variable y. The importance of the Backus-Gilbert approach is that it shows explicitly that there is a tradeoff between resolution and noise amplification and resolution: it is not possible to minimize both the spread and the noise simultaneously.

Suppose the independent variables we are retrieving are not values of a single function, but are different physical variables? For example, they could be the values of the sea surface temperature, wind speed, wind direction, ... I treated this problem in earlier chapters, but I did not consider any property of the inverse that is analogous to spread.

Let the independent variables be denoted by p_1', ..., p_m'. In order for this to make sense, we must multiply each p_i' by an appropriate constant h_i so that the physical units of the variables $p_i = h_i p_i'$ are all the same. Having done that, we see that, formally, there is no difference between the variables p_i defined here and the variables f_i as defined above to represent a discrete approximation to the function $f(y)$. Therefore, we should be able to define a generalized "spread" that describes, roughly speaking, how independent the retrieved values will be, and use the formalism of this chapter to trade this spread for noise amplification. While I don't know of anyone that has done this, there is no obvious reason why it should not be done. Other than the inherent difficulty of obtaining eigenfunctions of Q, that is.

References

Backus, G., and F. Gilbert, 'Numerical applications of a formalism for geophysical inverse problem,' *Geophysical Journal of the Royal Astronomical Society*, Vol. 13, pp 247-276 (1967).

Backus, G., and F. Gilbert, 'The resolving power of gross earth data,' *Geophysical Journal of the Royal Astronomical Society*, Vol. 16, pp 169-205 (1968).

Backus, G., and F. Gilbert, 'Uniqueness in the inversion of inaccurate gross earth data,' *Philosophical Transactions of the Royal Society of London*, Series A, Vol. 266, pp 123-192 (1970).

13. Convolution and Images

I do hate sums. There is no greater mistake than to call arithmetic an exact science. There are permutations and aberrations discernible to minds entirely noble like mine; subtle variations which ordinary accountants fail to discover; hidden laws of number which it requires a mind like mine to perceive. For instance, if you add a sum from the bottom up, and then from the top down, the result is always different.

Mrs. La Touche

Remote sensing measurements are often displayed in the form of images; *i.e.*, two-dimensional representations of the data, where the spatial coordinates in the image correspond to spatial coordinates in the scene being viewed (*e.g.*, the surface of the earth), and the brightness of each pixel corresponds to a measured radiant intensity. The resolution of such an image is determined by the nature of the instrument that was used to make the measurements (see Chapter 3).

The resolution determines our ability to discern fine detail in the image; often we want to improve the resolution of an image by manipulating the data in some way. For images that arise in remote sensing, it is usually the case that the image brightness is the convolution of the intensity of the scene being imaged and a function that characterizes the instrument that is used to make the measurements. Improving the resolution involves, essentially, trying to undo that convolution. We shall see that there are limits to how much we can improve the images; as one might expect, improved resolution comes only at the expense of increased noise. Although it might seem that this problem is unique to radiometry, it is also worth considering as an abstract mathematical inversion problem. I discussed the physical principles involved in Chapter 3; here, I shall consider the mathematical problem involved.

In this chapter I shall discuss two possible approaches to improving images. One follows the Fourier transform method of Chapter 10; the other follows the ideas of Chapter 12. I want this chapter to retain the mathematical abstractness of the other parts of the book, so I shall not dwell here on the properties of real radiometers or telescopes.

The mathematical nature of the problem is easily stated. I shall restrict the notation to a one-dimensional image for the sake of simplicity: it is easily ex-

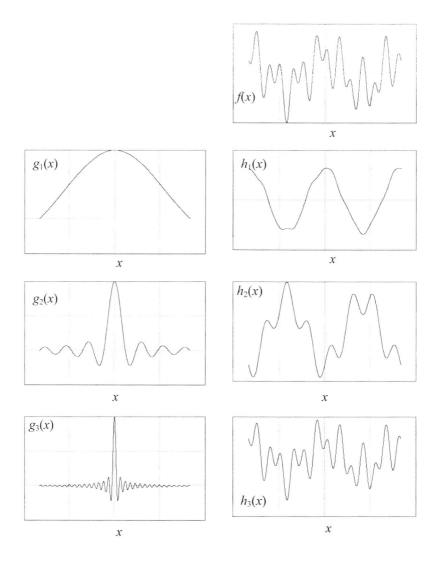

Figure 13-1. Convolution. The curve $f_1(x)$ (*upper right*) is the sum of three co-sines with relative frequencies of 2, 7, and 13. The curves g_1, g_2, and g_3 (*left*) are sinc functions of different widths. The curves on the right are $h_1 = f \star g_1$, $h_2 = f \star g_2$, and $h_3 = f \star g_3$.

tended to two or more dimensions for more realistic problems. First, however, I should mention some general properties of convolutions.

Suppose I have a function $f(x)$. Its spectrum will contain power over a range of frequencies, say, 0 to s_{max}. We say that f is *bandwidth limited*, which means that its spectrum is identically zero for frequencies $|s| > s_{max}$. Now suppose that we *convolve* f with some other function g that is also bandwidth limited: let $h = f \star g$ (*cf.* §8.2.5). In general, this process will remove high frequencies from f. Figure 13-1 shows a function f that is the sum of three co-sines with relative frequencies 2:7:13. The bottom row shows the convolution

$h_1 = f \bigstar g_1$, where g_1 is a narrow sinc function: because g_1 is very narrow, nothing much happens.

The row above this shows the convolution $h_1 = f \bigstar g_2$, where g_2 is wider than g_1. Here, we can see that the highest frequency has been removed from f. The result is a function h_2 that looks smoother than h_1. The row above that shows $h_3 = f \bigstar g_3$: because g_3 is still wider than g_2, h_3 is even smoother than h_2.

What happened to the high-frequency components that were filtered out of f by convolving it with g_2 and g_3? They were partly or completely removed from f. The *information* they contained is reduced or eliminated. This is what happens when we measure the distribution of intensity with a radiometer: the radiometer is only sensitive to a certain range of spatial frequencies, between, say, 0 and s_{max}. The scene itself might contain spatial frequencies higher than s_{max}, but they are attenuated or removed by the radiometer. The question I shall address in this chapter is this: given the convolution of some unknown function f with some known function g (*i.e.*, we measure $h = f \bigstar g$), what can we do to reconstruct the spatial frequencies that were removed from f by the convolution process? We call the process of undoing the effect of convolving f with g as *deconvolution*.

Exercise 13-1: Consider the Gaussian functions $f = \exp\{-a\pi x^2\}$ and $g = \exp\{-b\pi x^2\}$. What is $f \bigstar g$? How wide is it?

Exercise 13-2: We know that the Fourier transform of $\Pi(s)$ is sinc(x), where $\Pi(s) \equiv 1$, for $|s| = \frac{1}{2}$, and $= 0$ otherwise; and sinc$(x) = \sin(\pi x)/\pi x$. Use the similarity theorem (§8.2.3) to plot the Fourier transform of sinc(ax) for different values of a. Use this to show that, as $g_a(x) \equiv$ sinc(ax) get broader, the maximum spatial frequency, s_{max}, decreases.

13.1 Deconvolution in the absence of noise

Suppose that there is a spatially distributed quantity $B(x)$ whose value we want to know at each position x. We measure $T(x)$, which is the response of a measuring instrument—a radiometer or telescope, perhaps—to the field $B(x)$. In the case of interest here, $T(x)$ is the convolution of $B(x)$ with a "power pattern" $P(x)$ that is a fixed property of the instrument used to make the measurements. They are related by a convolution:

$$T(x) = \int_{-\infty}^{\infty} B(u)P(x-u)\mathrm{d}u, \tag{1}$$

Convolution

Scene		Instrument		Measurement
B(*x*)	★	*P*(*x*)	⇒	*T*(*x*)

Deconvolution

Measurement		Instrument		Estimated Image
T(*x*)	?	*P*(*x*)	⇒	*B*(*x*)

Figure 13-2. Convolution (top) and deconvolution (bottom).

or, in more succinct notation,

$$T = B \star P. \tag{2}$$

We measure $T(x)$ and we know $P(x)$: we want to infer the values of $B(x)$.

The power pattern $P(x)$ has two important properties that will affect the problem at hand. First, it is normalized so that

$$\int_{-\infty}^{\infty} P(x)\mathrm{d}x = 1. \tag{3}$$

Second, it is the Fourier transform of a function $p(u)$ that is identically zero outside of a finite interval; for the sake of concreteness, we can take it to be the interval [-1,1]. The *support* of a function is the part of the real line where that function is not identically zero. Here, the support of $p(u)$ is the interval [-1,1].

As an example, we have defined

$$\Pi(x) = 1 \qquad |x| \le \tfrac{1}{2}$$
$$= 0 \qquad |x| \ge \tfrac{1}{2}. \tag{4}$$

Let $p(u) = \Pi(u) \star \Pi(u)$, or $p(u) = 1 - |u|$, $|u| \le 1$, and $p(u) = 0$ otherwise. Then

$$P(x) = \int_{-\infty}^{\infty} e^{-2\pi iux} p(u)\,\mathrm{d}u = \int_{-1}^{1} e^{-2\pi iux}\left(1 - |u|\right)\mathrm{d}u = \operatorname{sinc}^2 x. \tag{5}$$

Any real measuring instrument has a finite aperture—a finite collecting area—and $p(u) = 0$ outside that aperture. Therefore there is no finite interval such that the corresponding power pattern $P(x)$ is identically zero outside of that interval. This property is quite general and it has an important effect on any possible efforts to improve the resolution of images $T(x)$. I show the situation schematically in Figure 13-2. The top panel shows that the measured quantity $T(x)$ is the convolution (★) of the brightness $B(x)$ with the instrument power pattern $P(x)$. The lower panel shows the inverse process: we want to operate on $T(x)$ in some as yet unknown way (?), taking $P(x)$ into account, to obtain the best possible estimate $\hat{B}(x)$ of the original brightness distribution.

In general, when we convolve one function with another one, the result is broader than either function was originally. Suppose that I measure $T(x)$; to what extent can I infer the details of the distribution of $B(x)$, given only the noisy measurements and a (perfect) knowledge of the function $P(x)$? We can use Fourier transforms to find out.

I shall recapitulate the discussion in Chapter 10. Let the Fourier transform of $T(x)$ be

$$t(u) = \int_{-\infty}^{\infty} T(x)e^{-2\pi iux}\,dx. \tag{6}$$

Similarly, let $p(u)$ and $b(u)$ be the Fourier transforms of $P(x)$ and $B(x)$, respectively. Using the convolution theorem (see §8.2.5),

$$t(u) = b(u)\,p(u). \tag{7}$$

Now certainly, if we know $P(x)$, we know $p(u)$—they are presumed to be constant properties that we can measure. We can also Fourier transform the measurements $T(x)$ to find $t(u)$. We could, *in the absence of any noise in the measurement*, find $\hat{b}(u)$ from

$$\hat{b}(u) = \frac{t(u)}{p(u)} \tag{8}$$

and then find $\hat{B}(x)$ from

$$\hat{B}(x) = \int_{-1}^{1} \frac{t(u)}{p(u)} e^{2\pi ixu}\,du. \tag{9}$$

Without noise, we would have undone the smoothing caused by the convolution in equation (1). The limits of integration are ± 1 because we assumed that $p(u) = 0$ outside the interval $(-1,1)$.

Exercise 13-3: Can we really do this? Think about it for a minute before you go on to the next section.

Exercise 13-4: Although I have used the convention that the radius of my aperture has a unit length, in reality, instruments come in different sizes. How does the size of the aperture affect the width of the power pattern $P(x)$? What about the electromagnetic wavelength λ?

13.1.1 Spatial filtering

If we want a physical realization of the deconvolution problem, we can view the convolution relation between T and B in the following manner. Because of the physical situation, the possible forms of $P(x)$ are limited to those that are the Fourier transform of a function that is identically zero outside the interval [-1,1]. Only the low-frequency parts of B are preserved by the convolution process. If x is a spatial coördinate, the Fourier transform of $B(x)$ is a function, $b(u)$, where u has the dimension of inverse length.[1] The variable u is often called a *spatial frequency* (at least, by radio astronomers and radar engineers); it is more often called a *wavenumber*. Either way, the effect of convolving B with P is to remove the higher spatial frequencies. In this sense, an antenna acts as a filter, removing higher spatial frequencies from the scene that it measures.

Schell (1969) made this point explicitly, and it is worth remembering. Regardless of the nomenclature, we can say colloquially that scanning with an antenna or radiometer removes the high-frequency components from a scene; these high-frequency components are gone forever and cannot be recovered. As we shall see presently, this is an overstatement: the higher-frequency components are selectively *attenuated*, and the greater the attenuation, the more noise will be introduced in reconstructing them.

13.2 Deconvolution in the presence of noise

I continue to recapitulate the discussion in Chapter 10 of the effects of instrument noise. Let $N(x)$ be a random variable with mean zero and variance σ_N^2. We measure $T'(x) = T(x) + N(x)$. Instead of equation (1), we have

$$T'(x) = \int_{-\infty}^{\infty} B(u)\,P(x-u)\,\mathrm{d}u + N(x) \tag{10}$$

or [letting $n(u)$ be the Fourier transform of $N(x)$]

$$t'(u) = b(u)p(u) + n(u). \tag{11}$$

Then, instead of equation (9), we would have

$$\hat{B}(x) = \int_{-1}^{1} \frac{[t(u) + n(u)]}{p(u)} e^{2\pi i x u}\,\mathrm{d}u. \tag{12}$$

---

[1] This is necessary because the product ux, which enters the exponential kernel of the Fourier transform, must be dimensionless.

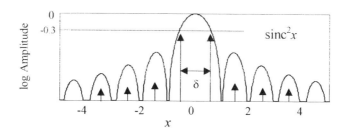

Figure 13-3. Power pattern of a uniformly illuminated aperture, plotted on a logarithmic scale. The nulls of the pattern at $x = \pm 1, 2,$ ...; the side lobes are maximum at $x = \pm 3/2, 5/2, \ldots$.

But $p(u)$ is zero outside $[-1,1]$. As we saw in Chapter 8, we make the smallest r.m.s. error if we approximate $b(u)$ by

$$b(u) = D(u)t(u), \tag{13}$$

where

$$D(u) = \frac{S_B(u)p(u)}{S_B(u)|p(u)|^2 + \sigma_N^2} \tag{14}$$

[compare this with equation (23) in Chapter 10]. Here, $S_B(u)$ is the spectrum of $B(x)$. Then the estimated brightness distribution will be

$$\hat{B}(x) = \int_{-1}^{1} \frac{S_B(u)\,p(u)\,t(u)}{S_B(u)|p(u)|^2 + \sigma_N^2} e^{2\pi iux}\,du. \tag{15}$$

This is one answer, but not the only reasonable one.

13.3 Numerical processing to improve the resolution

In the former discussion, we treated the spatial variable x as being continuous. Here, I shall treat it as being discrete. By this I mean that we have measured the values of $T(x)$ at the points $x_0, x_1, \ldots, x_n$; assume that they are evenly spaced with $x_{n+1} - x_n = d$. Say that $T(x)$ is a discretely-sampled image. Suppose I want to perform some linear transformation on the data to produce a more desirable image. What should I do? How, for that matter, do I define what the "more desirable" image is? In the last section, "more desirable" meant that we reconstructed the high-frequency components as well as possible.

In this section, I address two questions: *How can we reduce the effects of side lobes?* and *Can we improve the resolution of the image?*[2] These concepts both relate to the pattern $P(x)$ of the measuring instrument. For the sake of this argument, assume that $P(x)$ looks somewhat like Figure 13-3. The larger arrows point to the half-power points, or 3-dB points, of the pattern (since $\log_{10}2 = 0.3$ or 3 dB).[3] The full beamwidth δ is also shown. The smaller (vertical) arrows show the positions of the side lobes (see also Chapter 3 for a fuller discussion). The side lobes represent directions where the radiometer is sensitive to incident radiation; this sensitivity is unwanted as it degrades the resultant images.

There are two things to consider that will, in practice, sharply limit our ability to make either of these desired improvements. Firstly, any reduction of the sidelobe level will, in general, amplify the noise in the measurements. Secondly, reduction of the half-power beamwidth will probably increase the side-lobe levels and also increase the noise amplification.

To help us to understand the problem, I want to introduce the concept of an *effective pattern*, $Q(x)$. Suppose we transform the observed data $T(x_i)$ by replacing each value with a sum of the form

$$T'(x_i) = \sum_j m_{ij} T_j \qquad (16)$$

where the constants m_{ij} do not, in fact, depend on i: I shall write just m_j instead of m_{ij}. The resulting image is exactly the same as the one I would have gotten if I had been able to use a different pattern $Q(x)$ instead of $P(x)$.[4] It could turn out that, in an attempt to improve the resolution of the image, we inadvertently introduce large side lobes. We might not know that we had done so if we did not plot the effective pattern. In other words, I shall judge the improvement we have attained by comparing $P(x)$ with $Q(x)$. The advantages of doing this are, first, that I can make the comparison directly, without having to apply it to actual data, and that I can get a good visual impression from plots of $P(x)$ and $Q(x)$. I first dealt with this problem in Milman (1986); I give a simpler treatment here.

In the last section, I used Fourier transforms to process an image; here, I would like to operate on the image directly. One reason for doing this is to illustrate the effective power pattern that results from this method. Remember, however, that anything that can be accomplished in the image domain can also

---

[2] In the jargon of Chapter 3, we want to reduce the sensitivity to signals that are far from the boresight of the antenna and to reduce the half-power beamwidth of the antenna.

[3] 10 decibels (dB) equals a factor of 10.

[4] Some of the patterns $Q(x)$ are physically realizable, in that I could actually build an antenna with that power pattern. Others are not realizable: no real antenna could have that pattern. Either way, plotting $Q(x)$ will help us to understand the properties of the transformed image.

be accomplished in the Fourier transform domain, and *vice versa*. Sometimes, however, working in one domain can be more illuminating than the other.

Let us write the estimated value of B as a linear combination of the measured values $T(x)$:

$$\hat{B}(x) = \sum_k m_k T(x - kd).$$ (17)

Then, using equation (1),

$$\hat{B}(x) = \sum_k m_k \int_{-\infty}^{\infty} B(u) P(x - kd - u) du.$$ (18)

Therefore, if we define

$$Q(x) = \sum_{k=-n}^{n} m_k P(x - kd),$$ (19)

and reverse the order of summation and integration, equation (17) becomes

$$\hat{B}(x) = \int_{-\infty}^{\infty} B(u) Q(x - u) du,$$ (20)

in analogy with equation (1). Since $Q(x)$ must be normalized,

$$\int_{-\infty}^{\infty} Q(x) dx = 1 = \sum_k m_k \int_{-\infty}^{\infty} P(x - kd) dx.$$ (21)

Because of the normalization of $P(x)$,

$$\sum_k m_k = 1.$$ (22)

Therefore, we can always write a linear combination of the T's as the convolution of $T(x)$ with what I call an *effective pattern*, $Q(x)$. Every possible linear combination—any linear estimate $\hat{B}(x)$—corresponds to some effective pattern $Q(x)$. Does every possible $Q(x)$ correspond to a valid estimate of $B(x)$? Obviously not, since $Q(x)$ must be normalized to conserve power. Possible effective power patterns Q are limited by the nature of $P(x)$: since $P(x)$ is the Fourier transform of a function with finite support, it cannot be identically zero outside any finite interval; furthermore, $P(x) \geq 0$ (energy incident on the receiver increases the output; it can't suck power out of the receiver). Therefore, it seems reasonable to suppose that there is some minimum width that $Q(x)$ can have that depends on the width of $P(x)$.

13.3.1 Effects of noise

We shall see that the noise in the measurements limits our ability to make $Q(x)$ as narrow as we might desire. Let σ_N^2 be the r.m.s. level of noise in the measurements of the antenna temperatures. When we include noise we should write

$$\hat{B}(x) = \int_{-\infty}^{\infty} B(u) Q(x - u)\,du + N(x) \tag{23}$$

instead of equation (20), where $N(x)$ is a random variable with variance σ_N^2. Since the values of $N(x)$ are uncorrelated between different measurements,

$$\hat{B}(x) = \int_{-\infty}^{\infty} \sum_k m_k B(u)\, P(x - kd - u)\,du + \sum_k m_k N(x - kd). \tag{24}$$

If we made the same set of measurements over and over with the brightness distribution $B(x)$ unchanged, the variance of $\hat{B}(x_i)$ would be

$$\sigma_Q^2 = \sigma_N^2 \sum_k m_k^2. \tag{25}$$

That is, the noise associated with the effective pattern $Q(x)$ is α times the noise in the original measurements, where the noise amplification factor is

$$\alpha = \sum_k m_k^2. \tag{26}$$

The problem before us is, therefore, how can we choose a set of coefficients m_j, subject to the conditions that α is not too large, and that $\Sigma m_k = 1$.

Exercise 13-5: Show that, if every $m_j > 0$, the half-power beamwidth of Q is greater than that of P.

Exercise 13-6: Show that, if any of the m_j is < 0, then $\alpha > 1$.

Exercise 13-7: Show that, if there are M terms in the summation, the noise is minimum when every $m_k = 1/M$.

Exercise 13-8: $Q(x)$ was defined in equation (19). What is the Fourier transform of Q? Show that $Q(x) = P(x) \bigstar S(x)$, where

$$S(x) \equiv \sum_k m_k \delta(x - kd). \tag{27}$$

13.4 A Backus-Gilbert approach

In Chapter 12, I discussed the Backus-Gilbert approach to solving integral equations of this kind. Let us see how it would work here. To do this, let us introduce a spread function defined by

$$s_n = \int_{-\infty}^{\infty} x^{2n} Q^2(x) \, dx = \int_{-\infty}^{\infty} x^{2n} \left(\sum_k m_k P(x - kd) \right)^2 dx. \tag{28}$$

We might suppose that minimizing the spread will provide the best result. If σ_N^2 is the noise variance in the measurement of each value of T, the noise in

the linear combination $\sum_j m_j T(x - jd)$ will be as given in equation (25). The

parameter n is included in the exponent of x so that we can vary the weight given to the part of $Q(x)$ that is far from the center. Define

$$\chi_n^2(w) = (1 - w)s_n + wh\sigma_N^2, \tag{29}$$

where w is a number in the interval [0,1], and h is a constant chosen so that s_n and $h\sigma_N^2$ have the same units. We can think of $\chi_n^2(w)$ as being a generalized spread function. Different values of w correspond to giving the noise more or less weight relative to the spread. What choice might be appropriate for w will depend on the particular circumstances of the problem at hand.

To deal with the constraint that $\Sigma m_j = 1$, introduce the Lagrange multiplier λ and minimize

$$\left[(1 - w) \int_{-\infty}^{\infty} x^{2n} \left(\sum_k m_k P(x - kd) \right)^2 dx + wh\sigma_N^2 \sum_k m_k^2 + \lambda \left(\sum_k m_k - 1 \right) \right]$$

$$= (1 - w) \sum_{j,k} m_j m_k P_{jk} + wh\sigma_N^2 \sum_k m_k^2 + \lambda \left(\sum_k m_k - 1 \right). \tag{30}$$

In this last equation, I have switched the order of summation and integration and defined the quantities P_{jk} by

$$P_{jk} = \int_{-\infty}^{\infty} x^{2n} P(x - jd)\, P(x - kd)\, dx .$$ (31)

Note that the matrix $\mathbf{P}$, whose elements are P_{jk}, is symmetric and that the values of P_{jk} depend only on $|j - k|$ (for a given value of n).

The parameter w acts as a regularization parameter. If w is very small, little weight is given to the noise amplification and more weight given to making the spread as small as possible; the reverse is true when $w \to 1$.

We can rewrite equation (30) in matrix form. Define χ^2 by

$$\chi^2 = (1 - w)s + wh\sigma_Q^2 = (1 - w)\sum_{j,k} m_j P_{jk} m_k + wh\sigma_N^2 \sum_k m_k^2$$ (32)

or in matrix form,

$$\chi^2 = \mathbf{m} \cdot \left[(1 - w)\mathbf{P} + wh\sigma_N^2 \mathbf{I}\right] \mathbf{m} .$$ (33)

This is subject to the constraint that $\Sigma m_k = 1$. In these terms, we should recognize that this is a problem that we solved in §14.3. Here, the matrix $\mathbf{Q}$ in that section equals $(1 - w)\mathbf{P} + wh\sigma_N^2\mathbf{I}$. Then χ^2 is minimized for

$$\mathbf{m} = \frac{\mathbf{Q}^{-1}\mathbf{1}}{\mathbf{1}\cdot\mathbf{Q}^{-1}\mathbf{1}} = \frac{\left[(1 - w)\mathbf{P} + wh\sigma_N^2\mathbf{I}\right]^{-1}\mathbf{1}}{\mathbf{1}\cdot\left[(1 - w)\mathbf{P} + wh\sigma_N^2\mathbf{I}\right]^{-1}\mathbf{1}} ,$$ (34)

so

$$S(w) = \frac{1}{\mathbf{1}\cdot\mathbf{Q}^{-1}\mathbf{1}} = \frac{1}{\mathbf{1}\cdot\left[(1 - w)\mathbf{P} + wh\sigma_N^2\mathbf{I}\right]^{-1}\mathbf{1}} ,$$ (35)

where $\mathbf{1}$ is the vector $\mathrm{col}(1,\ldots 1)$. Note that $[(1 - w)\mathbf{P} + wh\sigma_N^2\mathbf{I}]$ is positive definite as long as $w < 1$. Because $\mathbf{P}$ is non-negative definite, none of its eigenvalues can be smaller than $wh\sigma_N^2$.

Exercise 13-9: Prove that $\mathbf{P}$ is non-negative definite.

As an example, we can consider the Gaussian power pattern

$$P(x) = \frac{1}{\sqrt{\pi}} e^{-x^2/2} .$$ (36)

This is convenient analytically, but we should be careful not to take it too seriously. The Fourier transform of a Gaussian is also a Gaussian, so the Fourier transform of $P(x)$ is *not* identically zero outside any finite interval. Therefore, this pattern decreases to zero faster with increasing x than any physically real-izable pattern. With this caveat, we can look at the deconvolution problem in detail.

Let the coefficients in equation (31) be

$$P_{jk} = \int_{-\infty}^{\infty} x^{2n} e^{-\frac{1}{2}\left[(x-j)^2 + (x-k)^2\right]} \, dx, \tag{37}$$

where I have allowed for the possibility that we will want to minimize the spread as defined with a factor x^4, x^6, ..., included to change the relative weight given to the center of the pattern. Figure 13-4 shows the results of modifying the power pattern, using 19 terms and spacing the measured values of T at an interval of 0.4. The value of w used was 0.01 in all three cases; the spread being defined relative to x^2, x^4, and x^8, respectively. Even with noise amplification factors of 1.4, 1.8, and 3.1, very little could be done to make the power pattern narrower; *i.e.*, to improve the resolution of the image.

As a second example, consider the power pattern

$$P(x) = \text{sinc}^2 x, \tag{38}$$

(which is the pattern of a uniformly illuminated antenna).[5] Now

$$P_{jk} = \int_{-\infty}^{\infty} x^2 \, \text{sinc}^2 \, \pi(x - j) \, \text{sinc}^2 \, \pi(x - k) \, dx; \tag{39}$$

note that, if we use x^4 instead of x^2 here, the integral will not converge.

Figure 13-5 shows the noise amplification α for different values of the parameter w. This shows the tradeoff between improved resolution and re-duced noise. Figure 13-6 shows the original $\text{sinc}^2 x$ power pattern and the ef-fective patterns for $w = 0.1$ and 0.2, which have $\alpha = 0.30$ and 0.21, respec-tively. We see that the noise is reduced, but the resolution sacrificed, as $w \to 1$.

These results show that there is very little that can be done to improve the resolution of the image given by equation (1). Once the data have been smoothed by performing a convolution like the one in equation (1), the detail that was lost cannot be recovered. More colloquially, you can't unscramble an omelet.

From the Fourier transform viewpoint, we can see why it is rarely possi-ble to make a substantial improvement in the power pattern. Because equation (39) represents a convolution, the Fourier transform $t(u)$ of $T(x)$ is zero wher-

---

[5] This is a more realistic case, since it is easily realized in practice. As a practical matter, an-tenna engineers usually design antennas so that the illumination of the primary reflector de-creases toward the edge, so that the side lobes are lower than this.

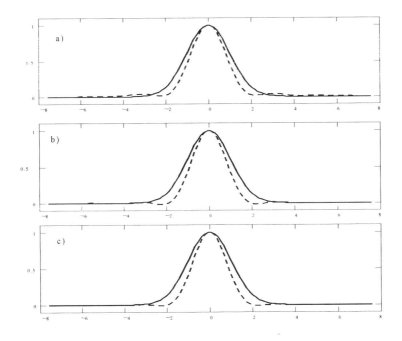

Figure 13-4. The original Gaussian power pattern (solid line) and the syn-
thesized pattern (dotted line). In a), I minimized $x^2 p(x)$; in b), $x^4 p(x)$; and in
c), $x^8 p(x)$. Note the negative side lobes at $x = \pm 2$, more pronounced in a)
than in b), and least-pronounced in c). In a), the noise amplification is 1.4;
in b), it is 1.8; and in c), it is 3.1. $w = 0.01$ in all three cases.

ever the Fourier transform $p(u)$ of $P(x)$ is zero. Remembering that $b(u)$ is the
Fourier transform of the brightness distribution,

$$t(u) \; = \; b(u)\, p(u) \tag{40}$$

implies that the components of $B(x)$ at frequencies where $p(u) = 0$ will be irre-
trievably lost. Since $p(u) = 0$ for $|u| > 1$, we cannot reconstruct those high-
frequency components of $B(x)$. While the analysis in the image domain gives
the appropriate results, and the computations are easier to do than they would
be in the Fourier domain, it is harder to see *why* the deconvolution problem is
usually next to impossible. If we concentrated only on the matrix algebra, we
might not realize what the limitations are to what we can accomplish.

13.5 Selecting a pattern

So far, we have considered only the spread and not the resolution and sidelobe
levels separately. There is yet another way to view this problem. Suppose that
$P(x)$ is the actual power pattern of the instrument, but we wish that it had been
some other pattern $Q(x)$ instead. Is there some way to transform the image so

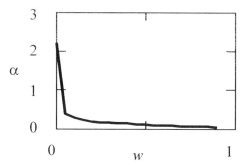

Figure 13-5. Noise amplification factor α for different values of the resolution parameter w and a $\text{sinc}^2 x$ power pattern.

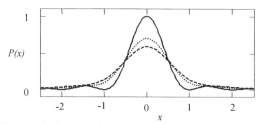

Figure 13-6. Effective patterns for modified $\text{sinc}^2 x$ power patterns; $w = 0.1$ (dotted line) and $w = 0.2$ (dashed line). The solid line shows the original $\text{sinc}^2 x$ power pattern.

that it becomes the image we would have gotten with an instrument that had the power pattern $Q(x)$? We can do that as follows.

Suppose we pick a pattern $Q(x)$ and we want to transform $T(x)$ by

$$T'(x) \equiv \sum_{k=-n}^{n} m_k T(x - kd),$$
(41)

where we assume that the coefficients m_j do not depend on x, and the measurements are made at points separated by d. To accomplish this, we need to pick the values of m_j that minimize

$$\chi^2 \equiv \int_{-\infty}^{\infty} \left[Q(x) - \sum_{k=-n}^{n} m_k P(x - kd) \right]^2 dx$$

$$= \int_{-\infty}^{\infty} Q^2(x) dx - 2 \sum_{k=-n}^{n} m_k \int_{-\infty}^{\infty} Q(x) P(x - kd) dx$$

$$+ \sum_{\substack{j=-n \\ k=-n}}^{n} m_j m_k \int_{-\infty}^{\infty} P(x-jd)P(x-kd)\,dx. \tag{42}$$

Let $Q = \int Q(x)dx$; define P_{ij} according to equation (8) with $n = 0$ and define

$$r_k \equiv \int_{-\infty}^{\infty} Q(x)P(x-kd)\,dx. \tag{43}$$

In vector notation,

$$\chi^2 = Q - 2\mathbf{m}\cdot\mathbf{r} + \mathbf{m}\cdot\mathbf{Pm}. \tag{44}$$

The coefficients m_j must still satisfy equation (22) to preserve the normalization. The noise amplification inherent in applying equation (21) is α defined in equation (26). As before, we want to minimize the quantity

$$(1-w)\chi^2 + wh\sigma_N^2|\mathbf{m}|^2 = (1-w)\left[Q - 2\mathbf{m}\cdot\mathbf{r} + \mathbf{m}\cdot\mathbf{Pm}\right] + wh\sigma_N^2\mathbf{m}\cdot\mathbf{m}$$

$$= (1-w)Q - 2(1-w)\mathbf{m}\cdot\mathbf{r} + \mathbf{m}\cdot\left[(1-w)\mathbf{P} + wh\sigma_N^2\mathbf{I}\right]\mathbf{m} \tag{45}$$

subject to the condition that $\mathbf{m}\cdot\mathbf{1} = 1$, where $\mathbf{1}$ is the vector $(1,1,\ldots,1)$. As before, w parameterizes the relative weight given to the noise amplification and the fit to the pattern $Q(x)$.

It is straightforward, given what we have already done, to evaluate

$$\frac{\partial}{\partial\mathbf{m}}\left\{(1-w)Q - 2(1-w)\mathbf{m}\cdot\mathbf{r} + \mathbf{m}\cdot\left[(1-w)\mathbf{P} + wh\sigma_N^2\mathbf{I}\right]\mathbf{m} - 2\lambda(\mathbf{m}\cdot\mathbf{1} - 1)\right\} = 0$$

$$= -2(1-w)\mathbf{r} + 2\left[(1-w)\mathbf{P} + wh\sigma_N^2\mathbf{I}\right]\mathbf{m} - 2\lambda\mathbf{1} \tag{46}$$

where -2λ is a Lagrange multiplier.[6] Rearranging this,

$$\mathbf{m} = \left[(1-w)\mathbf{P} + wh\sigma_N^2\mathbf{I}\right]^{-1}\left[(1-w)\mathbf{r} + \lambda\mathbf{1}\right]. \tag{47}$$

Since $\mathbf{m}\cdot\mathbf{1} = 1$, we can find λ from

$$1 = (1-w)\mathbf{1}\cdot\left[(1-w)\mathbf{P} + wh\sigma_N^2\mathbf{I}\right]^{-1}\mathbf{r} + \lambda\mathbf{1}\cdot\left[(1-w)\mathbf{P} + wh\sigma_N^2\mathbf{I}\right]^{-1}\mathbf{1}. \tag{48}$$

---

[6] The factor -2 is there for convenience; using -2λ instead of λ makes the algebra slightly simpler.

So

$$\lambda = \frac{1-(1-w)\mathbf{1}\cdot\left[(1-w)\mathbf{P}+wh\sigma_N^2\mathbf{I}\right]^{-1}\mathbf{r}}{\mathbf{1}\cdot\left[(1-w)\mathbf{P}+wh\sigma_N^2\mathbf{I}\right]^{-1}\mathbf{1}}.$$ (49)

This may be the best approach to modifying an image that is the result of a convolution. If you pick $Q(x)$ intelligently, it may be possible to approximate it adequately and still not amplify the noise too much.

Finally,

$$\mathbf{m} = (1-w)\left[(1-w)\mathbf{P}+wh\sigma_N^2\mathbf{I}\right]^{-1}\mathbf{r}$$

$$+\frac{1-(1-w)\mathbf{1}\cdot\left[(1-w)\mathbf{P}+wh\sigma_N^2\mathbf{I}\right]^{-1}\mathbf{r}}{\mathbf{1}\cdot\left[(1-w)\mathbf{P}+wh\sigma_N^2\mathbf{I}\right]^{-1}\mathbf{1}}\left[(1-w)\mathbf{P}+wh\sigma_N^2\mathbf{I}\right]^{-1}\mathbf{1}.$$ (50)

This is somewhat more complicated than we might wish. We will have to evaluate this numerically to get any real insight into how well we can fit a particular pattern $Q(x)$.

Exercise 13-10: Why is it not a good idea to pick $Q(x) = \Pi(x)$? Consider this problem in the Fourier transform domain. How are the Fourier components of $Q(x)$ related to the Fourier components of $P(x)$? What happens if there are high-frequency components in $Q(x)$ that are absent from $P(x)$?

Exercise 13-11: What are the eigenvalues of $[(1-w)\mathbf{P}+wh\sigma_N^2\mathbf{I}]$?

Exercise 13-12: Fix $\alpha = |\mathbf{m}|^2$ and minimize χ^2 subject to the constraints that $\mathbf{m}\cdot\mathbf{1} = 1$ and $|\mathbf{m}|^2 = \alpha$.

Not all possible patterns $Q(x)$ can be represented adequately as a weighted sum of P's. At this point, I encourage the reader who is interested to consider the following problem: suppose that $P(x) = \text{sinc}^2 x$. Consider some different possible forms for $Q(x)$ and calculate a curve of resolution *vs.* noise amplification. You should quickly get a good feeling for what patterns $Q(x)$ are possible and which are not. You should also not be surprised to find out that there is little, if anything, that you can do to improve the resolution of P significantly, although it is possible to reduce the level of the side lobes.

13.6 An integral-equation approach

We can look at this problem from yet another perspective. Suppose that there is a scene $f(x)$, which is a one-dimensional intensity distribution. Suppose that we know *a priori* that f is spatially limited; *i.e.*, there is an interval $[-x_0, x_0]$ such that $f = 0$ outside that interval. We call such a function *space limited*. Let $F(s)$ be the Fourier transform of f, and assume that $F(s)$ has been measured only within the interval $[-s_0, s_0]$. Since f is identically zero outside $[-x_0, x_0]$, its Fourier transform cannot be zero outside any finite interval; however, we have no knowledge of its value outside $[-s_0, s_0]$.

Suppose we measure $g(x)$ with an instrument that acts as a lowpass filter in the frequency domain: it limits our knowledge of $F(s)$ to the interval $[-s_0, s_0]$. In other words, $g(x)$ is bandwidth limited. This is a restatement of the problem we have been considering throughout this chapter, with the added restriction that $f(x)$ is space-limited. The problem is, given the measurement of $g(x)$, what is our best estimate of $f(x)$?

We can look at this as an integral-equation problem.[7] The bandwidth of f is $\Delta s = 2s_0$; we can call $\Delta x = 2x_0$ the *spatial width* of f. The problem can be characterized by the product $c = \Delta s \Delta x$. Since x and s are conjugate variables, xs is dimensionless. Furthermore, the Nyquist sampling interval is $\delta x = 1/\Delta s$. The number of points where we need to specify x, between $-x_0$ and $+x_0$, is $\Delta x / \delta x = \Delta s \Delta x$. For many applications, $c = \Delta s \Delta x \approx 10^3 - 10^4$.

We can write this as the integral equation

$$g(x) = \int_{-\infty}^{\infty} f(u)\mathrm{sinc}(x - u)\,du \tag{51}$$

There are two ways that we can solve equation (51): find its eigenvalues and eigenfunctions, or iteration. I shall describe each method briefly in turn.

13.6.1 Eigenfunctions

We can solve equation (51) by expressing f in a series expansion in the eigenfunctions of $\mathrm{sinc}(x)$. It turns out that the eigenfunctions of the integral equation

$$\lambda_m \phi_m(x) = \int_{-c}^{c} \phi_m(y)\mathrm{sinc}(x - y)\,dy \tag{52}$$

---

[7] Here, I follow the development of Rushforth and Harris (1968), and Rushforth and Frost (1980).

are *prolate spheroidal wave functions.*[8] Both the eigenvalues and the eigen-functions depend explicitly on the space-bandwidth product $c = \Delta s \Delta x$. If we limit our attention to cases where $c \gg 1$, the eigenvalues are[9]

$$1 - \lambda_n \approx \sqrt{\pi} 2^{3n+2} \frac{c^{n+\frac{1}{2}} e^{-2c}}{n!} \left[1 - \frac{6n^2 - 2n + 3}{32c} + O(c^{-2}) \right] \quad (53)$$

and the eigenvalues have the property that $\lambda_n \geq 0$ and

$$1 > \lambda_0 > \lambda_1 > \cdots > 0. \quad (54)$$

While this approach has some interesting possibilities, the prolate spheroidal wave functions are too complicated for this method to be practical.

13.6.2 Iteration

To begin, we need to define two operators. The first is the *bandwidth-limiting* operator B, which operates on $f(x)$ in the following manner. Calculate the $F(s)$, the Fourier transform of $f(x)$. Then, calculate the Fourier transform of $F(s)$ limited to the interval $[-s_0, s_0]$:

$$Bf(x) \equiv f_b(x) = \int_{-\infty}^{\infty} \Pi(s/s_0) e^{2\pi i x s} F(s) \, ds = \int_{-s_0}^{s_0} e^{2\pi i x s} F(s) \, ds \quad (55)$$

Second, define a *space-limiting* operator D, which operates on $f(x)$ and pro-duces $f_s(x) = \Pi(x/2)f(x)$. That is, it sets $f(x)$ to zero outside $[-x_0, x_0]$.

Using the convolution theorem, we can see that, $F(s)\Pi(s/s_0)$ is the Fourier transform of $f(x) \bigstar \text{sinc}(s_0 x) = f_b(x)$[10]. On the other hand, since we assumed that $f(x) = 0$ outside $[-x_0, x_0]$, the operator D does not affect f: $Df = f$. We can write equation (51) as $g(x) = Bf(x)$.[11] Since $Df = f$, $g = BDf$, and

$$g = Df - (I - B)Df = f - (I - B)Df. \quad (56)$$

We can use the iterative algorithm

$$f_{k+1} = D[g + (I - B)Df_k] \quad (57)$$

---

[8] See Matson (1994); Slepian and Pollak (1961); Landau and Pollak (1961); and Slepian and Sonnenblick (1965).

[9] Slepian and Sonnenblick (1965).

[10] Remember that $\bigstar$ denotes convolution and that $\text{sinc} \, x$ is the Fourier transform of $\Pi(x)$.

[11] This section follows Rushforth and Frost (1980), who quote Gerchberg (1974).

to get increasingly accurate estimates of f. We need to interpret this as meaning that first, we calculate $g + (I - B)Df_k$, and then operate on that with D to ensure that f_k is space-limited. The initial estimate is $f_1 = Dg$.

We can make use of the ϕ's without actually calculating them. Let $\phi_n(x)$ be the nth eigenfunction of sinc(x), so

$$\int_{-\infty}^{\infty} \text{sinc}(x - u)\phi_n(u)dx = \lambda_n\phi_n(x) \tag{58}$$

and

$$\int_{-\infty}^{\infty} \phi_n(x)\phi_m(x)dx = \delta_{nm}. \tag{59}$$

Finally, we can write

$$f(x) = \sum_{n=0}^{\infty} \alpha_n\phi_n(x). \tag{60}$$

The eigenfunctions $\phi_n(x)$ depend on $c = \Delta s\Delta x$, and they obey two orthogonality relationships:[12]

$$\int_{-\infty}^{\infty} \phi_n(x)\phi_m(x)dx = \delta_{nm} \tag{61}$$

and

$$\int_{-x_0}^{x_0} \phi_n(x)\phi_m(x)dx = \lambda_n\delta_{nm}. \tag{62}$$

We calculate the coefficients α_n from

$$\alpha_n = \frac{1}{\lambda_n}\int_{-x_0}^{x_0} f(x)\phi_n(x)dx. \tag{63}$$

If we operate on an eigenfunction I - B, we get

$$(I - B)D\phi_n(x) = D\phi_n(x) - \int_{-x_0}^{x_0} \phi_n(u)\text{sinc}(x - u)du = (1 - \lambda_n)D\phi_n(x). \tag{64}$$

If we substitute equation (60) into (51),

---

[12] See Slepian and Pollak (1961).

$$g(x) = \int\limits_{-x_0}^{x_0} f(u)\operatorname{sinc}(x - u)\,du$$

$$= \int\limits_{-x_0}^{x_0} \operatorname{sinc}(x - u)\sum_{n=0}^{\infty} \alpha_n \phi_n(u)\,du = \sum_{n=0}^{\infty} \alpha_n \lambda_n \phi_n(x). \tag{65}$$

Then, if we operate on g with equation (51),

$$f_2(x) = D\left(2g(x) - \int\limits_{-x_0}^{x_0} g(u)\operatorname{sinc}(x - u)\,du\right)$$

$$= D\sum_{n=0}^{\infty} \alpha_n \lambda_n \left[1 + (1 - \lambda_n)\right]\phi_n(x) \tag{66}$$

$$f_3(x) = D\sum_{n=0}^{\infty} \alpha_n \lambda_n \left[1 + (1 - \lambda_n) + (1 - \lambda_n)^2\right]\phi_n(x), \tag{67}$$

and so forth. Remember the equation

$$1 + (1 - \lambda) + (1 - \lambda)^2 + \cdots + (1 - \lambda)^n = \frac{1 - (1 - \lambda)^{n+1}}{\lambda}. \tag{68}$$

Then we can write

$$f_k(x) = D\sum_{n=0}^{\infty} \alpha_n \left[1 - (1 - \lambda_n)^{k-1}\right]\phi_n(x). \tag{69}$$

Therefore, in light of equation (54), $f_k \to f$ as $k \to \infty$. But, since $0 \le \lambda_n < 1$, the nth term in this iterative process converges more slowly than the $(n - 1)^{\text{st}}$. This shows that, even without calculating λ_n or ϕ_n, we can use our knowledge of the eigenvalues to show that this iterative procedure will converge.

Although the prolate spheroidal wave functions are rather complicated, there is a simple, but surprisingly accurate, approximation for the eigenvalues. The eigenvalues depend on c, although I haven't denoted it explicitly. Define the parameter $\hat{b}$ by[13]

---

[13] Slepian and Sonnenblick (1965), who cite Slepian, D., 'Some asymptotic expansions for prolate spheriodal wave functions,' *J. Math. and Phys.*, **44**, 99-140 (1965).

$$\hat{b}_{c,n} = \frac{(n+\frac{1}{2})\pi - 2c}{\gamma + 4\ln 2 + \ln c}, \tag{70}$$

where $\gamma \approx 0.5772156649\ldots$ is Euler's constant, and

$$\lambda_n \approx \frac{1}{1 + e^{\pi \hat{b}_{c,n}}}. \tag{71}$$

When $\hat{b}_{c,n}$ is small, $\lambda_n \approx 1$, while when $\hat{b}_{c,n}$ is large, $\lambda_n \approx 0$; if $\hat{b}_{c,n} = 0$, $\lambda_n = \frac{1}{2}$. So, for a fixed value of n, λ_n goes from 1 to 0 as c increases over a rather small range that depends on n. Another way to look at this is that, since $\hat{b}_{c,n} = 0$ when $(n + \frac{1}{2})\pi = 2c$, there are approximately $2c/\pi$ eigenvalues larger than zero. We can estimate the values of $2c/\pi$ components. However, since $c = \Delta x \Delta s = \Delta x/\delta x = M$, the number of components is approximately $2c/\pi$.

I included this example because it shows how we can use iteration and what we have learned about integral equations to solve a real remote-sensing problem: many of the threads that run through this book converge here. Note that I have not included the effects of noise here; I will leave it to the reader—as a kind of final exam—to answer the following questions:

Exercise 13-13: Suppose we have measured $g(x) = Bf(x) + n(x)$, where n represents Gaussian random noise with variance σ_N^2. What effect does the noise have?

Exercise 13-14: Can we usefully limit the noise amplification by deciding beforehand to iterate only n times?

Exercise 13-15: Could we devise an iteration scheme that converges to $1 / (\lambda + \sigma_N^2)$ instead of $1 / \lambda$?

Exercise 13-16: Are there other ways to solve this problem?

References

Gerchberg, R. W., 'Super-resolution through error-energy reduction,' *Opt. Acta*, **21**, 709-720 (1974).

Landau, H. J., and H. O. Pollak, 'Prolate spheriodal wave functions, Fourier analysis, and uncertainty—II', *Bell System Technical Journal*, **40**, 65-84 (1961).

Matson, C. L., 'Fourier spectrum extrapolation and enhancement using support constraints,' *IEEE Trans. Signal Processing*, **42**, 156-163 (1994).

Milman, A. S., 'Antenna Pattern Corrections for the Nimbus-7 SMMR,' *IEEE Trans. Geosci. & Remote Sensing,* **GE-24**, 212-219 (1986).

Rushforth, C. K., and R. L. Frost, 'Comparison of some algorithms for reconstructing space-limited images,' *J. Optical Society of America,* **70**, 1539-1544 (1980).

Rushforth, C. K., and R. W. Harris, 'Restoration, resolution, and noise,' *J. Optical Society of America,* **58**, 539-545 (1968).

Schell, A. C., 'The antenna as a spatial filter,' in *Antenna Theory*, 2nd ed., R. E. Collin and F. J. Zucker, Eds., New York: McGraw-Hill, pp 577-617 (1969).

Slepian, D., and E. Sonnenblick, 'Eigenvalues associated with prolate spheriodal wave functions of zero order,' *Bell System Technical Journal,* **44**, 1745-1759 (1965).

Slepian, D., and H. O. Pollak, 'Prolate spheroidal wave functions, Fourier analysis, and uncertainty—I', *Bell System Technical Journal,* **40**, 43-63 (1961).

14. Mathematical Appendix

The traditional mathematics professor of the popular legend is absentminded. He usually appears in public with a lost umbrella in each hand. He prefers to face the blackboard and to turn his back to the class. He writes a, he says b, he means c; but it should be d. Some of his sayings are handed down from generation to generation.

"In order to solve this differential equation you look at it till a solution occurs to you."

"This principle is so perfectly general that no particular application of it is possible."

"Geometry is the science of correct reasoning on incorrect figures."

"My method to overcome a difficulty is to go round it."

"What is the difference between method and device? A method is a device which you used twice."

George Polyá (1887- 1985)

I shall discuss several mathematical tools here because they are needed in different parts of the book, but do not fit neatly into any single chapter. One—Lagrange multipliers—is mentioned in most textbooks on advanced calculus, but bears repeating because it is central to one of the retrieval methods I want to discuss. Another—a method for dealing with derivatives of a scalar with respect to the components of a vector—is useful because it will allow us to do some minimization problems in our heads. Other topics include derivation of a matrix identity that we use in Chapter 12; a discussion of singular value decomposition of matrices; and empirical orthogonal functions.

14.1 Constrained extrema

Often, we need to find the extremum of an expression that is subject to some constraint. Suppose we want to find $f(\mathbf{x})$, a scalar function of the n-dimensional vector $\mathbf{x} = (x_1, \cdots, x_n)$, that satisfies certain constraints. Let E^n be the Euclidian vector space that contains these vectors. We want to minimize f, subject to the constraint $g(\mathbf{x}) = 0$, where $g(\mathbf{x})$ is also a scalar function of $\mathbf{x}$. Without the constraint, the extrema of $f(\mathbf{x})$ would be found by taking its derivative with respect to each component of $\mathbf{x}$ and setting each one equal to zero:

$$\frac{\partial f(\mathbf{x})}{\partial x_i} = 0 \qquad i = 1,\cdots,n.$$

In principle, we could use the constraint $g(\mathbf{x}) = 0$ to eliminate one variable from this set of equations: in practice, this is almost always impossible to accomplish. Instead, we use the following method.

The constraint $g(\mathbf{x}) = 0$ defines an n-1-dimensional surface in E^n. For instance, if $n = 2$, the function $g(\mathbf{x}) = x_1^2 + x_2^2 - 1$ defines a circle when $g(\mathbf{x}) = 0$. Suppose $\mathbf{x}_0$ is a point such that $g(\mathbf{x}_0) = 0$. Then let us move away from $\mathbf{x}_0$ by replacing x_1 with $x_1 + dx_1$, where dx_1 is an infinitesimal quantity. For the constraint to be satisfied, it is necessary that $x_2, \cdots, x_n$ change by an amount such that $g(\mathbf{x})$ is unchanged. If we expand $g(\mathbf{x})$ in a Taylor series and keep only the first-order terms, we get

$$g(\mathbf{x}) = g(\mathbf{x}_0) + \sum_{i=1}^{n} \frac{\partial g(\mathbf{x})}{\partial x_i} dx_i = 0. \qquad (2)$$

But since $g(\mathbf{x}_0) = 0$,

$$\sum_{i=1}^{n} \frac{\partial g(\mathbf{x})}{\partial x_i} dx_i = 0. \qquad (3)$$

Now it may seem plausible that we can satisfy both equations (1) and (3) by solving the set of equations

$$\frac{\partial f(\mathbf{x})}{\partial x_i} + \lambda \frac{\partial g(\mathbf{x})}{\partial x_i} = 0, \qquad i = 1,\cdots,n$$

$$g(\mathbf{x}) = 0. \qquad (4)$$

Here, I have introduced a new variable λ (called the *Lagrange*[1] *multiplier*).[2] Although we have formally added another variable, when this set of equations is satisfied, the value of λ itself will be unimportant—after all, it multiplies a quantity that we know must be zero. Another way to look at this is to realize that finding an extremum of f subject to $g = 0$ is equivalent to finding an ex-

---

[1] Joseph-Louis Lagrange (1736-1813) was a French mathematician. In 1788, he published *Méthode Analytique*, a textbook on celestial mechanics that laid the basis for all future work in mechanics. He lived through turbulent times. 'The revolution that began in 1789 pressed Lagrange into work on the committee to reform the metric system and then into teaching. When the great chemist Antoine-Laurent Lavoisier was guillotined, Lagrange commented: "It required only a moment to sever that head, and perhaps a century will not be sufficient to produce another like it"' (Encyclopædia Britannica CD ROM98).

[2] See, for example, Menke (1984, Appendix A.)

tremum of $f(\mathbf{x}) + \lambda g(\mathbf{x})$. This is plausible, but it requires proof. I offer a proof at the end of this section.

Let me illustrate this with a simple example. Let

$$f(x_1, x_2) = \alpha_1 x_1^2 + \alpha_2 x_2^2; \tag{5}$$

we want to find constants x_1 and x_2 that minimize the value of f, with the constraint that $x_1 + x_2 = 1$; *i.e.*, $g(x_1, x_2) \equiv x_1 + x_2 - 1 = 0$. (Without a constraint like this, of course, f is minimized by taking $x_1 = x_2 = 0$, which is not a very interesting or useful result.) This leads to three equations:

$$\frac{\partial}{\partial x_i}(f + \lambda g) = 2\alpha_i x_i + \lambda = 0 \qquad i = 1, 2 \tag{6}$$

and

$$x_1 + x_2 = 1. \tag{7}$$

The solution is

$$x_1 = \frac{\alpha_2}{\alpha_1 + \alpha_2} \quad and \quad x_2 = \frac{\alpha_1}{\alpha_1 + \alpha_2}. \tag{8}$$

As another example, suppose that we are given a function $q(x) > 0$ for x in [0,1], and $1/q(x)$ is absolutely integrable over that interval. We wish to find the function $f(x)$ that minimizes the quantity Q given by

$$Q = \int_0^1 f^2(x)q(x)dx, \tag{9}$$

subject to the constraint that

$$\int_0^1 f(x)dx = 1. \tag{10}$$

We will encounter a problem like this in Chapter 10. Now if the constraint equation (10) is satisfied,

$$Q = \int_0^1 f^2(x)q(x)dx + \lambda\left(\int_0^1 f(x)dx - 1\right). \tag{11}$$

The function $f(x)$ that minimizes it has the property that a small perturbation at any point x will not change the value of Q. Therefore, consider a small number $\varepsilon > 0$ and a function $h(x)$ that is arbitrary except that it is not identically zero and it is absolutely integrable over [0,1]. If we replace $f(x)$ with

$f(x) + \varepsilon h(x)$, the value of Q will be unchanged if f is indeed the desired function. To determine f, take the derivative of Q with respect to ε and set it equal to zero:

$$\frac{\partial}{\partial \varepsilon} \left\{ \int_0^1 [f(x) + \varepsilon h(x)]^2 q(x)\, dx \;+\; \lambda \left[\int_0^1 [f(x) + \varepsilon h(x)]\, dx \;-\; 1 \right] \right\} \Bigg|_{\varepsilon=0}$$

$$= \int_0^1 [2f(x)h(x)q(x) + \lambda h(x)]\, dx \;=\; 0. \tag{12}$$

Since we shall evaluate this expression in the limit as $\varepsilon \to 0$, we can ignore the term proportional to ε^2.

This equation must be true for any function $h(x)$ whatsoever, subject to the abovementioned limitations. This can only happen if the integrand is identically zero; *i.e.*,

$$2f(x)q(x) + \lambda \;=\; 0. \tag{13}$$

Therefore,

$$f(x) \;=\; -\frac{\lambda}{2q(x)}. \tag{14}$$

Using equation (10),

$$\int_0^1 \frac{dx}{q(x)} \;=\; -\frac{2}{\lambda}. \tag{15}$$

and

$$f(x) \;=\; \frac{1}{q(x) \displaystyle\int_0^1 \frac{dx'}{q(x')}}. \tag{16}$$

Exercise 14-1: Is this a reasonable answer? Can you find some functions $q(x)$ for which you can verify this result?

Exercise 14-2: Take the second derivative of Q and show that the extremum is indeed a minimum.

14.1.1 An application of Lagrange multipliers

Panofsky and Dutton[3] offer an application of Lagrange multipliers that is of interest here for two reasons. In addition to being an illustration of the mathematical method I am discussing here, it also explains an important physical phenomenon: why the distribution of particle speeds in a gas has a Gaussian distribution. It is of interest in remote sensing because the absorption coefficients of gases depend, in part, on the velocity distribution of the gas particles (*cf.* §2.5).

Consider the distribution of particle speeds v. Let $p(v)$ be the probability that a particle has a speed between v and $v + dv$. As for any probability distribution,

$$\int_{-\infty}^{\infty} p(v)\,dv = 1. \tag{17}$$

The *entropy* of a system is a measure of its capacity to do work. The Second Law of thermodynamics states that, in any isolated system, the entropy cannot decrease. Entropy is also a measure of the disorder of a system. As we know from everyday experience, a system will not go spontaneously from a state of disorder to an ordered state. After a party, the living room does not clean itself up. In thermodynamics, the entropy is defined by the relation

$$dS = \frac{dQ}{T}, \tag{18}$$

where S is the entropy; Q the work done by (or on) the system; and T is the temperature. Equation (18) states that an infinitesimal change in entropy corresponds to the change in work divided by the temperature.

Any closed system, left to its own devices, will reach a state where its entropy is maximum. In terms of the particles of a gas, we can write the entropy as being

$$S = -\int_{-\infty}^{\infty} p(v)\ln p(v)\,dv. \tag{19}$$

The gas has a fixed mean velocity; we assume that the gas is not moving with respect to the observer, so the mean velocity is zero. The thermal energy of the gas is

$$E_t = \frac{m}{2}\sum_i v_i^2, \tag{20}$$

---

[3] See Panofsky and Dutton (1984, pp 43-44).

where E_t is the total kinetic energy; m, the mass of each particle; v_i, the speed of the ith particle; and the index i runs over all of the particles in question. Assume that all of the particles have the same mass. Then in terms of probability distributions, the mean energy per particle is proportional to the variance of v:

$$E = \frac{m}{2} \int_{-\infty}^{\infty} v^2 p(v) \, dv . \tag{21}$$

Now suppose we are given a box with a gas inside. Suppose it contains N particles, and the total energy is E_t. What is the distribution of particle speeds? Or what function $p(v)$ will simultaneously maximize the entropy and satisfy the constraints that $\frac{1}{2}mv^2 = E = E_t/N$, and that $\int p(v) \, dv = 1$?

Let λ and ρ be scalars; we need to find the minimum of

$$S^* = \int_{-\infty}^{\infty} \left[p(v) \ln p(v) + \lambda p(v) + \rho v^2 p(v) \right] dv . \tag{22}$$

(The sign of $p \ln p$ is positive because, rather than find the maximum of the entropy S, we find the minimum of $-S$.) The terms in the integrand represent the entropy; normalization of the probability distribution; zero mean v; and the mean energy. We maximize the integral if we maximize the integrand for each value of v. Therefore, taking dS^*/dp,

$$0 = \ln p(v) + 1 + \lambda + \rho v^2 , \tag{23}$$

so

$$p(v) = e^{-\left(1 + \lambda + \rho v^2\right)} . \tag{24}$$

Now substitute $p(v)$ from equation (24) into equations (17) and (21) to eliminate λ and ρ. To do this, we will need to use the formulas

$$\int_{-\infty}^{\infty} e^{-ax^2} \, dx = \sqrt{\pi/a} \tag{25}$$

and

$$\int_{-\infty}^{\infty} x^2 e^{-ax^2} \, dx = \frac{3\sqrt{\pi}}{2} a^{-3/2} . \tag{26}$$

Substituting $p(v)$ into equation (17) yields

$$1 = e^{-(1+\lambda)} \int_{-\infty}^{\infty} \exp\{-\rho v^2\} dv = \sqrt{\frac{\pi}{\rho}}\, e^{-(1+\lambda)}. \tag{27}$$

Similarly, substituting equation (24) into equation (21) gives $\rho = \frac{1}{2}mE$. Therefore, the probability distribution is

$$p(v) = \sqrt{\frac{m}{2\pi E}}\, e^{-mv^2/2E}. \tag{28}$$

This is a Gaussian distribution; it is also one form of the central limit theorem (*cf.* §14.7).

Exercise 14-3: Provide the details that lead to equation (28).

Exercise 14-4: Derive equations (48) and (49), in §2.5.1, from equation (28).

14.1.2 Form of the Lagrange multiplier equation

I have written equations (8) and (4) in somewhat different forms. Suppose that the constraint has the form $g(\mathbf{x}) = 1$. We can write either

$$\frac{\partial f}{\partial x_i} + \lambda \frac{\partial g}{\partial x_i} = 0 \tag{29}$$

or

$$\frac{\partial f}{\partial x_i} + \lambda \frac{\partial}{\partial x_i}(g-1) = 0, \tag{30}$$

which emphasizes the point that λ multiplies a quantity that is necessarily zero. It is obvious that the two equations are equivalent, despite the somewhat different forms.

14.1.3 A proof of the Lagrange multiplier method

There are several ways to show that this method works. What follows is a geometrical interpretation that will provide added insight into Lagrange multipliers. In so doing, I will use the *directional derivative* of a vector function. The material that follows is usually covered in an advanced calculus course.

Let $f(\mathbf{x})$ be a scalar function of the *n*-dimensional vector $\mathbf{x}$. Let $\mathbf{u}$ be some other (*n*-dimensional) vector. Then the derivative of $f(\mathbf{x})$ in the direction $\mathbf{u}$ is

$$\frac{\partial f(\mathbf{x})}{\partial \mathbf{u}} \equiv \underset{\varepsilon \to 0}{Lim} \frac{f(\mathbf{x} + \varepsilon \mathbf{u}) - f(\mathbf{x})}{\varepsilon}. \tag{31}$$

If we expand $f(\mathbf{x})$ in a Taylor series [see equation (2)], we can see that

$$\frac{\partial f(\mathbf{x})}{\partial \mathbf{u}} = \sum_{i=1}^{n} u_i \frac{\partial f(\mathbf{x})}{\partial x_i}. \tag{32}$$

Now the constraint $g(\mathbf{x}) = 0$ defines an $n-1$-dimensional surface in E^n. The unit vector that is normal to that surface is the vector

$$\mathbf{n} = \frac{\nabla g(\mathbf{x})}{|\nabla g(\mathbf{x})|}, \tag{33}$$

where the *gradient* of $g(\mathbf{x})$ is the vector

$$\nabla g(\mathbf{x}) = \sum_{i=1}^{n} \frac{\partial g(\mathbf{x})}{\partial x_i} \mathbf{e}_i, \tag{34}$$

and where the vectors $\mathbf{e}_i$, $i = 1, \cdots, n$, are an orthonormal basis of E^n. Since

$$\mathbf{u} = \sum_{i=1}^{n} u_i \mathbf{e}_i, \tag{35}$$

we can write the directional derivative in the more compact manner

$$\frac{\partial f(\mathbf{x})}{\partial \mathbf{u}} = \mathbf{u} \cdot \nabla f(\mathbf{x}). \tag{36}$$

Consider some vector $\mathbf{u}$ that is tangent to the surface defined by $g(\mathbf{x}) = 0$. Since $\mathbf{n}$ is the unit normal to that surface, it is necessary that

$$\mathbf{u} \cdot \mathbf{n} = \frac{1}{|\nabla g(\mathbf{x})|} \sum_{i=1}^{n} u_i \frac{\partial g(\mathbf{x})}{\partial x_i} = 0. \tag{37}$$

But consider $f(\mathbf{x})$. If $\mathbf{x}$ is a point on the surface, then, to find an extremum of $f(\mathbf{x})$, we must move in a direction tangent to the surface. That is, when $f(\mathbf{x})$ is at an extremum, the directional derivative in any direction perpendicular to $\mathbf{n}$ must be zero. Therefore, for any vector $\mathbf{u}$ perpendicular to $\mathbf{n}$,

$$\frac{\partial f(\mathbf{x})}{\partial \mathbf{u}} = 0. \tag{38}$$

We can write each basis vector $\mathbf{e}_i$ as the sum of a vector parallel to $\mathbf{n}$ and a vector perpendicular to $\mathbf{n}$. The parallel component is $(\mathbf{e}_i \cdot \mathbf{n})\mathbf{n}$; the perpendicular component is

$$\mathbf{w}_i = \mathbf{e}_i - (\mathbf{e}_i \cdot \mathbf{n})\mathbf{n}; \tag{39}$$

since $\mathbf{n}$ is a unit vector, $\mathbf{n} \cdot \mathbf{n} = 1$, and the reader can verify that $\mathbf{w}_i \cdot \mathbf{n} = 0$. The derivative of $f(\mathbf{x})$ in the direction of $\mathbf{w}_i$ is

$$\frac{\partial f(\mathbf{x})}{\partial \mathbf{w}_i} = \mathbf{w}_i \cdot \nabla f(\mathbf{x}) = \left[\mathbf{e}_i - (\mathbf{e}_i \cdot \mathbf{n})\mathbf{n}\right] \cdot \nabla f(\mathbf{x}) = 0. \tag{40}$$

Since

$$\mathbf{e}_i \cdot \nabla f(\mathbf{x}) = \frac{\partial f(\mathbf{x})}{\partial x_i}, \tag{41}$$

$$\mathbf{e}_i \cdot \mathbf{n} = \frac{1}{|\nabla g(\mathbf{x})|} \frac{\partial g(\mathbf{x})}{\partial x_i}, \tag{42}$$

and

$$\mathbf{n} \cdot \nabla f(\mathbf{x}) = \frac{1}{|\nabla g(\mathbf{x})|} \sum_{j=1}^{n} \frac{\partial g(\mathbf{x})}{\partial x_j} \frac{\partial f(\mathbf{x})}{\partial x_j}, \tag{43}$$

we can write

$$\frac{\partial f(\mathbf{x})}{\partial \mathbf{w}_i} = \frac{\partial f(\mathbf{x})}{\partial x_i} + \lambda \frac{\partial g(\mathbf{x})}{\partial x_i} = 0 \tag{44}$$

for $i = 1, \ldots, N$. The parameter λ is given by

$$\lambda = \frac{-1}{|\nabla g(\mathbf{x})|^2} \sum_{j=1}^{n} \frac{\partial g(\mathbf{x})}{\partial x_j} \frac{\partial f(\mathbf{x})}{\partial x_j}. \tag{45}$$

This shows that solving the system of equations (4) is equivalent to finding the extrema of $f(\mathbf{x})$ subject to the constraint $g(\mathbf{x}) = 0$.

14.2 Vector derivatives of scalars

There are several ways to combine vectors and matrices to form scalars. For example, for any vector $\mathbf{v}$, $\mathbf{v} \cdot \mathbf{v} = v^2$ is a scalar. It is often necessary to take partial derivatives of such scalars with respect to the different components of one of the vectors. One way to do this is write the vector or matrix equation in terms of their components and then take the appropriate partial derivatives. The result will be a vector. It turns out that there is a way to take these derivatives by inspection, using rules that are almost the same as the rules of ordinary calculus (and therefore, they are easy to remember). The interested reader can consult Graybill,[4] who also shows how to find derivatives of a scalar with respect to a matrix.

Let us look at the following problem: we are given a vector $\mathbf{a}$ of length n, an $n \times n$ matrix $\mathbf{Q}$, and another vector $\mathbf{u}$, also of length n. Let $\mathbf{Q}$ and $\mathbf{u}$ be given; we want to find the vector $\mathbf{a}$ that simultaneously satisfies certain constraints and minimizes certain quantities. Let s be a scalar that is determined in some manner from $\mathbf{a}$, $\mathbf{Q}$, and $\mathbf{u}$. As an example, suppose that

$$s = \mathbf{a} \cdot \mathbf{u}. \tag{46}$$

Suppose we want to find the vector $\mathbf{a}$ that minimizes s. To do this, we could form the vector

$$\mathbf{x} = \frac{\partial s}{\partial \mathbf{a}} \equiv \mathrm{col}(\partial s / \partial a_1, \cdots, \partial s / \partial a_n). \tag{47}$$

Since

$$\mathbf{a} \cdot \mathbf{u} = \sum_{i=1}^{n} a_i u_i, \tag{48}$$

the vector of derivatives has the components

$$x_i = \frac{\partial s}{\partial a_i} = u_i. \tag{49}$$

Therefore, the simple result is that

$$\frac{\partial s}{\partial \mathbf{a}} = \frac{\partial}{\partial \mathbf{a}}(\mathbf{a} \cdot \mathbf{u}) = \mathbf{u}. \tag{50}$$

This looks just like elementary calculus: for scalars, $(d/dx)\, cx = c$, if c is constant. We should be emboldened to look at other possible scalars. For instance, let

$$s = \mathbf{a} \cdot \mathbf{a}. \tag{51}$$

Here, working things out component by component, we see that

---

[4] See Graybill (1969, §10.8).

$$\frac{\partial}{\partial \mathbf{a}}(\mathbf{a} \cdot \mathbf{a}) = \text{col}(\partial s/\partial a_1, \cdots, \partial s/\partial a_n) = 2\mathbf{a}. \tag{52}$$

This also looks familiar.

Let us try a harder one. Suppose

$$s = \mathbf{a} \cdot \mathbf{Qu}. \tag{53}$$

Then find the vector of derivatives from

$$\frac{\partial s}{\partial a_i} = \frac{\partial}{\partial a_i}\left(\sum_{\substack{k=1 \\ l=1}}^{n\mathclap{n}} a_k\, u_l\, q_{kl}\right)$$

$$= \sum_{l=1}^{n} u_l q_{il}. \tag{54}$$

This is the component form of the product $\mathbf{Qu}$. Therefore,

$$\frac{\partial}{\partial \mathbf{a}}(\mathbf{a} \cdot \mathbf{Qu}) = \text{col}(\partial s/\partial a_1, \cdots, \partial s/\partial a_n) = \mathbf{Qu}. \tag{55}$$

If, on the other hand,

$$s = \mathbf{u} \cdot \mathbf{Qa}, \tag{56}$$

then

$$\frac{\partial s}{\partial a_i} = \frac{\partial}{\partial a_i}\left(\sum_{\substack{k=1 \\ l=1}}^{n\mathclap{n}} a_l\, u_k\, q_{kl}\right) = \sum_{l=1}^{n} u_l q_{kl}. \tag{57}$$

This, in turn, is the component form of $\mathbf{Q^t u}$, so

$$\frac{\partial}{\partial \mathbf{a}}(\mathbf{u} \cdot \mathbf{Qa}) = \text{col}(\partial s/\partial a_1, \cdots, \partial s/\partial a_n) = \mathbf{Q^t u}. \tag{58}$$

Exercise 14-5: Show that $\mathbf{a} \cdot \mathbf{Qu} = \mathbf{u^t Q^t \cdot a}$.

Finally, suppose that

$$s = \mathbf{a} \cdot \mathbf{Qa}. \tag{59}$$

Then

$$\frac{\partial s}{\partial a_i} = \frac{\partial}{\partial a_i}\left(\sum_{\substack{k=1 \\ l=1}}^{n} a_k\, a_l\, q_{kl}\right) = \sum_{l=1}^{n} a_l\, q_{il} + \sum_{k=1}^{n} a_k\, q_{ki}. \qquad (60)$$

Since the first term in this equation is the component form of the vector $\mathbf{Qa}$ and the second term is $\mathbf{Q^t a}$,

$$\frac{\partial}{\partial \mathbf{a}}(\mathbf{a}\cdot\mathbf{Qa}) = \mathrm{col}(\partial s/\partial a_1,\cdots,\partial s/\partial a_n) = \mathbf{Qa} + \mathbf{Q^t a}. \qquad (61)$$

This is not quite the same as the corresponding result for scalars, since a scalar q is symmetric (since it has but one element), while a vector $\mathbf{Q}$ may or may not be. In the case where $\mathbf{Q}$ is symmetric,

$$\frac{\partial}{\partial \mathbf{a}}(\mathbf{a}\cdot\mathbf{Qa}) = \mathrm{col}(\partial s/\partial a_1,\cdots,\partial s/\partial a_n) = 2\mathbf{Qa}. \qquad (62)$$

This exhausts the possible combinations $\mathbf{a}$, $\mathbf{Q}$, and $\mathbf{u}$ that produce scalars. The reader may be interested in seeing if he can carry this idea further.

14.2.1 Derivative with respect to a scalar

Suppose the matrix $\mathbf{T}$ has components $t_{ij}(p)$ that each depend on a scalar p. Then the derivative $\partial \mathbf{T}/\partial p$ is also a matrix whose components are t'_{ij}. Denote this matrix by $\mathbf{T'}$. Let the scalar s be given by $s = \mathbf{a}\cdot\mathbf{Tb}$, where $\mathbf{a}$ and $\mathbf{b}$ are (constant) vectors. Then

$$\frac{\partial s}{\partial p} = \mathbf{a}\cdot\mathbf{T'b}. \qquad (63)$$

This is obvious if we write

$$s = \sum_{i,j} a_i t_{ij} b_j. \qquad (64)$$

14.3 A matrix identity

The following example will be of prime importance in Chapter 12, where I discuss the Backus-Gilbert retrieval method. Given a symmetric, nonsingular matrix $\mathbf{Q}$ and a vector $\mathbf{u}$, find a vector $\mathbf{a}$ such that $s = \mathbf{a}\cdot \mathbf{Qa}$ is minimized subject to the constraint that $\mathbf{a}\cdot \mathbf{u} = 1$. I have just consulted several papers on the subject; the authors all say something to the effect that "… it is obvious that the solution is…." Am I the only one to whom it is not obvious? Probably not.

First, note a few things about this problem. The constraint that $\mathbf{a} \cdot \mathbf{u} = 1$ means, among other things, that $\mathbf{a} = \mathbf{0}$ is not allowed as a solution. Furthermore, $\mathbf{a} \cdot \mathbf{Q} \mathbf{a} > 0$. To see this, just write it out in component notation. Therefore, this minimization problem is well posed.

We can use the method of Lagrange multipliers to solve it as I discussed in §14.1. Taking the derivative of s with respect to each a_i would give us n equations which we can write in vector form as

$$\frac{\partial s}{\partial \mathbf{a}} = 2\mathbf{Q}\mathbf{a} = \mathbf{0},$$ (65)

which could be satisfied if $\mathbf{a} = \mathbf{0}$. However, this would be inconsistent with the constraint that

$$\mathbf{a} \cdot \mathbf{u} - 1 = 0.$$ (66)

Also, we would be in the awkward position of having $n + 1$ equations but only n unknowns. To to solve this problem, introduce a Lagrange multiplier -2λ and write (the reason for the -2 will become obvious in a minute)

$$s = \mathbf{a} \cdot \mathbf{Q} \mathbf{a} - 2\lambda(\mathbf{a} \cdot \mathbf{u} - 1).$$ (67)

Because of the constraint, this does not affect the value of s, regardless of the value of λ. However, it does introduce the additional parameter needed to make the system of equations properly determined. Take the derivative of s in this equation; we can now solve the following system of equations:

$$\frac{\partial s}{\partial \mathbf{a}} = 2\mathbf{Q}\mathbf{a} - 2\lambda\mathbf{u} = \mathbf{0}.$$

$$\mathbf{a} \cdot \mathbf{u} - 1 = 0.$$ (68)

Therefore,

$$\mathbf{Q}\mathbf{a} = \lambda\mathbf{u},$$ (69)

or

$$\mathbf{a} = \lambda\mathbf{Q}^{-1}\mathbf{u}.$$ (70)

Take the dot product of each side with $\mathbf{u}$:

$$1 = \lambda\mathbf{u} \cdot \mathbf{Q}^{-1}\mathbf{u}.$$ (71)

Therefore, $\lambda = 1/(\mathbf{u} \cdot \mathbf{Q}^{-1} \mathbf{u})$ and

$$\mathbf{a} = \frac{\mathbf{Q}^{-1}\mathbf{u}}{\mathbf{u} \cdot \mathbf{Q}^{-1}\mathbf{u}}. \tag{72}$$

Also,

$$s = \mathbf{a} \cdot \mathbf{Q}\mathbf{a} = \frac{1}{\mathbf{u} \cdot \mathbf{Q}^{-1}\mathbf{u}}. \tag{73}$$

We shall use this result in Chapter 12.

14.4 Vector and matrix norms

The norm of a vector is its length. If $\mathbf{x}$ is a vector of length n, its norm is

$$|\mathbf{x}| = \sqrt{\sum_{i=1}^{n} x_i^2} = \sqrt{\mathbf{x} \cdot \mathbf{x}}. \tag{74}$$

Exercise 14-6: Show that $|a\mathbf{x}| = a|\mathbf{x}|$, where a is a scalar.

The norm of an $n \times n$ matrix $\mathbf{M}$ measures the size of the matrix, in the same way that the size of a vector $\mathbf{x}$ is measured by its length, $|\mathbf{x}|$. There are many different definitions of the norm of a matrix; the one I use here is defined to be

$$\|\mathbf{M}\| = \max\left\{ \frac{|\mathbf{M}\mathbf{x}|}{|\mathbf{x}|} \right\}. \tag{75}$$

where the maximum is taken over all (nonzero) n-dimensional vectors $\mathbf{x}$. Therefore

$$|\mathbf{M}\mathbf{x}| \le |\mathbf{x}| \cdot \|\mathbf{M}\|. \tag{76}$$

Note that $\|\mathbf{M}\| \ge 0$; if $\|\mathbf{M}\| = 0$, then $\mathbf{M} = \mathbf{0}$.

Exercise 14-7: Show that $\|a\mathbf{M}\| = |a| \cdot \|\mathbf{M}\|$, where a is a scalar.

Exercise 14-8: Show that $\|\mathbf{I}\| = 1$, where $\mathbf{I}$ is the identity matrix.

14.5 Inequalities

Take any two n-dimensional vectors $\mathbf{x}$ and $\mathbf{y}$, and any scalar λ. Then

$$|\mathbf{x} + \lambda\mathbf{y}|^2 = \sum_{i=1}^{n}(x_i + \lambda y_i)^2 = \sum_{i=1}^{n}x_i^2 + 2\lambda\sum_{i=1}^{n}x_i y_i + \lambda^2\sum_{i=1}^{n}y_i^2 \geq 0 \quad (77)$$

However, this is a polynomial in λ of the form $a\lambda^2 + b\lambda + c$; which is ≥ 0 for all values of λ if and only if $b^2 - 4ac \leq 0$. Therefore, if $|\mathbf{x} + \lambda\mathbf{y}|^2 \geq 0$ for all values of λ, it is necessary that

$$\left(\sum_{i=1}^{n}x_i y_i\right) \leq \left(\sum_{i=1}^{n}x_i^2\right)^{1/2}\left(\sum_{i=1}^{n}y_i^2\right)^{1/2}. \quad (78)$$

This is the Cauchy-Schwarz, or sometimes just the Schwarz, inequality.
   Another version of the Schwarz inequality is

$$\int|f(x)g(x)|dx \leq \left[\int|f(x)|^2 dx \int|g(x)|^2 dx\right]^{1/2}. \quad (79)$$

Equality holds if and only if $f(x) = g(x)$.

Exercise 14-9: Show that equation (79) follows directly from equation (78).

Exercise 14-10: Show that

$$\left|\int_{-\infty}^{\infty}f(x)g(x)dx\right| \leq \int_{-\infty}^{\infty}|f(x)g(x)|dx \leq \left[\int_{-\infty}^{\infty}|f^2(x)|dx\right]^{1/2}\left[\int_{-\infty}^{\infty}|g^2(x)|dx\right]^{1/2}. \quad (80)$$

14.5.1 The Triangle Inequality

Consider $|\mathbf{x} + \mathbf{y}|^2$:

$$|\mathbf{x} + \mathbf{y}|^2 = \sum_{i=1}^{n}x_i^2 + 2\sum_{i=1}^{n}x_i y_i + \sum_{i=1}^{n}y_i^2 \leq \sum_{i=1}^{n}x_i^2 + 2\left|\sum_{i=1}^{n}x_i y_i\right| + \sum_{i=1}^{n}y_i^2$$

$$\leq \sum_{i=1}^{n}x_i^2 + 2\sqrt{\sum_{i=1}^{n}x_i^2}\sqrt{\sum_{i=1}^{n}y_i^2} + \sum_{i=1}^{n}y_i^2 = (|\mathbf{x}| + |\mathbf{y}|)^2. \quad (81)$$

Taking the square root of each side, this proves the triangle inequality

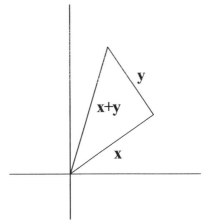

Figure 14-1. The triangle inequality.

$$|\mathbf{x} + \mathbf{y}| \leq |\mathbf{x}| + |\mathbf{y}|. \tag{82}$$

This is called the triangle inequality because it is equivalent to the geometrical proposition that the length of one side of a triangle must be $\leq$ the sum of the lengths of the other two sides. See Figure 14-1.

Exercise 14-11: Show that, for any vector $\mathbf{x}$ of length m and $m \times m$ matrix $\mathbf{M}$,

$$\mathbf{x} \cdot \mathbf{M} \mathbf{x} \leq \|\mathbf{M}\| \mathbf{x} \cdot \mathbf{x}. \tag{83}$$

Exercise 14-12: Supose $\mathbf{M}$ is a square, symmetric, positive-definite matrix. Show that $\|\mathbf{M}\| = \lambda_1$, where λ_1 is the largest eigenvalue, and show that $\|\mathbf{M}^{-1}\| = 1/\lambda_n$, where λ_n is the smallest eigenvalue.

14.6 Two useful relationships from statistics

There are two relationships from statistics that are sometimes useful to know. I shall simply state them here.[5] While they are not used elsewhere in this book, it may help the reader at some time to know of their existence.

1. Let x_1, x_2, x_3, and x_4 be four Gaussian random variables with zero mean. The average value $<x_1 x_2 x_3 x_4>$ is given by

$$\left\langle x_1 x_2 x_3 x_4 \right\rangle = \left\langle x_1 x_2 \right\rangle \left\langle x_3 x_4 \right\rangle + \left\langle x_1 x_3 \right\rangle \left\langle x_2 x_4 \right\rangle + \left\langle x_1 x_4 \right\rangle \left\langle x_2 x_3 \right\rangle. \tag{84}$$

Notice the pattern of the subscripts on the r.h.s.

2. Let z be a real Gaussian random variable with zero mean. Then

---

[5] A good source for a proof is Davenport and Root (1958) or Goodman (1985).

$$\left\langle e^{iz} \right\rangle = e^{-\frac{1}{2}\left\langle z^2 \right\rangle}. \tag{85}$$

14.7 Law of large numbers and the central limit theorem

The normal, or Gaussian, distribution plays an important role in many areas of physics because many different variables have that distribution. We saw in §14.1.1 how to use Lagrange multipliers to show that the particle speeds in a gas are normally distributed. The central limit theorem says that, under rather general conditions, many other random processes are also normally distributed.

To begin, suppose there is a set of n random variables X_1, ..., X_n, that all have the same probability distribution. Suppose that they are also statistically independent. The mean of these variables is

$$S_n \equiv \frac{1}{n} \sum_i^n X_i. \tag{86}$$

Also, suppose that the mean $\mu_i = \langle X_i \rangle$ of each random variable is finite. Then the *Law of large numbers* says that[6]

$$\lim_{n \to \infty} S_n = \lim_{n \to \infty} \frac{1}{n} \sum_{i=1}^n \mu_i. \tag{87}$$

There are three ways to define convergence of random variables.[7] S_n converges *quadratically* if

$$\lim_{n \to \infty} \mathrm{var}\left(S_n - \frac{1}{n} \sum_1^n \mu_i \right) = 0. \tag{88}$$

Alternatively, S_n converges *in probability* if, in the limit $n \to \infty$,

$$\mathrm{Pr}\left(\left| S_n - \frac{1}{n} \sum_1^n \mu_i \right| > \varepsilon \right) = 0, \tag{89}$$

where ε is any small positive number. Here, $\mathrm{Pr}(x)$ means the probability of x. Finally, S_n converges *with probability 1* if, in the limit $n \to \infty$,

---

[6] See, for example, Parzen (1960, p 417 *ff*).
[7] See Parzen (1960, pp 414-415).

$$\Pr\left(S_n - \frac{1}{n}\sum_1^n \mu_i = 0\right) = 1.\tag{90}$$

There are different necessary and sufficient conditions for the law of large numbers to hold, depending on which meaning of convergence we have in mind.

For our purposes, we need only to consider the following case. Suppose that all of the X_i are uncorrelated, have the same (finite) mean, and have uniformly bounded variances. By this, I mean that there exists a constant M_X, such that $\text{var}(X_i) < M_X$ for each $i = 1, ..., n$. Then S_n converges to $(1/n)\Sigma\mu_i$ in all three of the senses defined above.

Consider now a fixed value of n. Suppose that $<X_i> = \mu$ and $\text{var}(X_i) = \sigma^2$ are both finite and independent of i. Furthermore, assume that the variables are statistically independent. Let

$$Z_n = \frac{S_n - n\mu}{\sqrt{n}\sigma}.\tag{91}$$

Then the *Central limit theorem* states that Z_n is normally distributed with variance 1 and mean 0.[8] That is, if I add a large number of random variables that are independent and identically distributed, the sum is normally distributed, regardless of the distributions of the original variables.

A proof of the central limit theorem would take us too far afield, but Panofsky and Dutton[9] give the following example. Suppose that X_i is a random variable that can only take the values 1 and 0, with probabilities p and $q = 1 - p$, respectively. Define S_n according to equation (86). Then $<S_n> = p$.

The variance of S_n is

$$\left\langle\left(S_n - \langle S_n\rangle\right)^2\right\rangle = \left\langle\left(S_n - p\right)^2\right\rangle = \left\langle\left(X_1 - p\right)^2\right\rangle\tag{92}$$

since the variables X_i are uncorrelated, which implies that

$$\text{var}\left(X_1 + \cdots + X_n\right) = n\,\text{var}\left(X_1\right).\tag{93}$$

But

$$\left\langle\left(X_1 - p\right)^2\right\rangle = p\left(1 - p\right)^2 + q\left(0 - p\right)^2;\tag{94}$$

i.e., $X_1 - p = 1 - p$ with probability p, and $X_1 - p = 0 - p$ with probability $q = 1 - p$. Therefore,

---

[8] See Parzen (1960, p 430 *ff*).

[9] See Panofsky and Dutton (1984, p 41 *ff*).

$$\left\langle \left(S_n - \langle S_n \rangle \right)^2 \right\rangle = p(1-p).$$

In the same way, we can evaluate the higher moments of S_n, i.e.,

$$\mu_k \equiv \left\langle \left(S_n - \langle S_n \rangle \right)^k \right\rangle. \tag{95}$$

Now consider the moments of a Gaussian distribution:

$$m_k \equiv \frac{1}{\sqrt{2\pi}\sigma} \int_{-\infty}^{\infty} x^k \, e^{-\frac{1}{2}(x - \langle x \rangle)^2 / \sigma^2} \, dx. \tag{96}$$

If $\mu_k = m_k$ for each $k = 1, ..., \infty$, then the two distributions are the same.

Exercise 14-13: Show that S_n is Gaussian distributed by evaluating the moments μ_n and m_n for $\sigma^2 = p(1 - p)$ and $<x> = p$.

This is an illustration of the central limit theorem. Even though each individual X_i can take on only the values 0 or 1, the mean of n of the X_i's is a normally distributed random variable. There are several versions of the central limit theorem, depending on what assumptions are made about the finiteness of the moments μ_k and the independence of the different random variables X_i. The reader can consult Parzen[10] for statements and proofs of the theorems.

14.8 Hilbert spaces

When we are looking for the solutions to a mathematical problem, we always must keep in mind the questions, "Do solutions necessarily exist?" and "What kind of object might be a solution?" In Chapter 7, the solutions were vectors. In other contexts, the solutions will be functions $f(x)$. Just as vectors are inhabitants of vector spaces, there is a kind of space inhabited by functions. It is in this space that we will find solutions to integral equations—when the solutions exist. I discuss integral equations in Chapter 10. But before we enter this space, let us look at another kind of mathematical problem for a moment.

Consider, for a moment, polynomial equations and the field of rational numbers. Let $p(x)$ be a polynomial of degree n with integer coefficients. In some cases, the equation $p(x) = 0$ will have n solutions that are rational numbers. However, most of these equations will have irrational solutions. The simplest such equation is $x^2 = 2$. Pythagoras—or someone of his school—proved that $\sqrt{2}$ is irrational long ago, much to the dismay of all concerned. So when we look for solutions to polynomial equations, we need a field that in-

---

[10] (1960, Chapter 10).

cludes more than just the rational numbers, even though the coefficients are integers.

Vector spaces like E^n, the Euclidian n-dimensional vector space, should be familiar to the reader. The elements of E^3 are common 3-dimensional vectors that we might envision as arrows of different lengths. Another kind of space that will be useful in Chapter 10 is a *Hilbert space*, which is a space whose elements are complex functions, rather than vectors. We shall see that Hilbert spaces are very much like the vector spaces I discussed in Chapter 5. There is an important difference: while a vector space will usually have a dimension n that is finite, the dimension of a Hilbert space may be infinite.

Throughout this section, $f(x)$ and $g(x)$ will be two complex functions that obey the condition that

$$\int_{-\infty}^{\infty} |f(x)|^2 \, dx < \infty, \tag{97}$$

and similarly for $g(x)$. In other words, they are square integrable. For physical problems, this is not an unreasonable limitation: it is physically impossible to have an infinite amount of anything. Such functions can be elements of a Hilbert space. The definition of a Hilbert space also uses the notion of the *inner product* of two functions.[11]

A *space* is set of objects: vectors, functions, points, other sets, or whatever. For our purposes, we limit ourselves to spaces that have addition and scalar multiplication defined in the usual ways. Also, we will assume that any space is closed, in the sense that, if $\mathbf{x}$ and $\mathbf{y}$ are any two elements of the space and a is any scalar, then $\mathbf{x} + \mathbf{y}$ and $a\mathbf{x}$ are also elements of that space.

14.8.1 Inner product

Just as we have defined the dot—or inner—product of two vectors, we can define the inner product of two complex functions to be

$$(f, g) \equiv \int_a^b f(y) g^*(y) \, dy \tag{98}$$

where $g^*(x)$ is the complex conjugate of $g(x)$. Here, a and b are constants such that $-\infty < a < b < \infty$. The inner product has the properties that

$$(f, g) = (g, f)^*, \tag{99}$$

and that for any complex scalar α,

---

[11] The reader can find a good mathematical introduction to the subject in Young (1988).

$$(\alpha f, g) \; = \; \alpha(f,g) \quad and \quad (f, \alpha g) \; = \; \alpha^*(f,g). \tag{100}$$

The inner product of any function with itself is > 0 unless that function is identically equal to zero:

$$(f,f) \; = \; 0 \; \Rightarrow \; f \; \equiv \; 0. \tag{101}$$

The parallel between the inner product defined here and the dot (or inner) product of two vectors is obvious. If $(f,g) = 0$, the two functions are *orthogonal*.[12] If there is a set of functions $\{\phi_n(x)\}$, it is *orthonormal* if, for any i and j,

$$\left(\phi_i, \phi_j\right) \; = \; \begin{cases} 1 & i = j \\ 0 & \text{otherwise} \end{cases}. \tag{102}$$

If a vector space has dimension n, there are (at most) n vectors that are mutually orthogonal.[13] However, there can be an infinite number of different functions $f_j(x)$ that are mutually orthogonal. In this sense, a Hilbert space has a dimension that is infinite.

Exercise 14-14: Show that, if $f_n(x) = \sin(nx)$, f_n and $f_n{}'$ are orthogonal over the interval $(0, 2\pi)$ if $n \neq n'$.

14.8.2 Function norm

There are many ways to define the *norm* of a function. The norm of a vector is defined in equation (74). It is equal to the length or magnitude of the vector. I should state, for the sake of completeness, that if α is a complex number, then its norm is

$$|\alpha| \; = \; \sqrt{\alpha \alpha^*}. \tag{103}$$

The corresponding norm of a function $f(x)$ is

$$\|f\|^2 \; = \; (f,f) \; = \; \int_a^b f(y)\, f^*(y)\, \mathrm{d}y. \tag{104}$$

There are many other ways to define norms. But any norm of a function of any $f(x)$ must have the following properties:

1. $\|f(x)\| \geq 0$. If $\|f(x)\| = 0$, then $f(x) \equiv 0$.

---

[12] Just as two vectors whose dot product $= 0$ are orthogonal.

[13] *Mutually orthogonal* means that, given a set of vectors $\mathbf{e}_j, j = 1, \cdots, n$, that $\mathbf{e}_i \cdot \mathbf{e}_j = 0$ for $i \neq j$.

2. $\|\alpha f(x)\| = |\alpha| \|f(x)\|$, where α is a scalar.

3. $\|f(x) + g(x)\| \le \|f(x)\| + \|g(x)\|$.

14.8.3 Metric

A *metric* is a measure of the distance between two elements in a space. One common metric is the Euclidean distance between two points in E^n:

$$d(\mathbf{x},\mathbf{y}) = |\mathbf{x} - \mathbf{y}| = \sqrt{\sum_{i=1}^{n}(x_i - y_i)^2}. \qquad (105)$$

There are many other metrics. One example is the so-called *Manhattan metric*

$$d_M(\mathbf{x},\mathbf{y}) = \sum_{i=1}^{n}|x_i - y_i|. \qquad (106)$$

Exercise 14-15: Why is this called the Manhattan metric? (Hint: the streets in Manhattan are laid out in a regular grid pattern.)

Any metric must have the following properties:

1. $d(\mathbf{x},\mathbf{y}) > 0$ if $\mathbf{x} \ne \mathbf{y}$; $d(\mathbf{x},\mathbf{x}) = 0$.

2. $d(\mathbf{x},\mathbf{y}) \le d(\mathbf{x},\mathbf{z}) + d(\mathbf{z},\mathbf{y})$ for any vector $\mathbf{z}$.

This is a different form of the triangle inequality, since if $\mathbf{x}$, $\mathbf{y}$, and $\mathbf{z}$ are vectors in E^3, this is just the statement that the sum of any two sides of a triangle cannot be less than the length of the third side.

Exercise 14-16: Prove that the Euclidean and Manhattan metrics have these properties.

14.8.4 Cauchy sequences

We need one more concept, that of *completeness*. First, we define a *Cauchy sequence* to be a sequence of elements of a metric space $\mathscr{S}$. Let $x_1, x_2, \ldots \in \mathscr{S}$ be an infinite sequence of elements. It is a Cauchy sequence if for any $\varepsilon > 0$, where ε is understood to be a very small positive number, there is a number N such that

$$d(\mathbf{x}_j,\mathbf{x}_k) < \varepsilon, \quad \text{for all} \quad j,k > N. \qquad (107)$$

The sequence x_1, x_2, ... approaches x as a limit if, for every $\varepsilon > 0$, there is a corresponding N such that

$$d\left(x_j, x\right) < \varepsilon \quad \text{for all} \quad j > N. \tag{108}$$

A space $\mathscr{S}$ is *complete* if every Cauchy sequence in $\mathscr{S}$ approaches a limit in $\mathscr{S}$.

Example: The set of rational numbers is also a space. The natural metric is $d(x,y) = |x - y|$. However, this space is not complete. Consider the sequence of numbers 3., 3.1, 3.14, 3.141, 3.1415, ... Each number is one decimal place closer to π. This sequence approaches a limit, but that limit is not a rational number.

Exercise 14-17: Prove that this is a Cauchy sequence and that it converges to π. Does it converge in the space of rational + irrational numbers?

Formally, a Hilbert space is a complete space that has an inner product and a metric defined on it. The very notion of a convergent series requires that there be a metric: otherwise, there would be no way to say whether or not the distance between points is $< \varepsilon$. The point of completeness is crucial to our study of Fourier series and integral equations, because otherwise we would have no assurance that any series would converge to a limit.

In a Hilbert space, complex functions play roles that are analogous to vectors in vector spaces. There is also the analog of a matrix. For every function of two variables $K(x,y)$ there is an operator, denoted simply by K, defined by

$$Kf = \int_a^b K(x,y) f(y) \, dy; \tag{109}$$

therefore

$$(Kf, g) = \int_a^b \int_a^b K(x,y) f(y) g(x) \, dx \, dy. \tag{110}$$

This is similar to $\mathbf{f} \cdot \mathbf{Mg}$, where $\mathbf{M}$ is a matrix.

14.8.5 $L^2[0,1]$

Consider the space of all complex functions that are defined on the interval $[0,1]$ and that are square-integrable; *i.e.*,

$$\int_0^1 |f(x)|^2 \, dx < \infty. \tag{111}$$

The inner product was defined by equation (98); here, $a = 0$ and $b = 1$. Note that these functions do not have to be continuous. This is a Hilbert space that is usually denoted $L^2[0,1]$. The restriction to functions defined on [0,1] is not a real limitation for physical problems, where we can never make measurements over an infinite time or space interval. If a function $g(x)$ is defined on the interval $[a,b]$, where $b > a$, then making the substitution $(x - a) / (b - a) \to x$ produces a new function defined on [0,1].

I want to show that $L^2[0,1]$ is like a vector space, and, in particular, that there is a set of basis functions that have the property that any function $f(x)$ can be represented as a linear combination of these basis functions. Although this discussion could be generalized to any Hilbert space, I will use the notation of complex functions here.

Suppose that there is a complete sequence of orthonormal functions $\{\phi_n(x)\}$, $n = 0, 1, \ldots$ defined on this interval. Here, *complete* means that if there is a function $f(x)$ that is orthogonal to every one of the $\phi_n(x)$, then $f(x) = 0$ everywhere. Furthermore, consider the *complex linear span* of $\{\phi_n(x)\}$, which is defined to be the set of all functions $f_k(x)$ of the form

$$f_k(x) = \sum_{n=1}^{k} \lambda_n \phi_n(x), \tag{112}$$

where the λ_n are complex numbers. There is a theorem[14] that says that if $\{\phi_n(x)\}$ is complete, then the space containing all the $f_k(x)$'s is equal to $L^2[0,1]$.

If the set $\{\phi_n(x)\}$ is complete, then any function can be expressed as a linear combination of the ϕ_j's:

$$f(x) = \sum_i a_i \phi_i(x). \tag{113}$$

In other words, any element of $L^2[0,1]$ is a linear combination of the functions $\{\phi_n(x)\}$.

Since the functions $\phi_n(x)$ are assumed to be orthonormal, the coefficients a_i can be found from

---

[14] See Young (1988) Chapter 4.

$$(f, \phi_i) = \sum_{k=1}^{m} a_i(\phi_k, \phi_i) = a_i. \tag{114}$$

The upshot of all of this is that we can call $\{\phi_n(x)\}$ a *basis* of $L^2[0,1]$ in that any element of $L^2[0,1]$ is a linear combination of $\{\phi_n(x)\}$. This is the basis of a Taylor series or Fourier series representation of a function. Now recall that the basis of an n-dimensional vector space is a set of n orthogonal vectors $\{\mathbf{u}_i\}$ and that any vector in that space can be written as a linear sum of those vectors: *i.e.*, any vector $\mathbf{x}$ is

$$\mathbf{x} = \sum_{i=0}^{n} \alpha_i \mathbf{u}_i. \tag{115}$$

where the α_i's are scalars. Recall also that

$$\alpha_i = \mathbf{x} \cdot \mathbf{u}_i. \tag{116}$$

Therefore, we are tempted to identify the basis functions of $L^2[0,1]$ with the basis of a vector space. However, there is one important difference: while the vector spaces that we are familiar with all have finite dimensions, the number of basis functions of $L^2[0,1]$ is infinite.

Exercise 14-18: The Taylor series of $f(x)$ is the representation

$$f(x) = f(x_0) + \sum_{n=1}^{\infty} \frac{(x - x_0)^n}{n!} f^{(n)}(x), \tag{117}$$

where x_0 is some particular value of x, and $f^{(n)}(x)$ denotes the nth derivative of f. Clearly it is defined only for continuous functions. What is the relationship between the set $\{(x - x_0)^n\}$ and an orthonormal basis of the space of all continuous functions on $[0,1]$?

14.8.6 The Inverse of a linear function

Let $\mathbf{A}$ be a linear function that maps $\mathcal{E}$ to $\mathcal{E}$ where $\mathcal{E}$ is a Hilbert space. That is, for every $\mathbf{x} \in \mathcal{E}$, there is a $\mathbf{y} \in \mathcal{E}$ such that $\mathbf{y} = \mathbf{A}\mathbf{x}$. If $\|\mathbf{A}\| < 1$, then[15]

$$(\mathbf{I} - \mathbf{A})^{-1} = \sum_{n=0}^{\infty} \mathbf{A}^n. \tag{118}$$

---

[15] See Young (1988).

Here, $\mathbf{A}^n$ means $\mathbf{A} \cdot \mathbf{A} \cdot \; \cdots \; \mathbf{A}$ n times. So if $\mathbf{y} = \mathbf{A}\mathbf{x}$, then $\mathbf{A}\mathbf{y} = \mathbf{A}\mathbf{A}\mathbf{x} = \mathbf{A}^2\mathbf{x}$, and so forth. $\mathbf{A}^0 = \mathbf{I}$, the identity operator.[16] This is an extension of the theorem that states that, for $|r| < 1$,

$$\sum_{n=0}^{\infty} r^n = \frac{1}{1-r}. \tag{119}$$

Here is a heuristic proof of equation (118).

Let $\mathbf{S}$ be a linear transformation from $\mathcal{E}$ to $\mathcal{E}$ such that

$$\mathbf{S}\mathbf{x} = \lim_{N\to\infty} \sum_{n=0}^{N} \mathbf{A}^n \mathbf{x} \tag{120}$$

assuming for the time being that that limit exists. Then

$$(\mathbf{I} - \mathbf{A})\mathbf{S}\mathbf{x} = \lim_{N\to\infty} \left(\sum_{n=0}^{N} \mathbf{A}^n \mathbf{x} - \sum_{n=0}^{N} \mathbf{A}^{n+1} \mathbf{x} \right)$$

$$= \lim_{N\to\infty} \left(\sum_{n=0}^{N} \mathbf{A}^n \mathbf{x} - \sum_{n=1}^{N+1} \mathbf{A}^n \mathbf{x} \right) = \mathbf{A}^0 \mathbf{x} - \lim_{N\to\infty} \mathbf{A}^N \mathbf{x} = \mathbf{x}. \tag{121}$$

In this last equation, I have used the property that $\|\mathbf{A}^N\| = \|\mathbf{A}\|^N$ and $\|\mathbf{A}\| < 1$, so

$$\lim_{N\to\infty} \|\mathbf{A}^N \mathbf{x}\| = \mathbf{0}. \tag{122}$$

Therefore,

$$\mathbf{S}\mathbf{x} = (\mathbf{I} - \mathbf{A})^{-1}\mathbf{x}, \tag{123}$$

establishing equation (118).

Exercise 14-19: Show that the sum in equation (120) converges.

Corollary: Let $\mathbf{B} = \mathbf{I} - \mathbf{A}$. Then $\mathbf{A} = \mathbf{I} - \mathbf{B}$. If $\|\mathbf{I} - \mathbf{B}\| < 1$,

$$\mathbf{B}^{-1} = \sum_{n=0}^{\infty} (\mathbf{I} - \mathbf{B})^n. \tag{124}$$

---

[16] The proof of this theorem can be found in Young (1988).

14.9 Orthogonal expansions

We often want to express a function $f(x)$ as a series. It is especially useful if we have a set of functions that are orthonormal. To be specific, suppose there is a set of functions ϕ_i, $i = 0, 1, \ldots$. They have the property that

$$\int_a^b \phi_i(x)\phi_j(x)dx = \delta_{ij}. \tag{125}$$

where $\delta_{ij} = 1$ if $i = j$ and $\delta_{ij} = 0$ otherwise. The constants a and b are real, with the obvious restriction that $a \neq b$. Now we want to represent $f(x)$ as the series

$$f(x) = \sum_i c_i\,\phi_i(x). \tag{126}$$

This series represents f in the interval (a,b). We can find the coefficients c_i as follows.

Multiply both sides of equation (126) by ϕ_j and integrate from a to b:

$$\int_a^b f(x)\phi_j(x)dx = \int_a^b \sum_i c_i\,\phi_i(x)\phi_j(x)\,dx$$

$$= \sum_i c_i \int_a^b \phi_i(x)\,\phi_j(x)\,dx. \tag{127}$$

Using the orthogonality relationship, this reduces to

$$\int_a^b f(x)\phi_j(x)\,dx = c_j. \tag{128}$$

Therefore, it is easy (at least formally) to calculate the appropriate coefficients, given the function $f(x)$.

There is a further property that is sometimes useful. Suppose we have calculated c_i, $i = 0, \ldots, n$. Since ϕ_{n+1} is orthogonal to ϕ_i, $i = 0, \ldots, n$, ϕ_{n+1} is orthogonal to the sum

$$f_n(x) \equiv \sum_{i=0}^{n} c_i\phi_i(x). \tag{129}$$

Therefore, when the functions $\phi_i(x)$ are orthonormal (and only when they are orthogonal, at least, if not orthonormal), it is possible, given an approximation f_n to f with $n+1$ terms, it is possible to add the $n+2^{nd}$ term without having to re-calculate the preceding n terms.

14.10 Orthonormalization

Suppose that we have a set of functions ψ_k, $k = 1, 2, \ldots$, that are linearly independent but not necessarily orthonormal. We can always find a new set of functions that span the same space and *are* orthonormal by using the Gram-Schmidt orthonormalization method. Suppose that the ψ_k, $k = 1, 2, \ldots$, are not orthonormal. Take ψ_1 and form

$$\psi_1' = \frac{\psi_1}{|\psi_1|}. \tag{130}$$

Then take ψ_2 and form

$$\psi_2' = \frac{\psi_2 - (\psi_2, \psi_1)}{|\psi_2 - (\psi_2, \psi_1)|}. \tag{131}$$

Continuing in the obvious manner, we get $\{\psi_k'\}$, which is an orthonormal set.

Exercise 14-20: Fill in the steps to show how Gram-Schmidt works. This is discussed in any book on modern algebra.

14.11 Summing series

Many series have a form similar to that of E_1, the first exponential integral:

$$E_1(x) \equiv \int_x^\infty \frac{e^{-t}}{t}\,dt = \int_1^\infty \frac{e^{-xt}}{t}\,dt$$

$$= -\gamma - \ln x - \sum_{n=1}^\infty \frac{(-1)^n x^n}{n\,n!} \quad (x > 0), \tag{132}$$

where $\gamma = 0.577215665\ldots$ is Euler's constant.

Calculating the sum in equation (132) can be a problem. If we simply evaluate each term in turn, the round-off errors will dominate if x is not very small. To avoid this problem, we can regroup the terms as follows. Write

$$\sum_{n=1}^\infty \frac{(-1)^n x^n}{n \cdot n!} = -\frac{x}{1} + \frac{x^2}{2 \cdot 2!} - \frac{x^3}{3 \cdot 3!} + \frac{x^4}{4 \cdot 4!} - \cdots$$

$$= -x\left(1 - x\left(\frac{1}{2\cdot2!} - x\left(\frac{1}{3\cdot3!} - \frac{x}{4\cdot4!} + \cdots\right)\right)\right)$$

$$= -x\left(1 - \frac{x}{2}\left(\frac{1}{2} - \frac{x}{3}\left(\frac{1}{3} - \frac{x}{4\cdot4} + \cdots\right)\right)\right). \tag{133}$$

The round-off error is minimized by performing the calculation starting with the *innermost* term. This method works well, in general, for series with terms proportional to $x^n/n!$.

14.12 Clenshaw's algorithm

A similar idea, due to Clenshaw, helps us to sum certain series.[17] Suppose we want to evaluate

$$y(x) = \sum_{n=1}^{N} a_n f_n(x), \tag{134}$$

where $f_n(x)$ is a sequence of functions and the a_n are known coefficients. Suppose further that the functions f_n obey the recursion relation

$$f_{n+1}(x) = c_n(x)f_n(x) + d_n(x)f_{n-1}(x). \tag{135}$$

Note that the coefficients c_n and d_n can depend on x. This situation is fairly common, where we want to write a function as a weighted sum over some orthogonal functions—polynomials, perhaps, or Bessel functions. Most of these sets of functions obey a recursion relation.

The obvious way to evaluate equation (134) is to evaluate $a_0 f_0(x)$, add $a_1 f_1(x)$, and so forth. This is not, however, the most efficient way to do things, especially if evaluating $f_n(x)$ is complicated. Instead, we can do the following. Let $y_{N+1} = y_{N+2} = 0$. Then evaluate

$$y_{n-1} = y_n c_{n-1}(x) + y_{n+1}d_n(x) + a_{n-1} \tag{136}$$

for $n = N$, $N-1$, ..., 0. We can write the sum as follows. Note that I have not denoted how c_n, d_n, and f_n depend on x explicitly. First, solve equation (136) for a_{n-1}:

$$a_{n-1} = y_{n-1} - y_n c_{n-1}(x) - y_{n+1}d_n(x). \tag{137}$$

Then $y(x) = \Sigma a_n f_n(x)$ is given by the lengthy expression

---

[17] See Clenshaw (1962), and Press *et al.* (1992, pp 176-177).

$$= f_0(y_0 - c_0 y_1 - d_1 y_2)$$

$$+ f_1(y_1 - c_1 y_2 - d_2 y_3)$$

$$+ f_2(y_2 - c_2 y_3 - d_3 y_4)$$

$$+ f_3(y_3 - c_3 y_4 - d_4 y_5) + \dots \qquad (138)$$

Now consider the factor that multiplies y_3: it is

$$f_3 - c_2 f_2 - d_2 f_1 = 0. \qquad (139)$$

In fact, all of the terms cancel for $k > 1$. Therefore, the sum is simply

$$\sum_{n=0}^{N} a_n f_n(x) = (y_0 - c_0 y_1) f_0(x) + y_1 f_1(x). \qquad (140)$$

We only need to evaluate $f_0(x)$ and $f_1(x)$—not all N of the functions $f_n(x)$—to evaluate the sum. The usefulness of this method is not primarily that it saves computation, but that it uses the recursion relation, which is exact, in the computation: this reduces the round-off errors. This method is almost always stable in the downward direction. If it is unstable in this direction, because the lowest-order terms are almost equal, a similar method can be derived that works in the upward direction.

Exercise 14-21: Extend this line of reasoning to the case where there is a four-term recursion relation for a_n [*i.e.*, equation (135) has an additional term $e_n f_{n-2}$].

14.12.1 Power series

We can also extend these ideas to summation of a power series. Suppose we express a function $S(t)$ as a power series in t:

$$S(t) = \sum_{k=0}^{\infty} a_k t^k, \qquad (141)$$

where the coefficients a_k obey the recursion relation

$$a_{k+1} = \alpha_k a_k + \beta_k a_{k-1}. \qquad (142)$$

Define a new set of constants b_k such that

$$b_k \equiv a_k t^k. \tag{143}$$

Then $S(t)$ is just the sum of the coefficients b_k. The recursion relation between the coefficients b_k is

$$b_{k+1} = t\alpha_k b_k + t^2 \beta_k b_{k-1}. \tag{144}$$

Now, choose N to be the number of terms to evaluate. Define a series of constants y_k by $y_{N+1} = y_{N+2} = 0$, and, for $k \leq N$,

$$y_k = t\alpha_k y_{k+1} + t^2 \beta_{k+1} y_{k+2} + 1. \tag{145}$$

Starting with $k = N$, evaluate $y_N, y_{N-1}, \ldots, y_0$. Now rewrite equation (145) as

$$1 = y_k - t\alpha_k y_{k+1} - t^2 \beta_{k+1} y_{k+2}. \tag{146}$$

The last few equations, for $y_0, y_1, \ldots$, are

$$1 = y_0 - t\alpha_0 y_1 - t^2 \beta_1 y_2, \tag{147}$$

$$1 = y_1 - t\alpha_1 y_2 - t^2 \beta_2 y_3, \tag{148}$$

$$1 = y_2 - t\alpha_2 y_3 - t^2 \beta_3 y_4, \tag{149}$$

$$1 = y_3 - t\alpha_3 y_4 - t^2 \beta_4 y_5, \ldots. \tag{150}$$

Multiply equation (147) by b_0; equation (148), by b_1; equation (149), by b_2; and so forth. Then sum the left- and right-hand sides of these equations. Summing the left-hand side yields $S = \Sigma b_k$. Now consider the contribution that the coefficients that multiply y_3 make to the sum of the right-hand sides, which is

$$b_3 - t\alpha_2 b_2 - t^2 \beta_2 b_1 = 0. \tag{151}$$

Similarly, we see that all the terms cancel that multiply y_k for every $k > 1$. Therefore, the only terms that remain in the sum are

$$S(t) = \sum_{k=0}^{\infty} b_k = y_1 b_1 + (y_0 - t\alpha_0 y_1) b_0. \tag{152}$$

14.13 Proof by induction

The method of proof by induction may be unfamiliar to some readers. It is often used to prove that a *series* of statements are true. I use this method in Chapter 11, for example, to show that a certain set of equations that contain an integer parameter k are all true. Because the method of proof by induction can seem to be obscure—it may look like we are claiming to prove the truth of a statement by simply assuming that it is so—I shall give a short explanation here.

Suppose we have a set of statements S_n, $n = 0, 1, \ldots$. In particular, assume that there is an infinite number of such statements. We wish to prove that they are all true. Proof by induction requires two steps.

1. Prove that S_0 is true. How this is accomplished obviously depends on what the statement actually is. This step is often called the *induction hypothesis*.

2. Prove that, for any (positive) value of n, if S_n is true, then S_{n+1} must also be true. I shall denote this compactly by writing $S_n \rightarrow S_{n+1}$.

3. If both parts have been proved, then every statement S_n is true. Note that the procedure requires that every statement be associated with a different positive integer, and that every positive integer be associated with exactly one statement.

One needs to be careful in using inductive proofs. The proof that $S_n \rightarrow S_{n+1}$ must be true for *every* positive n. I know of one seeming paradox—a supposed proof that all horses have six legs—that arises from a sequence of statements where $S_n \rightarrow S_{n+1}$ is indeed true for every value of n—except for $n = 1$.

We can see why induction provides a rigorous proof as follows. Suppose that S_0 is true and that $S_n \rightarrow S_{n+1}$ for all $n \geq 0$. But suppose that at least one of the statements S_n is false. S_0 cannot be false, since we base the method of induction on the proposition that we can prove that S_0 is true. So there must be an $m > 0$ such that S_m is false, but S_n is true for every $n < m$. In other words, the set of all integers for which S_n is false must have a greatest lower bound. But S_m cannot be false, because $S_{m-1} \rightarrow S_m$, and S_{m-1} is true. And if there can be no m such that S_m is false, then all of the statements S_n must be true.

References

Clenshaw, C. W., *Mathematical Tables*, Vol. 5, National Physical Laboratory, H. M. Stationery Office, London (1962).

Davenport, W. B., and R. L. Root, *An Introduction to the Theory of Random Signals and Noise*, McGraw-Hill, New York (1958).

Goodman, J. W., *Statistical Optics*, John Wiley & Sons, New York (1985).

Index

$\in$ 104, 344
$<\cdot>_t$ 116
$<\cdot>_x$ 116
$(\underline{\mathbf{B}}\mathbf{x})\ \mathbf{x}$ 173
$<x>$ 102
$\{\}$ 248
$\{x_i\}$ 247
$\mathbf{A}$ 147
$a(u)$ 78
$\mathbf{A}^\dagger$ 169
$\mathbf{A}^{-1}$ 109
$\mathbf{AS}_x\mathbf{A}^t + \mathbf{N}$ 154
$\mathbf{A}^t$ 105
$\mathcal{A}_{ul}$ 55
$\mathbf{B}$ 172, 173
$\mathcal{B}_{lu}$ 54
$B(x)$ 77
$B(\lambda,T)$ 38, 85
$\mathcal{B}_{ul}$ 55
$C(\tau)$ 71
col 105
$\mathrm{col}(x_1, x_2, \ldots)$ 11
$\mathrm{cov}(x_i, x_j)$ 100
$\mathrm{d}(f_j,f_k)$ 248
$\mathbf{D}^\dagger$ 125
det 114
DFT 198
diag 113, 114
$\mathrm{E}[f(x)]$ 102
$\mathbf{E}^n$ 147
$E_n(z)$ 89
e_p 44
E_u 53
$e_\lambda(\theta)$ 90
$\mathcal{F}$ 248
FFT 198, 203
f_v 24
$G(r)$ 82
$H(x)$ 182
$H(\omega)$ 233
I_λ 37
I_v 41
$J_0(x)$ 191
$\mathcal{K}$ 108
$K(\mu,p)$ 94
$K^\dagger$ 264
$\mathbf{n}$ 149

$\mathcal{N}$ 108, 133
NEΔT .... *See* noise-equivalent ΔT
$\mathbf{nn}^t$ 154
P 73
$P(x)$ 77, 346
$p(z)$ 92
$\mathbf{Q}$ 144
R 51, 148
r_h 45
r_p 44
r_v 45
s_c 199
$\mathrm{sinc}x$ 79
s_{max} 199
$\mathbf{S}_x$ 118
$\mathbf{T}$ 144
$T(\tau_\lambda)$ 87
T_a 74
T_B 41
tr 117
$\mathbf{U}$ 111
$\mathcal{V}$ 104
$\mathrm{var}(x_i)$ 100
WSS 217
$\mathcal{X}$ 251
$\mathbf{x}\cdot\mathbf{y}$ 106
$\mathbf{x}^t$ 105
$\mathbf{xy}^t$ 115
ΔE 52
Δt 60
ΔE_{ul} 53
$\star$ 188
$\Pi(x)$ 190, 346
Σ_{tot} 117
α 255
$\alpha_{\beta ij}$ 172
β 24
β_{ijk} 172
χ^2 .... 154, 258
χ_h 16
χ_{ij}^2 144
$\delta(x)$ 181
δ-function 181
δ_{ij} 263
$\delta\Omega$ 74
ε_N^2 149
ε_x^2 148

ε_λ 39, 86
κ_λ 39, 86
λ 76, 119
λ_i 119
λ_{max} 42
μ 86
μ_i 132
ρ 46
$\rho(z)$ 92
ρ_{ij} 117
σ^2 100
σ_N^2 145
σ_x^2 102
σ_{xy}^2 103
τ_λ 46, 86

A

absorption coefficient 39, 45, 54
    depends on temperature   57
absorption coefficients 62, 171
adjoint
    definition   264
algebraic multiplicity 122
algorithm 12, 19, 99, 100, 144, 146, 156
    optimal   19
aliasing 199
antenna temperature 74
apparent temperature .... See brightness
    temperature
asymmetric kernel 265
autocorrelation function 78, 216, 217, 218,
219, 221, 222, 223, 228, 229, 230
    and stationarity   231
    and Wiener-Khintchine theorem   220
    power pattern   79
    summary   224

B

Babbage, C. 279
Backus and Gilbert 324, 329, 330, 353
bandwidth limited function................... 344
basis functions 248
Bessel functions........................... 191, 288
black body 38, 42, 57, 62
    enclosure   44
black body radiation 43
blackbody enclosure 48
Boltzmann's constant 70
boundedness condition 250
brightness 78
brightness temperature 41
broadening, Doppler.......................... 58, 61
broadening, pressure.................. 58, 60, 61

C

calibration.. 67
Cauchy sequence 388
Cauchy-Schwarz inequality.......... 118, 381
central limit theorem 102, 385
Central limit theorem 384
Chahine, M................................. 311, 314
Chahine's method............................... 317
Churchill, W....................................... 323
Clenshaw's algorithm.......................... 395
collisional excitation 60
collisional transition 57
complete space 389
completeness 388
complex linear span............................. 390
complex permittivity 33
computer
    human   13
constraints .... *See* Lagrange Multipliers
continuum............................ 31, 34, 53, 62
convergence... 285
convergence factor 295, 296
convolution.. 77
    and Fourier transform   204
convolution integrals........................... 251
convolution theorem..................... 187, 252
Cooley-Tukey algorithm 179
Corollary 1 292
Corollary 2 293
Corollary 3 293
Corollary 4 294
correlation coefficient 117
cosines
    product of two   240
covariance 99, 103, 114, 117, 130, 132,
133, 273
covariance matrix 100, 114, 115, 127
    and eigenvalues   159
covariance matrix, and noise 133, 154
cross spectrum
    defined   223
cross-correlation
    defined   223
    propagating waves   238
cross-periodogram
    defined   223
cutoff frequency 253

D

deconvolution...................................... 345
    and noise   348
derivative
    Fourier transform   191
determinant.. 114

DFT 226
Diadochus, P..7
diagonal matrix 113
discrete Fourier transform 198
E
effective pattern 351, 356
   defined   350
eigenfunction expansion 265
   integral equations   262
eigenfunction solution
   asymmetric kernel   265
eigenfunctions
   and resolution   360
   finding   269
   Hilbert space   262
   of a kernel of an integral equation 263
   orthogonal   263
   symmetric kernels   263
eigenvalues 119, 132
   of a kernel of an integral equation 263
eigenvectors.................................119, 141
Einstein coefficients 57
Einstein, A..53
El Niño 9
emission coefficient................................39
emissivity........................... 31, 45, 63, 64
   ocean surface   63
empirical orthogonal functions.............129
energy levels......................................51, 53
energy of a photon 53
entropy..371
equation of radiative transfer47, 85, 87, 91,
94, 312
equivalence, matrix............................... 113
ergodic..218
error variance.......... 16, 274, 282, 291, 298
evaluation set ....See regression
even function 188
existence
   of solutions   259
exponential integral 89
F
fast Fourier transform 203
FFT ...._See_ fast Fourier transform
FFT algorithm.......................................179
filters...233
Fleming, H..............................26, 290, 297
Fontenelle, B. 147
Fourier series 179, 192
   complex   197
   derived from Fourier transform   195
Fourier series and discrete Fourier
  transform 198

Fourier transform.................................. 184
   and convolution   187
   definition   183
Fourier transform coloring book 207
Fourier transform, 2-D 190
fraction of the variance explained . 24, 131,
273, 274
Fredholm equation................................311
   first kind   249
   second kind   250
Fredholm equation of the first kind 338
Fredholm equations 249
Fresnel reflection coefficients 45
G
Gaussian
   Fourier transform of   185
   uncertainty principle   202
Gaussian distribution............................ 102
Gaussian noise.......................................149
generalized functions.............................181
geometric multiplicity 122
geometric series 195
Gram-Schmidt orthonormalizaton........394
ground truth 144
H
hammer..245
Hankel transform 206, 207
Hermitian
   definition   189
Hilbert space 247, 248, 250, 251, 262, 386,
387, 389, 390, 391
Hilbert spaces 385
homogeneous.........................._See_ stationary
Huygens, C...53
hydrogen.....................................50, 51, 61
hydrometeors 33
hydrostatic equilibrium....................92, 93
I
induction...398
information .. 103, 114, 129, 156, 162, 200,
228, 326, 340
   independent pieces of   21, 273
information content 271
infrared 9, 62, 311, 313
infrared radiation 33
inhomogeneous atmosphere 91
inner product 386
integral
   Fourier transforms   191
integral equations
   Fourier-transform method   251
integral equations 247
   and matrix equations   251

and resolution   336
    ways to solve   248
intensity 37
inverse, matrix 109
Inversion.. 297
ionization potential 52
iteration . 26, 138, 171, 177, 248, 290, 295,
    297, 310, 313
    and resolution   361
    equivalent to matrix inverse   295
iterative method...........................291, 317
    regularized   298
K
kernel...........................311, 312, 315, 319
    asymmetric   265
    degenerate   264
kernel, symmetric 263
kinetic energy 58
Kirchoff, G. 40
Kirchoff's law 30, 39, 40, 41
Kirchoff's Law 44
Koestler, A. 215
$L^2[0,1]$ 389
L
lagged product See autocorrelation function
Lagrange multiplier 337, 368
    defined   367
Lanczos, C... 195
Lang, A.. 143
Laplace transform................................ 205
laser 57
law of large numbers 383
least squares................................ 150, 152
Lemma 1.. 292
Lemma 2.. 292
Liou, K.-N. 311, 314
load
    calibration   69
logistic equation 287
Lorentz line shape 61
M
matrix equations
    compared with integral   248
matrix inverse
    equivalent to iteration   295
matrix iteration 290
matrix method
    compared with integral equations   261
matrix, non-negative definite............... 132
matrix-inverse....................290, 291, 298
Maxwell, J. C. 29
mean 101
mean values

assumed zero   19
metric 388
microwave radiation.............................. 33
N
Newman, J.. 247
Newton, I..................................... 53, 215
Newton's method 280
Newtonian reflector.............................. 76
noise 149, 171
    amplification factor   255
    and effective pattern   352
    definition   14
    in integral equations   253
    included in solution   254
    tradeoff with resolution   256, 330, 332
noise amplification 152, 331, 358, 359
noise amplification factor............. 149, 352
noise amplification factor (α)....... 330, 332
    and spread   335
    and spread   330
noise amplification factor (α_j) 328
noise covariance matrix........................ 329
noise power 253
noise-equivalent ΔT............................... 72
noiselike process 237, 239
nondefective
    matrix   122
non-derogatory matrix.......................... 122
nonlinear equations 171, 172, 174, 310
nonlinear problems.............................. 171
non-inear systems................................. 25
non-negative definite............................ 145
nonsymmetric matrices 121
norm, function 387
norm, matrix....................................... 380
norm, vector 380
null space..... 108, 109, 110, 111, 150, 168,
    260
Nyquist frequency 82, 199
Nyquist sampling interval 83, 200, 360
Nyquist, H. 71, 73, 199
O
odd function 188
optical depth.............................. 46, 48, 86
optical thickness.................................... 46
orthogonal expansion 393
orthogonal functions.................... 387, 393
orthogonal matrix 111
orthogonal transformation 112
orthonormal basis 107
orthonormalization 394
outer product 106
outer product (matrix) 115